# The Shorebird Guide

# The Shorebird Guide

MICHAEL O'BRIEN
RICHARD CROSSLEY
KEVIN KARLSON

Photographs by
Richard Crossley
Kevin Karlson
Mike Danzenbaker
Arthur Morris
and other noted photographers

Silhouettes by
Michael O'Brien

Maps by Kenn Kaufman

Houghton Mifflin Company
Boston New York

Visit our Web site: www.houghtonmifflinbooks.com

Library of Congress Cataloging-in-Publication Data

Crossley, Richard.
The shorebird guide / Richard Crossley, Kevin Karlson, Michael O'Brien.
p. cm.
Includes bibliographical references (p. ) and index.
ISBN-13: 978-0-618-43294-3
ISBN-10: 0-618-43294-9
1. Shore birds—Identification. 2. Shore birds—Pictorial works.
I. Karlson, Kevin. II. O'Brien, Michael, date. III. Title.
QL696.C4C76 2006
598.3'3—dc22 2005020996

Book design by Anne Chalmers

Printed in China

SCP 10 9 8 7

# CONTENTS

**PREFACE:**

# A Simplified Approach to Shorebird Identification

A CLOUD OF BIRDS bursts from a seemingly deserted mudflat. With uncanny synchrony, several thousand shorebirds swirl about, twisting and turning as one entity—truly one of nature's awe-inspiring spectacles. The realization that many of these birds are in the midst of an 8,000-mile journey adds to a fascination with the shorebird family shared by many birders. Shorebirds are among the most visible and easily observed members of the avian world. With their worldwide distribution and propensity to feed and rest on open beaches, mudflats, and fields, they may be enjoyed by anyone who looks for them.

## THE PROBLEM

With only about 217 species worldwide and approximately 50 species found breeding regularly in North America, shorebirds make up a small percentage of the world's more than 8,000 bird species. Since most North American birders encounter only about 35 to 40 shorebird species in a single year, one would think that the identification of these visible birds would pose few problems. However, shorebirds often present seemingly insurmountable identification problems to a large number of birders. Why is this? Distance is part of the problem. Shorebirds are frequently located on inaccessible mudflats hundreds of yards away. Even with a spotting scope, the images are often small and details difficult to see.

When observers are fortunate enough to have close views, they are presented with a complex variety of plumages. While some shorebirds change little in appearance throughout the year, many species look very different at different seasons or at various life stages. Furthermore,

**Winter Dunlin flock, Jan., N.J.**

the transition between these plumages creates a seemingly endless variation of appearances in some species—so much so that a casual observer looking at an August flock of Western Sandpipers, for example, might assume that a juvenile, a first-summer, a breeding adult, and a molting adult are actually four different species. With so much seasonal and individual variation in plumage, it is nearly impossible to come up with a set of plumage-based "field marks" that always work. To complicate things, much of the current shorebird literature is illustrated entirely with detailed paintings or close-up photographs that do not reflect real-life field conditions. Furthermore, these texts frequently focus on minutiae without giving proper emphasis to the fundamentals—the basic framework for identification.

## THE SOLUTION

When advanced birders look at a flock of shorebirds, they are often able to identify the vast majority of birds with a quick binocular scan. How do they do this? Experts have an easier time identifying shorebirds because they are using an easier identification method. Instead of skipping straight to plumage details, they base their identifications, first and fore-

**Bolivar Flats, Apr., Tex.** A group of birders watches shorebirds.

most, on relative size, structure, behavior, and voice. They start with the basics. All these characteristics are far less variable than plumage details and are therefore an easier, more reliable *starting point* for identification. In fact, virtually any shorebird can be identified solely on the basis of these fundamental characteristics. That's not to say that plumage details are not important to identification—they very often are. But in order to get off on the right track, the identification process must begin with the fundamentals.

This holistic style of identification (which has its roots in *jizz* birding, named after "General Impression of Size and Shape" used by the military) is by no means limited to shorebirds. It applies to all birds. By simply focusing on those characters that are least variable and most visible in the field, not only will your identifications become more accurate, but you'll also identify a higher percentage of what you see. You may practice your skills anywhere, even in your backyard, while taking a walk or looking out your office window. It is especially instructive to practice identifying birds by naked eye, without the aid of binoculars. Such exercises help hone your skill at evaluating size, structure, and behavior. With more practice, you'll be amazed at how many birds you will identify almost subconsciously before even beginning your critical thought process. As you store more visual impressions in your mind, not only will your skill level increase, but birding will also become more fun and more rewarding.

# ACKNOWLEDGMENTS

The unique format and approach of *The Shorebird Guide* is a result of many years of field study by the authors combined with the influence of many talented people, both from the birding community and elsewhere. Some of these were our mentors, providing guidance and instruction during the formative years. Others were sounding boards for our ideas, especially workshop participants, who helped us to learn how to *teach* shorebird identification. Every question, blank look, and understanding smile helped form our teaching styles and design this book. To all of the above, thank you.

Several people deserve special mention for their extraordinary help with the physical preparation of the book: Lloyd Spitalnik and Scott Elowitz, for their guidance and technical support during the design stage; our agent, Russ Galen, who helped shape our idea into a presentable package; Debra Crossley and Dale Rosselet, who added important technical support during the page-layout process; Steve Howell, who helped us track down several hard-to-find photos; Kenn Kaufman, who graciously provided the range maps; George Myers, who helped with preparation of the silhouette pages; and Yoshimitsu Shigeta, for acting as a liaison to several Japanese photographers whose work was invaluable to the Rarities photo section. For access to museum skins or library resources, we thank Jim Dean (Smithsonian Institution), David Mizrahi (New Jersey Audubon Center for Research and Education), Matt Sharp (Academy of Natural Sciences), and Doug Wechsler (Visual Resources for Ornithology). We also thank Paul Lehman and Tom Parsons for use of their extensive personal libraries, and Pete Dunne, who provided encouragement and advice in the early stages of this project.

Countless birders and nonbirders encountered over the years have provided lively discussions and insights that influenced our approach to this work. Space precludes our naming them all, but each has helped to improve the quality of this guide. For answering queries, providing research materials, or various other forms of assistance, we thank Jonathan Alderfer, Per Alstrom, Lee Amery, Jessie Barry, Robert Behrstock, Wes Biggs, Adrian Binns, Keith Brady, Ned Brinkley, Jeff Bouton, Paul Buckley, Bob Carlough, Daniel Chan, Paul Cook, Cameron Cox,

Brian and Margaret Crossley, Heidi Cummings, the late Tom Davis, Jorge Montejo Diaz, Chris Dooley, Greg Downing, Jon Dunn, Pete Dunne, Gail Dwyer, Vince Elia, Graham Etherington, Sean Farrel, Shawneen Finnegan, David Foster, Russell Fraker, Mark Garland, Graham Gordon, Patti Hodgetts, Craig Hohenberger, Paul Holt, Julian Hough, Steve Howell, Kwong Kit Hui, George Jett, Margaret Karlson, Paul Kerlinger, Chris Kisiel, Richard Lanctot, Sheila Lego, Paul Lehman, Tony Leukering, Cyndi Loeper, Barbara McGlauglin, Clive Minton, Steve Mlodinow, Arthur Morris, Stephen Moss, Killian Mularney, Marleen Murgitroyde, Larry Niles, Evan Obercian, John O'Brien, Paul O'Brien, Laura X. Payne, Don Riepe, Gary Rosenberg, Dave Rudholm, Will Russell, David Sibley, Brian Small, Lloyd and Sandy Spitalnik, Brian Sullivan, Clay and Pat Sutton, Roy Sutton, Dave Tetlow, Laurilee Thompson, Pavel Tomkovich, Declan Troy, Chris Vogel, Dick Walton, Nils Warnock, Sophie Webb, the late Claudia Wilds, Chris Wood, Louise Zemaitis, and Jim Zipp.

This book would not be as visually appealing without contributions by the following photographers: Goran Altstedt, Lee Amery, Yuri Aztukhin, Nigel Blake, the late Don Bleitz, Lysle Brinker, Richard Brooks, P. A. Buckley, Milo Burcham, Daniel Chan, Richard J. Chandler, Robin Chittenden, Lyann Comrack, Mike Danzenbaker, Don Des Jardin, Paul Doherty, Linda Dunne, Scott Elowitz, Ian Fisher, Paul J. Fusco, Hugh Gallagher, James A. Galletto, Brian M. Guzzetti, Martin Hale, Brandon Holden, Julian Hough, Steve Howell, Barry Hughes, Kwong Kit Hui, Himaru Iozawa, Tsutomu Ishikawa, George Jett, Aaron Lang, Gordon Langsbury, Vernon Laux, Tony Leukering, Tim Loseby, Bruce MacTavish, Eric McCabe, Mike McKavett, Clive Minton, Arthur Morris, Alan Murphy, Jari Peltomaki, Simon Perkins, Richard Revels, Wayne Richardson, Jim Rosso, Robert Royse, Bill Schmoker, Yoshimitsu Shigeta, Brian Small, Brian Sullivan, Ruth Sullivan, David Tipling, Ray Tipper, Arnoud B. van den Berg, Shuji Watanabe, Chris Wood, Steve Young, and Jim Zipp. Their beautiful, interesting, and illustrative photos are the heart of this guide. We thank them for their fine work.

Special thanks to Lisa White, our editor, who worked tirelessly to keep all three authors on track. Thanks also to Beth Fuller, Beth Kluckhohn, David Pritchard, Hugh Willoughby, Shelley Berg, and Mimi Assad, and book designer Anne Chalmers, for all the hard work behind the scenes during the production of our guide.

Finally, and most importantly, we thank our families for the support they gave and sacrifices they endured during the lengthy process of preparing this book. Undying thanks to Debra, Samantha, and Sophie Crossley, Dale Rosselet, Alec and Bradley Smith, and Louise Zemaitis.

**Top: American Golden-Plover on nest, June, Manitoba**

**Bottom: Buff-breasted Sandpiper nest, June, Alaska**

# Introduction

## WHAT ARE SHOREBIRDS?

Shorebirds are a varied group of wading birds in the order Charadriiformes, which also includes skuas, jaegers, gulls, terns, skimmers, and alcids. They are distributed nearly throughout the world, and many are long-distance migrants. Most shorebirds (also known as "waders") spend most of their time near water, though many species prefer upland pastures, plowed fields, and even forest. Most shorebirds feed on a variety of invertebrates, including worms, insects, small crustaceans, and mollusks. Some also feed on small fish or amphibians and occasionally seeds or berries. They typically nest on the ground in a small scrape in sand or gravel or a small depression in grass, tundra, or leaf-litter. Young are precocial and leave the nest soon after hatching, somewhat protected from the elements by downy feathers. Chicks are cared for by one or both parents, depending on the species. There are 12 families of shorebirds worldwide, 7 of which occur in North America. Of these, only 4 occur regularly. Shorebirds vary widely in size, leg length, bill shape, and feeding behavior—all adaptations to take advantage of different food sources and habitats.

### North American Shorebird Families

#### Thick-knees: Burhinidae (page 319)

A unique family of large upright shorebirds with large eyes, short thick bills, and long legs with swollen ankle ("knee") joints. Primarily nocturnal. They occur in dry open country with scattered vegetation. When alarmed they will usually run or crouch rather than fly, using their cryptic patterning as camouflage. They eat primarily invertebrates, especially insects, as well as small vertebrates and sometimes eggs. Thick-knees are widely distributed in tropical and temperate regions. Nine species are recognized worldwide, and one has occurred accidentally in North America north of the United States/Mexico border.

#### Lapwings and Plovers: Charadriidae (pages 319–42)

Members of this cosmopolitan group are small- to medium-sized and typically upright in posture. They have large eyes, short, often thick bills, short necks, and medium-long legs. Plovers have long, narrow, pointed wings, and lapwings have broader wings with rounded tips. Lapwings and plovers are primarily visual feeders, exhibiting a characteristic robin-like feeding style in which they walk (or run) in a slightly crouched position, then stop in a more upright stance to scan for food. Flocks tend to scatter while feeding, unlike many other shorebirds, which may feed in tight packs. Food items, including a wide variety of invertebrates, seeds, and berries, are plucked from the ground when seen. Lapwings and plovers also use the ultrasensitive soles of their feet to detect movements of subsurface prey items. They use a wide variety of open habitats, including tidal flats, shorelines, pastures, agricultural fields, and tundra. About 66 species occur worldwide, 14 in North America, including 4 that are rare or accidental.

### Oystercatchers: Haematopodidae (pages 342–46)

Large, bulky, predominantly black-and-white shorebirds with loud piping calls. All are similar in appearance, with laterally compressed orange bills used to pry open oysters, mussels, and other shellfish. Cosmopolitan in distribution, they occur primarily in coastal areas and are either non-migratory or short-distance migrants. About 10 species are recognized worldwide, 2 occurring regularly in North America and another as a vagrant.

### Stilts and Avocets: Recurvirostridae (pages 346–49)

Large slender shorebirds with very long legs and long slim bills. Most numerous in shallow inland lakes and ponds and quiet coastal lagoons, where they are typically seen wading in shallow water. They feed primarily on a variety of invertebrates and small vertebrates that are picked off the surface of the water or mud. Avocets frequently feed by skimming the water surface with a distinctive side-to-side swipe of the bill. With partially webbed toes, avocets sometimes swim to pursue food. Stilts and avocets are widespread in most temperate and tropical regions. About 9 species occur worldwide, with 2 occurring regularly in North America and another as a vagrant.

### Jaçanas: Jaçanidae (pages 349–50)

A unique family of shorebirds with short bills, broad rounded wings, and long legs with extremely long toes, enabling them to walk on floating vegetation. They occur in shallow freshwater wetlands with abundant floating vegetation, where they eat primarily insects, snails, and small fish. Widespread in tropical and subtropical regions. Eight species occur worldwide, with 1 species occurring as a vagrant and rare breeder in North America.

### Sandpipers, Phalaropes, and Allies: Scolopacidae (pages 350–451)

The largest and most widespread group of shorebirds. Highly variable in structure, ranging from small to large, slim to bulky, and short- to long-legged. Bills are usually slender but range from short to long and may be straight, decurved, or upturned. Most species have relatively long pointed wings designed for long-distance migration, but woodcocks have distinctly rounded wings. All but Sanderling share a rudimentary hind toe. Scolopacids are found in a wide variety of open habitats, mostly near water but often in pastures or agricultural fields and sometimes in forests. Food items include a wide variety of invertebrates, especially worms, as well as small fish, fruit, and seeds. Although some species locate their food visually, many find their food through tactile clues while probing in the mud with supersensitive bill-tips. About 87 species are recognized worldwide, with 43 species occurring regularly in North America and another 20 occurring accidentally. One other, the Eskimo Curlew, was historically abundant in North America but is now probably extinct.

### Coursers and Pratincoles: Glareolidae (pages 451–52)

A unique family of small- to medium-sized shorebirds with long pointed wings, short decurved bills, and relatively short necks. Coursers are fast-running, ploverlike birds with long legs, large eyes, and cryptic patterning. They occur in arid habitats where several species are primarily nocturnal. Pratincoles are ternlike in appearance and behavior with extremely long wings, forked tails, and short legs. They are primarily aerial feeders, hawking for insects in graceful ternlike flight. Widespread in tropical and temperate regions of the Old World, there are about 17 species of coursers and pratincoles worldwide. One species of pratincole has occurred as a vagrant in North America.

## SHOREBIRD MIGRATION

Many shorebirds travel thousands of miles in their biannual migrations, some crossing several continents and the open ocean. Although some species, such as American Oystercatcher, travel only short distances each year, others, such as Bar-tailed Godwit, make marathon nonstop flights of up to 6,000 miles. Such long-distance travel demands enormous amounts of energy and is fraught with risk: predators, bad weather, and exhaustion take thousands of migrants each year. Yet shorebird migration has evolved over tens of thousands of years to improve each species' chances for survival.

What makes such long-distance travel worth the risk? The need to reproduce and seasonal variability in resources are the key forces guiding migration. Food resources change through the year, and migration is essentially a movement from one seasonally abundant food source to another. During migration, most shorebirds stop at specific food-rich stop-over sites in order to molt or to accumulate enough body fat to fuel the rest of their travels. At these critical stop-over sites, many birds need to double their weight before moving on. For Arctic-nesting species these fat reserves also need to sustain birds through the initial stages of courtship and breeding in a landscape dominated by snow and ice. Although particularly rich feeding areas attract many species of shorebirds, each species has evolved to fill a unique ecological niche; hence each has its own unique structure, feeding style, breeding and wintering areas, migration routes, and stop-over sites.

Weather plays a major role in shorebird migration. Because some species travel long distances nonstop, often over inhospitable habitats, it is essential that they have favorable weather to help them along. Most shorebirds fly at speeds between 30 and 50 mph, but with a tail wind they can make much better time, increasing their chances for survival. Peak movements typically occur with, or just in advance of, frontal systems.

Most shorebird migration takes place at night, with noisy flocks often taking off around sunset, particularly during high tide. Most species make limited diurnal movements, especially in the morning, however, and long overwater travel necessitates diurnal migration in many species. Also, a few species (such as American Oystercatcher) do most of their migration during the day. Although most shorebirds normally migrate at relatively high altitudes, up to 20,000 feet or more, under cloudy or stormy weather they often fly much lower, making them more detectable by birders. Some shorter-distance migrants such as Killdeer and American Oystercatcher do most of their migration at lower altitudes.

An important aspect to understand about fall migration of shorebirds is its protracted nature. In many species, failed breeders depart a week or more before successful breeders, adult females up to a week before adult males, and adult males a week or more before juveniles. As a result, the first adults may arrive in a given area up to a month before the first juveniles. Furthermore, subadults of some species skip the first one or more breeding seasons and spend those summers either on the wintering grounds or at stop-over sites partway toward the breeding grounds.

These birds often migrate later in spring and earlier in fall than adults.

For birders, one of the most exciting aspects of shorebird migration is the chance to see a rarity. With such long-distance movements, it takes only a small initial error in navigation to bring a shorebird far out of its normal range. In general, juvenile birds are more likely to wander out of range than adults. Adult birds, which have made the trip before, can use visual clues to help them navigate. Without adult birds to follow, the first fall migration of a shorebird is guided by a rather imprecise innate knowledge of the approximate distance and direction of their intended wintering grounds. Wherever they end up, they will imprint on that location and, if they survive, will likely return there year after year. By this mechanism, shorebirds may be better able to adapt to a changing environment and more readily find new wintering and stop-over areas if old ones become less favorable. Many adult vagrants are, most likely, birds that wandered off course as juveniles and have habituated to a different migration route and wintering area than most of their clan.

## THREATS TO SHOREBIRD POPULATIONS

Like many living things, shorebirds are resilient creatures with a capacity to adapt to most challenges that the natural world can throw at them. In fact, these natural challenges have made shorebirds what they are today. Unfortunately, human-induced changes to the natural world are coming along at a faster rate than shorebirds can adapt to, causing serious declines in many species. Some species, such as the Eskimo Curlew, have likely been lost to us forever.

Habitat loss is a widespread problem for shorebirds. In the United States, about 50 percent of natural wetlands have been filled or drained, and we continue to lose about 35 square miles of wetland each year. Native grasslands have suffered even greater losses, resulting in severely restricted habitat for "grasspipers" passing through or nesting in the prairies.

Pollution poses another major threat to shorebirds. Solid waste and chemicals dumped into estuaries where shorebirds feed threaten to contaminate or destroy critical feeding areas. Oil spills have even more dramatic implications. A single oil spill at the wrong place and wrong time could threaten an entire population of shorebirds.

A perpetual and ever-increasing problem is direct competition with humans. The beaches and shorelines so critical as feeding or nesting habitat for shorebirds are also among our most popular recreational spaces. Although shorebirds often attempt to squeeze in around us, they inevitably get flushed, wasting limited feeding time and burning valuable energy needed for their long migrations. Humans also compete with shorebirds for food resources. Horseshoe Crab eggs provide a critical food source for several species of migrating shorebirds along the shores of Delaware Bay. Red Knots, in particular, must eat an estimated 18,000 Horseshoe Crab eggs per day and double their body weight in order to make the long flight to the Arctic and breed successfully. In recent years, adult Horseshoe Crabs have been used as cheap bait for eel and whelk fisheries. Overharvesting of this resource has caused drastic declines in Horseshoe Crabs and, in turn, similar declines in Red Knot and other species.

**Oil drilling on the Texas coast, April.** Oil drilling in shorebird-sensitive regions such as the Gulf Coast not only disturbs critical foraging and nesting habitats but also increases the risk of potentially devastating oil spills.

**Horseshoe Crab harvesting along Delaware Bayshore, May, Thompson's Beach, N.J.** Horseshoe Crabs are used as bait for eel and whelk fisheries, in direct competition with Red Knots and other shorebirds which depend on Horseshoe Crab eggs as their primary food source during spring migration.

Perhaps the greatest threat to shorebird populations is climate change. Tundra, where the majority of shorebirds breed, is the only major environment whose range is completely unable to shift northward in response to global warming. The most optimistic models predict that tundra habitat will be cut in half by the end of the twenty-first century because of a northward shift of boreal forest. Furthermore, sea-level rise will affect shoreline habitats in the Arctic and farther south. More frequent severe storms and flooding may heavily impact birds breeding in coastal lowlands such as coastal tundra, salt marshes, and beaches.

Although the prognosis may sound bleak, there is still hope. Research is under way to learn more about specific needs of shorebirds and how to stabilize or increase their populations. One of their most obvious needs is a safe haven where food is abundant and disturbance is minimal. Several conservation organizations are working hard to protect such important areas. Supporting these organizations is the best way to help ensure that future generations will be able to enjoy shorebirds as we do today.

## HOW TO IDENTIFY SHOREBIRDS: A SIMPLIFIED APPROACH

When in the field looking at a flock of shorebirds (or any birds, for that matter), an experienced birder will first take note of a species or object that he or she recognizes, such as a gull or egret or, better yet, one of the shorebirds themselves. This provides a solid reference point for judging size. With a handle on how big the various birds are, the birder will then notice overall structure of the birds (how they are proportioned, how long their legs and bills are, and so on) and how the birds are feeding or walking. For the most experienced birders, many of these observations are made subconsciously in just a few seconds, paving the way for the more conscious observation of plumage or structural details, voice, and so forth. With just a quick look, the experienced birder can form a surprisingly accurate first impression based simply on size, structure, and behavior. Because much of this process takes place at the subconscious level, it may be helpful to review the key steps in order to train yourself to be more effective at identifying shorebirds. The points that follow are listed in approximate order of importance.

**Relative size:** At the risk of stating the obvious, relative size is probably the single most important piece of information needed when trying to identify any bird. At the same time, it may be one of the most difficult characteristics to judge accurately. Fortunately, shorebirds tend to form mixed-species flocks, and where they occur, there are often other birds such as herons or gulls (or inanimate objects such as a soda can, shell, or discarded sandal) to compare them to. The key to judging relative size is having a known-identity bird available for direct comparison. If you see a distant shorebird standing next to a Ring-billed Gull, and the shorebird is only slightly smaller, you are probably looking at a large shorebird such as a godwit or a Willet. Or while looking at a bird that you think is a Baird's Sandpiper, if a Semipalmated Sandpiper walks by, and the Semipalmated is larger, you might want to reconsider your identification. One important aspect to consider when judging size is optical size illusion.

**Sanderling (left) and Long-billed Curlew, Jan., Tex.** Sometimes differences in size and structure are striking.

**White-rumped and Semipalmated Sandpipers, June, N.J.** Relatively subtle details of size and structure are sometimes useful for identification. Notice the White-rump's slightly larger size and longer wingtips.

When two same-sized birds are seen through a telescope, with one bird closer and the other farther away, your mind will interpret the far bird as being larger because of the telescope's shallower depth of field.

**Structure:** One of the most fundamental steps in identifying a shorebird is to take note of its basic structure. At a minimum this includes looking at how long its legs are and what shape and how long or thick its bill is. Features such as bill length are usually best judged by comparing them to another part of the bird such as the head or legs. (In general, we refer to a "medium-length" bill as one that is about equal to or slightly longer than the head depth, while a "short bill" is shorter and a "long" bill longer.) As you gain more experience, other aspects of structure will become more useful in a relative sense. For example, some birds look slim, others look chunky, some small-headed or long-necked, some pot-bellied or hunch-backed, and so forth. On a finer scale, some *details* of structure can be very helpful. For example, it is often very useful to look at where the primary tips fall relative to the tips of the tertials (known as the "primary projection") or where the primary tips fall relative to the tail tip (known as the "wing point"). Differences in primary projection and wing point can be very useful in distinguishing between similar species such as American and Pacific Golden-Plovers. Fine details of bill structure can be helpful as well, such as the typically bulbous tip on a Semipalmated Sandpiper's bill or the long nasal groove on a Wandering Tattler's bill.

Structural characteristics are perhaps most easily learned if the evolutionary forces that created them are understood. Shorebirds have evolved a wide array of bill shapes to best capture their preferred prey items. Differences in wing shape have evolved to suit different flight requirements. Longer, more pointed wings create less drag, allowing the faster flight needed by long-distance migrants (such as Baird's Sandpiper). Shorter, more rounded wings provide better maneuverability, a luxury limited to shorter-distance migrants. Likewise, longer-distance migrants tend to look thick-chested because of their larger flight muscles and correspondingly larger breastbones compared with those of shorter-distance migrants.

It is important to be cautious when evaluating some aspects of structure. A bird's shape may vary considerably under different behavioral or weather conditions. For example, when it is windy and cold, a bird may stand with its feathers fluffed up, neck pulled in, and bill pointed straight into the wind. At other times, the same bird in a more alert or active posture may have its feathers sleeked down and neck stretched out, giving it a much slimmer, longer-necked appearance. With practice, a keen observer will learn to associate a variety of shapes with each species.

Some of the more subtle aspects of structure are fully appreciated only after watching a bird for a prolonged period of time. As with using head shape to identify a female goldeneye or scaup, the distinctiveness of a shorebird's shape (such as that of a dowitcher or a yellowlegs) may be fully appreciated only after studying the posture that it assumes most of the time or when in a relaxed pose. For that reason, we find studying birds in the field to be much more informative than studying photos, which all too often catch birds in atypical postures. (In this guide we have tried, as much as possible, to use photos that capture "typical" poses.)

**Behavior:** The way a bird behaves—how it walks, how it feeds, how it flies, the habitat it selects—is genetically programmed and subject to minimal variation. While each species has its own unique behaviors, each also shares some traits with related species. Foraging motions often provide the most useful clues. For example, the two dowitcher species exhibit a distinctive sewing-machine-like probing behavior shared only by snipes, Dunlins, and a few other species. Most plovers exhibit a characteristic robinlike walk-stop-pluck feeding style. If you see this behavior from a distant bird, you may not know the species, but you will likely recognize it as a plover. In some cases, a bird's behavior may be enough to identify it. For example, when a Spotted Sandpiper takes off and flies low over the water to an adjacent shoreline, its vibrant, stiff-winged flight is virtually unique among shorebirds. Only the closely related Common Sandpiper, a vagrant to North America, exhibits a similar flight style. The two yellowlegs share many traits, but Greaters often exhibit a characteristic feeding style in which they dash about chasing small fish. Lessers stick to a more methodical feeding style.

Another characteristic that is closely related to both structure and feeding behavior is body stance. Some species, such as American Golden-Plover, tend to stand fairly upright, while others, such as Ruddy Turnstone, tend to look more hunched. Likewise some species, such as Solitary Sandpiper, tend to hold their bills nearly level, while others, such as Stilt Sandpiper, tend to hold their bills tilted sharply downward. On a more subtle level, Least Sandpipers often hold their bodies in a more crouched position than Semipalmated or Western Sandpipers. While these generalizations may not always work, they add to your mind's overall "typical" impression for each species and will usually lead you in the right direction.

**Impression of basic color patterns:** When viewing a bird at a distance where plumage details are not visible, you may still gain an impression of its overall color pattern. For example, a nonbreeding Dunlin looks dingy brownish gray on the head, breast, and back, with a contrasting white belly. By comparison, a nonbreeding Western Sandpiper looks paler and cleaner gray above, with a whiter breast and more white in the

**Sanderling, Sept., N.J.** Sanderlings are often "chased by waves" as they actively pick and shallowly probe in wet sand.

**Long-billed Dowitcher, Apr., Tex.** Dowitchers probe deeply and rapidly, "sewing-machine" style.

**Red Phalarope, Feb., N.J.** Phalaropes spin in the water, creating a vortex which draws nutrients to the surface.

throat and face. Sometimes these basic color patterns will serve only to narrow down your choices, but sometimes they will be sufficient to make an identification. For example, if you see a distant dowitcher in summer and it shows a distinct orange wash on the neck and a contrasting whitish belly, it will be a Short-billed. No plumage of Long-billed Dowitcher matches that pattern.

**Voice:** Most shorebirds are highly vocal. They give often elaborate flight songs on the breeding grounds, and most species also give distinctive flight calls and other vocalizations year round. With practice, you'll find that voice can be just as useful as appearance in shorebird identification. In some cases voice can be one of the best ways to distinguish similar species. Semipalmated and Common Ringed Plovers, for example, are often more easily distinguished by voice than by appearance. Also, when a shorebird is flying high overhead, it can often be more readily identified by its call than by its appearance. It should be noted, however, that vocalizations are variable and that some variations are less distinctive than others. Caution is always warranted when identifying a bird based solely on voice.

**Plumage details:** Using plumage details to identify shorebirds is standard practice among birders and is the method emphasized in most field guides. To use plumage details successfully, however, you need to have a good understanding of shorebird topography, molt patterns, and the effects of wear on feathers. When identifying a shorebird based on plumage details, it is essential to know, first, what plumage the bird is in, and second, what feather groups you are looking at. (See Shorebird Topography, Shorebird Molt, and Shorebird Aging below.)

**Lighting conditions:** It is important to bear in mind that lighting conditions may have a major influence on how a bird looks in the field. In everyday life we constantly interpret what we see based on how known-identity objects appear. For example, when we see a white building in bright midday sun, the sunlit area may appear white and the shaded area deep blue. On a cloudy day the same building may appear dull grayish white with indistinct shadows, while at sunrise or sunset it may appear deep pink or orange. Yet in all cases our minds automatically interpret the whole building as being white. While it is easy to understand that the brightly lit side of a bird will appear lighter and more washed-out than the side in shade, some effects of lighting are more subtle. For example, the apparent color of an object differs subtly depending on the color of its surroundings. The shaded side of that white building may appear tinted with green if next to trees or a lawn but tinted with red if next to a red building. Reflected light has the same effect on birds. Also, an object against a dark background appears lighter than the same object against a light background. It is second nature for people to make these interpretations of familiar objects that they see, and experienced birders do the same with birds.

**Variation:** Birds are variable—more so than any book will ever be able to portray. Some species have several morphologically distinct subspecies. Dunlin, for example, has 6 subspecies recognized worldwide. Many species also show some degree of sexual dimorphism. An extreme case is the Ruff, in which males are substantially larger and much more elaborately patterned than females. Other species, such as Black-necked Stilt, show only subtle sexual dimorphism, and some species show no dimorphism at all. All species show some degree of individual variation. A certain amount of individual variation is predictable, but there is always the possibility of seeing a shorebird that simply does not fit the mold. Although most aberrant birds should be identifiable based on a careful evaluation of size, structure, behavior, and voice, some birds are best left unidentified.

**Probability:** Before even seeing a shorebird, it is extremely helpful to know what species are likely to be present at a given location at a given time of year. Furthermore, it is helpful to know if that species is likely to be present in large numbers or more likely found singly among large flocks of other birds. For example, a large flock of small shorebirds on the Delaware Bay shoreline in December is likely to be either Sanderlings or Dunlins, whereas a similar flock at the same location in May probably includes some Sanderlings or Dunlins but may be mostly Semipalmated Sandpipers. Likewise, a flock of dowitchers on a lakeshore in Colorado in August is more likely to be Long-billeds, whereas a flock of dowitchers on a tidal flat in Maine on the same date is more likely to be Short-billeds. While probability should never be the sole basis for an identification, it should always be a consideration.

## SHOREBIRD TOPOGRAPHY

Plumage patterns almost always follow the contours of different feather groups. Although it is not necessary to know all the different feather groups in order to identify shorebirds, learning the major feather groups can significantly increase your understanding of why a bird looks the way it does. If you have trouble learning all the feather groups, you should at least try to learn the three most obvious ones on the upperparts: the scapulars, tertials, and wing coverts. These are the feathers that tend to have the most visible patterns and are the most important for identification. Also, it is extremely helpful to be able to distinguish among the tertials, primaries, and tail. The relative positions of these feathers may be very important for identifying some species.

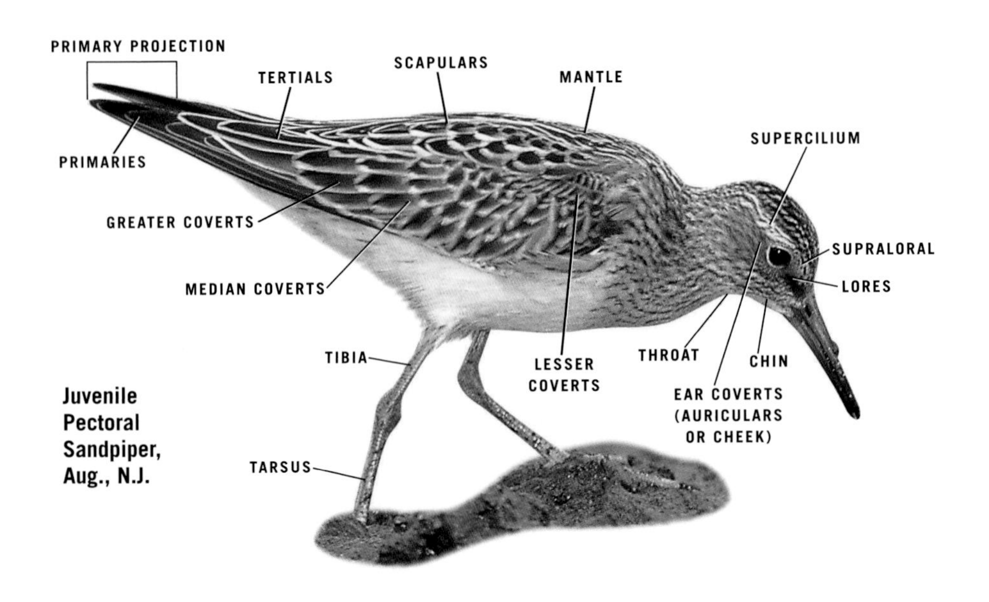


**Juvenile Pectoral Sandpiper, Aug., N.J.**

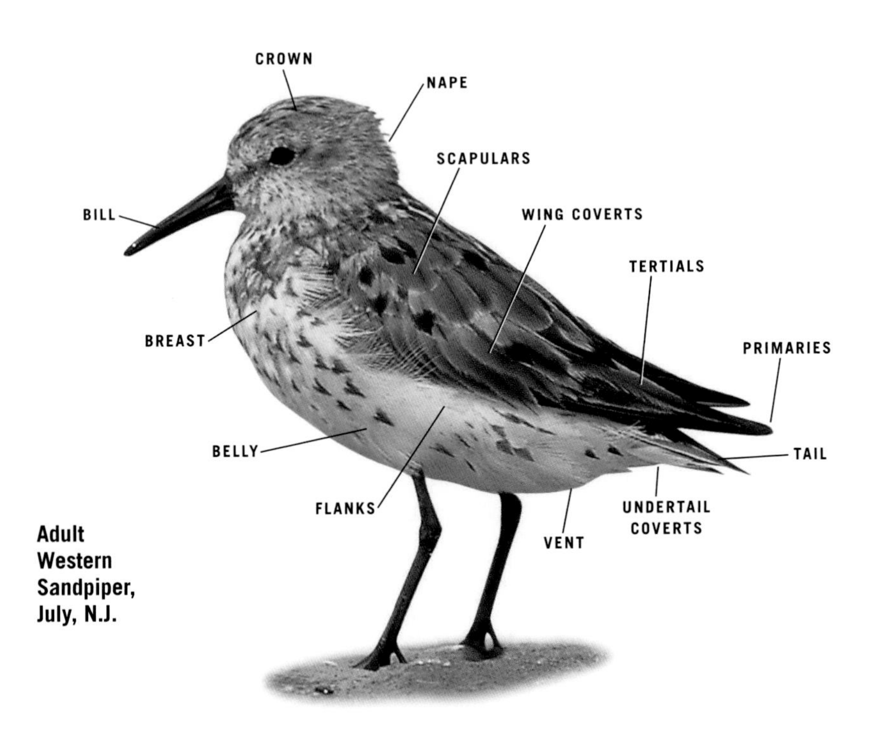


**Adult Western Sandpiper, July, N.J.**

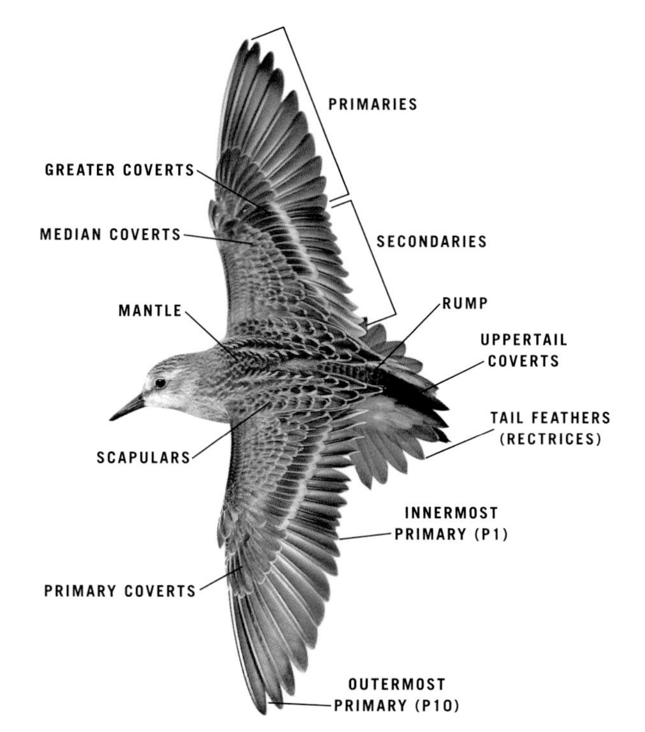


**Juvenile Semipalmated Sandpiper, Aug., N.Y.**

## SHOREBIRD MOLT

Over time, feathers wear out and need to be replaced. Molt is the cyclic replacement of these feathers. Most shorebirds go through two molts each year. After breeding they undergo a complete molt during which they replace all of their feathers. This molt produces nonbreeding (basic) plumage and is called the "prebasic molt." In spring they undergo another partial molt of primarily head and body feathers that produces breeding (alternate) plumage. This molt is termed the "prealternate molt."

Molt is a physically demanding process and occurs only in areas with rich food resources. It does not typically coincide with other physically demanding activities such as breeding and active migration. Although most shorebirds follow the same basic molt sequence, each species has evolved its own slightly different schedule depending on how far it migrates and where it stops to feed along the way. As a general rule, the longer-distance migrants do most of their molting near the wintering grounds, saving energy for migration. Shorter-distance migrants, on the other hand, often molt on or near the breeding grounds or during migration at favored stop-over sites before continuing on. Many species combine both strategies, beginning a molt on the breeding grounds, continuing it at a migratory stop-over site, and completing it on the wintering grounds.

The timing of a molt and the region in which a molt takes place can be very useful identification clues. It is particularly important to look for

**Molting adult Least Sandpiper, Aug., Calif.**
• Primary molt progresses in the order in which the primaries are numbered, from innermost (p1) to outermost (p10). Note the contrast between worn (paler and browner) outer primaries and fresh (darker, grayer) inner primaries. A similar contrast can be seen in the primary coverts, which are molted in the same sequence.

flight feather molt (molt of the primaries and secondaries), as this tends to be the most visible and least variable in terms of timing. Flight feather molt is very conspicuous. When a bird is undergoing flight feather molt, large gaps can be seen in the wings as it flies or stretches. Flight feather molt takes place in an orderly, predictable manner. It begins with the inner primaries and progresses outward. After the first few primaries have been replaced, the secondary molt begins and typically progresses in the reverse order, beginning with the outer secondaries. In some species the inner primaries are replaced while a bird is near the breeding grounds; the molt is then suspended so that the bird has a full set of flight feathers for migration. Such birds complete their molt on the wintering grounds. Region and timing of flight feather molt can be especially useful in distinguishing several similar species pairs such as American and Pacific Golden-Plovers, Greater and Lesser Yellowlegs, Long-billed and Short-billed Dowitchers, and Semipalmated and Western Sandpipers.

It should be noted that molt timing does vary within a species. For example, failed breeders typically begin to molt earlier than successful breeders. Nonbreeding subadults begin their molt even earlier, sometimes in a seamless continuation of the previous molt. Also, different populations may be on different schedules depending on where they breed, winter, and stop over during migration. For example, a Least Sandpiper undergoing primary molt in August on the Atlantic Coast is probably going to winter nearby, whereas one showing no primary molt at the same location and date may be headed to South American wintering grounds. It should also be noted that occasionally there will be aberrant birds that, for one reason or another, are just off schedule.

## SHOREBIRD AGING

When making a plumage-based identification of a shorebird, it is usually necessary to first determine what plumage the bird is in. This requires an understanding of basic molt patterns of a shorebird and a vocabulary to describe these patterns.

There is some confusion surrounding age and molt terminology, in part because several different systems are presently in use. In this guide, we use what is known as the *life-year system.* This system describes the appearance of a bird at different seasons or life stages but does not necessarily correspond to when the bird molts in a new set of feathers. In other words, it refers to what a bird looks like in the summer (breeding season) or winter (nonbreeding season). A bird in its first year of life, from hatching to when it is one year old, is called a *first year* or *first life year.* During this first year of a shorebird's life, however, there are usually three distinct stages (generally corresponding to distinct plumages)—*juvenile, first winter,* and *first summer*—and we use these terms in this book. In some species there are recognizable *second winter* or *second summer* stages (plumages). In most species, however, after the first life year, the only recognizable stages (plumages) are *breeding* and *nonbreeding.* (Note that seasonal references used in the life-year system refer to Northern Hemisphere seasons.)

Another system, known as the *Humphrey-Parkes system,* is the one used to describe plumages based on the molt by which they were acquired, with no reference to time of year. For example, all birds go through one complete molt (a molt during which all feathers are replaced) each year after the breeding season. This molt is called the *prebasic molt* and results in *basic plumage.* Some birds, including shorebirds, undergo a second partial molt (typically including a variable number of head and body feathers) in spring, the *prealternate molt,* which results in *alternate plumage.* Basic and alternate plumages are analogous to nonbreeding and breeding stages (plumages) in life-year terms. Other terms are used to describe immature plumages.

In this guide we use the life-year system for two reasons. First, the life-year system is familiar to most of us and therefore convenient to use. Second, in the case of shorebirds, it usually corresponds well to molt timing. In the aging summary below, we use the life-year term first, then the Humphrey-Parkes term in parentheses for reference.

One of the key elements in aging a shorebird (or any bird) is looking for contrasts in the degree of feather wear. When feathers wear, they become paler and usually browner overall and are typically frayed and chipped along the edges (with the stronger shaft often remaining as a

**Juvenile Western Sandpiper, Aug., N.J.**
uniformly fresh plumage with relatively small scapulars and wing coverts

**Molting juvenile Western Sandpiper, Sept., N.J.**
a few replaced scapulars look larger and plain gray next to juvenal feathers

**First-winter Western Sandpiper, Sept., N.J.**
worn juvenal coverts and tertials look brownish-tinged and contrast with fresh scapulars and mantle

**First-summer Western Sandpiper, June, N.J.**
a few "breeding-type" feathers have molted in, but most have a nonbreeding appearance

**Nonbreeding Western Sandpiper, Feb., Calif.**
nonbreeding birds show no contrast between coverts and scapulars

**Molting adult Western Sandpiper, Apr., Tex.**
in early spring, the first few signs of breeding plumage appear

**Breeding Western Sandpiper, May, Alaska**
full breeding plumage is bright and striking, but worn coverts are retained from winter

**Molting adult Western Sandpiper, Aug., N.J.**
molting birds in late summer often look shabby; note fresh nonbreeding scapulars and worn wings

**Juvenile Greater Yellowlegs, Aug., N.J.**
uniformly fresh plumage with relatively small scapulars and coverts

**First-winter Greater Yellowlegs, Nov., N.J.**
worn juvenal coverts and tertials look brownish-tinged and contrast with fresh scapulars and mantle

**First-summer (left) and breeding adult Greater Yellowlegs, Apr., Tex.**
many one-year-old birds remain in mostly nonbreeding plumage through the summer; these birds often remain on the wintering grounds or at stop-over sites through the breeding season

**Nonbreeding Greater Yellowlegs, Nov., Tex.**
nonbreeding birds show no contrast between coverts and scapulars

**Molting adult Greater Yellowlegs, Apr., N.J.**
in early spring, the first few signs of breeding plumage appear

**Breeding Greater Yellowlegs, May, N.J.**
full breeding birds are heavily patterned throughout upperparts and breast

**Molting adult Greater Yellowlegs, Aug., N.J.**
molting birds in late summer often look shabby; note fresh nonbreeding scapulars

pointed feather tip). Worn feathers often look more drooped in appearance as they become more loosely attached before dropping. The contrast between a worn feather and a fresh, crisp-edged feather can be striking. One thing to keep in mind is that patterned feathers wear less evenly than unpatterned feathers because dark-pigmented areas are more resistant to wear than light-pigmented areas. For example, a dark feather with white notches along the edge will wear down to a dark feather with a saw-toothed worn edge. Likewise, a dark feather with a prominent white tip when fresh will wear down to an all-dark feather. This is important to keep in mind because wear can significantly change the overall appearance of a bird. For example, an adult Least Sandpiper in April or May looks brown above with black spots. This is due to the pattern on the scapulars, which are blackish with brown edges and grayish tips. By July, however, when these feathers are worn, the grayish tips have disappeared and the brown edges become narrower, making the upperparts as a whole look blackish brown. The more one understands how feather wear changes the appearance of a bird, the easier it will be to interpret much of the variation seen in the field.

**Juvenile (first basic plumage):** When a shorebird fledges, it is wearing its juvenal plumage and is referred to as a juvenile. Because all these feathers grow in at the same time (the only time in a shorebird's life that this happens), the uniform condition of this plumage is distinctive. Juvenal plumage is characterized by smaller scapulars and wing coverts than those of subsequent plumages. These feathers are also weaker than all subsequent feathers, so they wear more quickly. Although exact feather patterns vary widely from one species to the next, most juvenal scapulars and coverts show some type of bold pattern, often including contrasting pale fringes or pale notches along the feather's edge. The combination of uniform condition, small feathers, and pale fringes frequently gives a juvenile shorebird a scaly overall appearance. In late summer, when the first southbound juveniles appear in their pristine, crisply patterned plumage with each feather in perfect arrangement, they often look strikingly different from the worn, often molting adults. Later in the season, as adults molt into fresh nonbreeding plumage and juveniles become more worn, the small feather size, uniformity of plumage condition, and juvenal feather patterning become the most useful aging criteria. How long a shorebird holds its juvenal plumage varies from species to species. Some species, such as Dunlin, begin to molt out of juvenal plumage while still near the breeding grounds, while others, such as Baird's Sandpiper, hold virtually all of their juvenal feathers until they arrive at their wintering quarters in South America. It may be helpful to note that juveniles tend to migrate as much as a month later than adults in the fall, so in any given region the first juveniles typically arrive when adults are peaking in numbers.

**First winter (formative plumage; acquired by the preformative or post-juvenal molt):** At some point during fall and winter, juvenile shorebirds will molt into first-winter plumage. This molt typically involves replacement of most head and body feathers (though lower back and rump feathers are often retained), some wing coverts and tail feathers, and, for some species, a variable number of primaries (usually just outers). Head and body plumage is usually molted by mid- to late fall, with primaries molted later. The newly molted-in feathers are essentially the same as those of adult nonbreeding plumage, so most first-winter birds closely resemble nonbreeding adults. For most species, many juvenal wing coverts are retained, resulting in a usually subtle contrast between fresh, more plainly patterned scapulars and worn, juvenal-patterned wing coverts. For some species, such as Willet, that replace most of the wing coverts as well, it is often nearly impossible to distinguish a first-winter from a nonbreeding adult. In other species, such as Mountain Plover, full juvenal plumage is retained through the first winter, so there is no first-winter plumage.

**First summer (first alternate plumage; acquired by the first prealternate molt):** This is a highly variable plumage, acquired through a partial molt (late winter to spring) of a variable number of head and body feathers. With smaller shorebirds, this plumage may be essentially the same as either breeding or nonbreeding adults or anywhere in between. With larger shorebirds, first-summer plumage is typically closer to that of nonbreeding adults. First-summer birds are often recogniz-

**First-summer Curlew Sandpiper, Aug., N.J.** Primary molt during the first winter of a shorebird's life varies greatly between species and individuals. Many molt no primaries at all during their first winter, while a few molt all primaries. A third pattern is for only the outer primaries to be molted, as occurred on this one-year-old Curlew Sandpiper. Adults never show this pattern. Worn inner primaries are pale and tattered with yellowed shafts compared to relatively fresh outers.

able by a variable number of extremely worn wing coverts and flight feathers retained from juvenal plumage. Where these birds spend their first summer is also variable. Some stay on the wintering grounds, while others migrate partway or all the way to the breeding grounds. As a rule, the farther north a bird migrates, the more adultlike its plumage will be because of increased sexual hormone levels.

**Second winter (second basic plumage; acquired by a prebasic molt):** This plumage is acquired through a complete molt after the breeding season (early summer to early winter). This molt typically takes place a few weeks earlier than in adult birds. Second-winter plumage is essentially indistinguishable from adult nonbreeding plumage.

**Second summer (second alternate plumage; acquired by a prealternate molt):** This is a variable plumage, acquired through a partial molt (late winter to spring) of a variable number of head and body feathers. With smaller shorebirds, this plumage is essentially indistinguishable from adult breeding plumage. With larger shorebirds, second-summer plumage is typically intermediate between adult breeding and nonbreeding plumages. Although many birds in their second summer go to the breeding grounds to mate, others migrate only partway north.

**Adult nonbreeding (definitive basic plumage; acquired by a prebasic molt):** This plumage is acquired through a complete molt after the breeding season (midsummer to midwinter). In most species, nonbreeding plumage is plainer and paler than breeding plumage, though some species show little or no change in appearance.

**Adult breeding (definitive alternate plumage; acquired by a prealternate molt):** This plumage is acquired through a partial molt (late winter to spring) of most or all head and body feathers and a variable number of wing coverts and tertials. In most species, breeding plumage is characterized by relatively boldly patterned, colorful head and body plumage, while at least some wing coverts are contrastingly dull and worn. In some species there is little difference between breeding and nonbreeding plumages.

**Transition:** This is a generic term for the transition between any two plumages. Birds in transitional plumages may show any combination of old, worn feathers from one plumage and new, fresh feathers from the next. It is in these transitional plumages that molt patterns become useful for identification.

# How to Use This Guide

## SPECIES PHOTOS

In this guide we present an approach to bird identification quite different from that found in other books. Rather than relying solely on stunning portraits, we present a more real-life image of each species, including distant birds, mixed-species flocks, and varied lighting. Along with full-frame portraits, each gallery should convey a much more realistic impression of each species and place it in context with the real world. After all, this is how we see birds in the field.

All identification information is covered in the photographic section. Each species account begins with a brief text describing fundamentals of size, structure, and behavior—characteristics that apply to all individuals regardless of age or plumage. *This critical information should be considered when viewing all photos.* Status is also briefly described, along with a range map (see below) to give a sense of where and when each species is likely to occur. Definitions of a variety of terms used throughout the book may be found in the glossary.

The gallery of photos was carefully selected. The first photo is simply an impression photo. It offers a distant view, often of a flock, designed to portray a typical, real-life field impression of the species, ideally including preferred habitat and representative feeding behaviors. No comments accompany this photo, but information on date and location may be found in the appendix.

Subsequent photos progress chronologically through the life cycle of the bird, beginning with juvenal plumage and continuing through subadult to adult appearances. These comprehensive layouts also include transitional plumages and the progressive stages of plumage wear. Group photos are interspersed throughout and often contain similar species. These photos are particularly important because they depict variation in plumage and posture within a species and allow for direct comparison with other shorebirds on the same date and under the same conditions. Such comparisons help form the basis for the evaluation of relative size and structure that is so crucial to the identification process.

Each photo (except the impression photo) is accompanied by a caption highlighting selected identification or aging criteria. Fundamental plumage characters that apply to all ages are mentioned in the juvenile caption, thus covering all the basics on the first page. Subsequent pages reinforce some of these points and highlight other details pertinent to each photo, including comparisons with other species. With photos of more distant birds, general patterns visible at long range are noted.

Flight photos are situated at the end of each account and are accompanied by descriptions of the most common flight calls (vocalizations are described more thoroughly in the text section of the book). Many photo captions are intended as quizzes, so are posed as questions. Answers to these quiz photos may be found in the appendix. Full-page photos have no captions in the photographic section, but brief captions are located in the appendix as well.

The Species Photos section is divided into two parts. The first covers domestic species—those occurring commonly in the Lower 48 states. The second depicts vagrants and regional specialties. The latter group includes a few species that breed regularly in Alaska or northern Canada but migrate and winter primarily in the Old World. Vagrants receive more limited coverage than domestic species but are covered in a similar format, minus the range map and the impression photo.

To make full use of the photographic portion of this book, it is important for the reader to be familiar with information provided in the introduction. In particular, the reader should have a basic familiarity with the different families of shorebirds. This will help in finding the right section of the book when trying to identify an unfamiliar shorebird. Also, when trying to make a plumage-based identification, an understanding of shorebird topography and plumages is critical.

### Range Maps

The maps show breeding range (red), winter range (blue), and migratory range (gray). Within this color scheme, a darker shade indicates where the species is more common and a lighter shade where it is less common.

## SPECIES ACCOUNTS

The text portion of this book provides a concise summary of status, taxonomy, behavior, migration, molt, and vocalizations. The text in this guide is designed to be supplementary information to assist with various aspects of field identification. As in the Species Photos, vagrants and regional specialties receive more limited treatment than domestic species but are covered in the same format.

### Status

A brief description of North American and worldwide distribution and abundance is given, including, with many species, a population estimate.

### Taxonomy

If any subspecies are recognized, there is a brief discussion of breeding and (when known) nonbreeding distribution. If not covered in the photographic section, field identification of subspecies is briefly discussed.

### Behavior

In this section there are brief descriptions of preferred habitat at different seasons, typical feeding style, primary foods taken, flocking behavior, and other habits pertinent to identification. There is also a synopsis of breeding behavior including egg dates, incubation time, fledging period, and a description of courtship displays.

### Migration

Timing and routes of migration are described for both spring and fall. When known, the differential timing and distribution of adult and juvenile migrations are noted.

### Molt

Timing and extent of molt is described separately for adults and for birds in their first life year (and second life year in a few cases). In our molt discussions, we have purposefully not used Humphrey-Parkes molt terminology in order to avoid inevitable confusion between the two different systems of plumage terminology. Instead, we simply describe which feathers are replaced and when, without actually naming the molt.

### Vocalizations

A brief description and phonetic representation are given of the songs of domestic species and the primary calls of all species. Note, however, that there may be many variations to each call type as well as "hybrid" calls or other odd vocalizations. It is not our intent to portray the full repertoire of each species. Rather, we wish to highlight the most commonly heard vocalizations—those most useful for identification.

Although voice descriptions are frequently awkward and of limited value, we hope that they will steer the listener in the right direction and enable easier recall of vocalizations once heard. Each letter in a phonetic representation is intended to describe a specific sound that would be made by speaking the English language. In general, consonants indicate how hard or soft a certain part of the sound is, and vowels denote the relative pitch or change in pitch. Longer sounds are indicated by stringing several vowels together, and various combinations of letters may be used to convey different overtones (again, use the English language as a guide). Distinctly higher volume or a more emphatic section of a song or call is indicated by capital letters.

# Species Photos

# Domestic Species

# Black-bellied Plover

***Pluvialis squatarola*** p. 320

**Size:** **L 10¾–12 in. (27–30 cm); WS 29–32 in. (73–81 cm); WT 105–320 g**
**about the size of Greater Yellowlegs; slightly larger than American Golden-Plover**

**Structure:** **compared to American Golden-Plover, chunkier and less attenuated with thicker neck and chest, larger head, heavier bill; primaries project slightly past tail tip**

**Behavior:** **run-stop-pluck feeding style; beaches, mudflats, plowed fields; nests on tundra**

**Status:** **common in coastal areas; scarcer inland than American Golden-Plover**

**1 Juvenile with Dunlin, Oct., N.J.**
- juvenile and non-breeding show bland face-pattern with prominent dark eye; white belly and under-tail coverts
- aged by streaked breast, crisply check-ered upperparts with dark-centered feathers; fresh juveniles may be evenly gold-washed
- may sometimes retain full juvenal plumage until spring

**2 Nonbreeding, Jan., Fla.**
• breast more smudged and upperparts more softly patterned than juvenile (with pale-centered feathers and dark central shaft)
• primary molt is still under way: note fresh (blackish) inner primaries contrasting with worn (brown) outers

**3 Juvenile, nonbreeding, and molting adult with other species, Oct., N.J.**
• Which is which, and what are the most striking differences?
• What other species are visible in this photo? Focus on size and structure.

**4 Juvenile (left) with American Golden-Plover and Killdeer, Oct., N.J.**
• compared to golden-plover, note bulkier appearance, larger head and bill, thicker neck, blander face-pattern, whiter belly and undertail coverts

**6 Breeding female (right) and males with first summer (second from left) and Red Knot, May, N.J.**
• males show whiter crown; paler, grayer back than females
• first summer resembles nonbreeding adult (but shows worn retained juvenal wing coverts)

**7 Breeding male, July, Alaska**
• mostly white rump and tail in all plumages
• raised wing and spread tail are part of nest distraction displays

**8 Molting adults and first summers with American Oystercatcher (2), Western Willet (2), and Ruddy Turnstone (4), Aug., N.J.**
• late-summer flocks often show a wide range of plumages; spring birds have a similar appearance
• contrast between white vent and dark belly of breeding adult is striking even at long range

**9 Juveniles, Dec., N.J.**
- all plumages show black axillaries ("wing-pits"); mostly white rump and tail; white wing stripe
- flight call is a far-carrying whistle, *plEE-uu-ee*

**10 Jan., Fla.**
- looking at silhouettes is a great way to learn structure

**11 Breeding adults with Short-billed Dowitcher, Dunlin, Ruddy Turnstone, Red Knot, and Semipalmated Sandpiper, May, N.J.**
- stocky body and long pointed wings; black belly contrasting with white underwing; undertail coverts and tail visible from long range
- Can you identify all the birds?

# American Golden-Plover

***Pluvialis dominica*** **p. 323**

**Size: L 9½–11¼ in. (24–28 cm); WS 23½–28¾ in. (59–72 cm); WT 100–200 g**
**slightly smaller than Black-bellied Plover**

**Structure: compared to Black-bellied, slimmer and more attenuated with smaller head and bill, slimmer neck and chest, longer legs; usually four primaries project past tertials**

**Behavior: run-stop-pluck feeding style; pastures, plowed fields, mudflats; nests on tundra**

**Status: locally common in Great Plains and Texas coast in spring; uncommon in Northeast in fall; rare in much of West; much scarcer along coast than Black-bellied Plover**

**1 Juvenile, Sept., N.J.**
- in all plumages looks slim and attenuated with small dovelike head; long primaries (well past tail tip)
- juvenile and non-breeding show bold supercilium set off by dark cap, thin dark eye-line; dirty under-parts
- aged by uniformly fresh plumage with neatly mottled breast and crisply checkered upperparts
- brightest juveniles may be more gold-washed; brightest on crown, mantle, and rump (but not on face, neck, or breast as in Pacific Golden-Plover)

**2 Nonbreeding, Mar., Tex.**
• from juvenile by more smudged breast and more softly patterned upperparts with paler feather centers
• spring migrants hold mostly nonbreeding plumage until late April, then molt very quickly to breeding plumage

**3 Nonbreeding with Black-bellied Plovers and Laughing Gull, Apr., Tex.**
• compared to Black-bellied, note smaller size and slimmer, more attenuated shape; smaller head and bill; darker and browner overall with duller underparts, darker cap, bolder supercilium
• a few dark spots are beginning to appear on the underparts

**4 Breeding male, June, Manitoba**
• contrasting dark cap; wide white patch at breast-sides
• plain wings common to all plumages
• adults perform distraction display when threat is sensed near their nest site

**6 Breeding male, June, Alaska**
• dark overall; broad white patch at breast-side; underparts black including flanks and undertail coverts
• moist grassy tundra is preferred nest habitat

**7 Breeding female, June, Alaska**
• female duller than male, often with pale mottling on face, flanks, and undertail coverts

**8 Molting adult, late Aug., N.J.**
• molting adults often retain extensive dark smudge behind eye, setting off whitish forehead and supercilium

**9 Molting adults with Black-bellied Plover and Buff-breasted Sandpipers, Sept., N.J.**
• slimmer, smaller-headed, and browner than Black-bellied with dark cap set off by whitish supercilium
• Which one is the Black-bellied Plover?

**10 Molting adult (center) with Black-bellied Plover and Ruddy Turnstone, Aug., N.Y.**
• smaller head and bill, slimmer chest than Black-bellied; darker overall with sharply defined cap and supercilium
• this Black-bellied is particularly small

**11 Breeding (top center), with Black-bellied Plovers, late Aug., N.J.**
• slimmer, browner than Black-bellied with all-dark rump and tail, less conspicuous wing stripe

**12 Molting adult, Aug., N.J.**
• gray-brown underwing; indistinct wing stripe; slight toe projection
• flight call is an urgent whistled *pleE-dle;* much higher than Black-bellied

# Pacific Golden-Plover

***Pluvialis fulva*** **p. 325**

Size: **L 9¼–10½ in. (23–26 cm); WS 21½–24½ in. (54–61 cm); WT 100–200 g same as American Golden-Plover**

Structure: **recalls Black-bellied Plover; compared to American Golden-Plover, more front-heavy and upright with rounder body, larger head, heavier chest, often larger bill and longer legs, shorter wings; typically shows three primaries past tertials (variable)**

Behavior: **run-stop-pluck feeding style; lawns, pastures, grassy headlands, rocky shores, beaches; nests on tundra**

Status: **uncommon during migration and winter along Pacific Coast; rare elsewhere**

**1 Juvenile, Sept., Japan**

- in all plumages note Black-bellied Plover-like structure; typically shows three primaries past tertials (variable)
- juvenile and nonbreeding show dark eye isolated on pale face
- aged by uniformly fresh plumage with neatly mottled breast and crisply checkered upperparts
- juvenile with larger gold spots above; sparser, more contrasting markings on head and underparts than juvenile American; usually bright gold overall including face, nape, and breast; duller birds brightest on head; often pale, pink-tinged legs and bill-base

**2 Juveniles with juvenile American (right), Sept., Alaska**
• paler than American with gold-washed head and neck, more contrastingly streaked breast; supercilium less contrasting, palest above (not behind) eye

**3 Nonbreeding, Jan., Sri Lanka**
• gold-washed face; scattered bright gold markings above; pale face with isolated dark eye and cheek-spot
• long tertials almost reach tail tip (variable); short primary projection
• bill size variable, but this bird has distinctively large bill
• differs from juvenile by more smudged breast, less neatly checkered upperparts

**4 Nonbreeding (left and center right) with Black-bellied Plovers, Feb., Calif.**
• smaller and brighter than Black-bellied with relatively smaller head and bill, longer legs, darker cap; gold-washed head and upperparts
• drooping supercilium creates broader dark nape than American Golden-Plover when viewed from behind

**5 Nonbreeding, Sept., Hawaii**
- supercilium brightest above eye, drooping behind; face washed with yellow; large gold spots on crown and upperparts
- often stops with foot lifted up

**6 Breeding male, June, Alaska**
- even-width white stripe along neck and flanks; scapulars often more golden than coverts
- male more solidly black below than female

**7 Molting adult with Greater Sand-Plover, Curlew, and Broad-billed Sandpipers, Apr., Hong Kong**
- even-width white stripe along neck and flanks, extending to undertail coverts; large gold spots on upperparts including crown
- heavier-chested, less attenuated, and more upright than American Golden-Plover

**8 Breeding, Mar., Hawaii**
- even-width white stripe along neck and flanks with minimal widening at breast-sides
- supercilium droops behind eye, resulting in relatively broad dark nape line
- relatively large gold spots on upperparts, including crown
- short wings and heavy chest

**9 Molting adult, Apr., Japan**
- golden head and upperparts with usually larger spots than American; "ghost" of white flank stripe; drooping supercilium results in broader dark nape
- less attenuated, more front-heavy, often more upright than American; long tertials almost reach tail tip; short primary projection

**10 Molting adult American Golden-Plover, Sept., N.J.**
- duller, darker overall than Pacific with usually smaller spots; "ghost" of white breast patches; flared supercilium results in narrower dark nape
- more attenuated, less upright than Pacific; short tertials fall well short of tail tip; longer primary projection

**11 Nonbreeding, Feb., Calif.**
- all-dark rump and tail; plain wings with indistinct wing stripe
- heavier chest than American; pale face with isolated dark eye; often longer toe projection
- flight call is a Semipalmated Plover–like *ku-EEid;* lower pitched than American

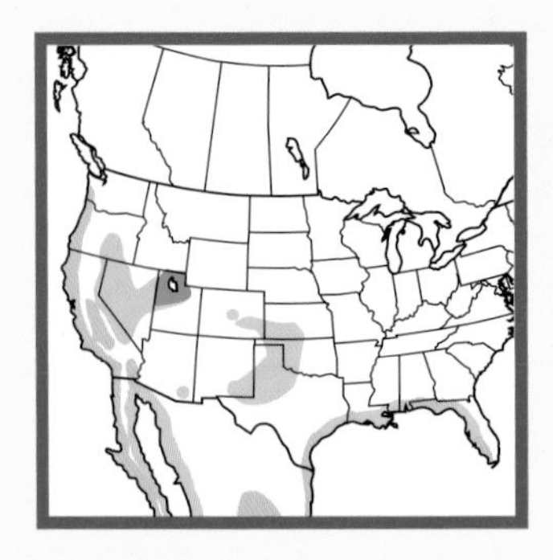

# Snowy Plover

***Charadrius alexandrinus*** **p. 330**

**Size:** **L 6–6¾ in. (15–17 cm); WS 16½–18½ in. (41–46 cm); WT 34–58 g**
**slightly smaller than Semipalmated or Piping Plovers**

**Structure:** **rounded and front-heavy; compared to Piping looks larger-headed and less attenuated with slimmer bill, longer legs**

**Behavior:** **run-stop-pluck feeding style; runs faster and farther than other plovers; beaches, salt flats**

**Status:** **uncommon and local; coastal populations threatened**

**1 Juvenile/first winter, early Nov., Calif.**

- all plumages show grayish (often pinkish-tinged) legs, slim black bill, breast-side patches; usually slightly darker upperparts and darker cheeks than Piping Plover
- most head and body feathers (including scapulars) have been replaced; aged with close view by retained juvenal coverts and tertials, which are slightly brownish-tinged and have well-defined buff fringes; coverts contrast subtly with grayer, gray-fringed scapulars

**2 Nonbreeding male, Feb., Calif.**
- slim, dark, pointed bill always distinctive
- breast-band always incomplete
- all non-juveniles show unpatterned upperparts, but difficult to distinguish adult from first winter
- males in winter often show darker cheeks and breast-patches but sexes often similar
- western *C. a. nivosus* shows variably dark loral stripe and bit of dark feathering at bill-base

**3 Nonbreeding, Dec., Fla.**
- *C. a. tenuirostris* of e. Gulf Coast averages paler upperparts; usually clean whitish lores
- colored leg-bands help with population studies

**4 Nonbreeding, Feb., Calif.**
- western *nivosus* averages darker upperparts; variably dark lores (more extensive in nonbreeding)

**5 Nonbreeding (left) with Sanderling (right) and Western Sandpiper, Feb., Calif.**
- dark cheeks often prominent at long range
- roosting birds often hide behind tide-strewn debris

**6 Breeding male (left) and female, Apr., Tex.**
• breeding male shows blacker head and breast markings; female duller but variable
• many individuals show a pale cinnamon wash on crown and nape

**7 Feb., Calif.**
• potbellied and long-legged with stubby rear end, slim bill

**8 Nonbreeding with Least Sandpiper (lower right), Feb., Calif.**
• front-heavy look; dark tail with white edges; thin, contrasting wing stripe
• flight call is a low, burry *drrrp*

**9 Breeding male, Apr., Tex.**
• in courtship flight, gives a drawn-out, rising *toorrEET* call

# Wilson's Plover

***Charadrius wilsonia*** **p. 331**

**Size: L 6½–8 in. (16–20 cm); WS 15½–19½ in. (39–49 cm); WT 55–70 g**
**slightly larger than Semipalmated Plover; smaller than Killdeer**

**Structure: front-heavy with large head and bill; thick neck**

**Behavior: run-stop-scan feeding style; often runs with head low; upright stance when alarmed; sandflats and beaches, especially near vegetation or debris**

**Status: fairly common but strictly coastal; rare stray north to New England and s. California**

**1 First winter, July, Fla.**
• all plumages show long, heavy, all-black bill and dull fleshy-pink legs
• pale, neatly fringed median coverts are characteristic of juvenal plumage; many scapulars have been replaced by grayer nonbreeding feathers
• Fiddler Crabs provide a major food source

**2 Nonbreeding (right) with Semipalmated Plover, Jan., Fla.**
• brown face-pattern and breast-band like juvenile and breeding female; breast-band may be broken or complete
• What differences do you see from Semipalmated?

**3 Molting male, Dec., Fla.**
• long heavy black bill always distinctive
• the black face- and breast-pattern of breeding plumage is already well developed at this early date

**4 Nonbreeding (right foreground) with Piping (left) and Black-bellied Plovers and Laughing Gull, Jan., Fla.**
• size, structure, and head-pattern are most useful characters at long range
• Wilson's often roosts next to driftwood or other debris

**5 Breeding male at nest, Apr., Tex.**
• darker lores and fore-crown than female; always a complete black breast-band

**6 Nonbreeding *C. w. beldingi* (left) with Semipalmated (center) and Collared Plovers, Jan., Puerto Vallarta, Mex.**
• Pacific Coast *beldingi* is smaller and darker (same shade as Semipalmated) with broader dark lores and shorter, narrower supercilium

**7 Female Wilson's Plover, May, Fla.**
• short "sandplover" wing stripe on inner primaries
• flight calls include a short *pip* or *pi-dit* and a sharp rising *pweet*

# Semipalmated Plover

***Charadrius semipalmatus*** **p. 334**

**Size:** **L 6¾–7¼ in. (17–19 cm); WS 17¼–20¾ in. (43–52 cm); WT 31–69 g**
**intermediate between "peep" and Sanderling**

**Structure:** **chunky but attenuated; rounded head, short thick neck, stubby bill**

**Behavior:** **run-stop-pluck feeding style; forages in scattered flocks; beaches, mudflats, plowed fields; nests on tundra**

**Status:** **common in coastal areas; locally common inland**

**1 Juvenile, late July, N.Y.**
• all plumages show orange-yellow legs, stubby bill, dark breast-band, and dark brown upperparts
• aged by uniformly fresh plumage with neat pale fringing and dark subterminal lines on upperparts; pale fringing often visible on crown
• "foot-trembling" feeding method evident here
• named for partial webbing between middle and outer toes, visible in the field from close range

**2 Juvenile, Aug., N.Y.**
• bill mostly dark with limited orange base in juvenile and nonbreeding
• pale-fringed crown visible at close range

**3 Juvenile (left) and breeding adult, late July, N.Y.**
• frequent squabbles while feeding
• juvenile aged by neat pale fringing on upperparts

**4 First winter, Feb., Calif.**
• aged by subtly pale-fringed coverts contrasting with darker, unpatterned scapulars

**5 Nonbreeding, Dec., Fla.**
• differs from juvenile by uniformly unpatterned upperparts
• breast-band sometimes incomplete in nonbreeding and juvenile

**6 Breeding male on nest with chicks, early July, Alaska**
• black mask and breast-band; extensive orange bill-base indicate breeding plumage
• male shows more solidly black facial markings than female
• often nests in open, barren habitats, using muted plumage as camouflage

**7 Breeding, Aug., N.J.**
• some birds (for example, second from right) show less black on face; these may be females or first-summer males
• much variation in breast-band width depending on pose and individual variation

**8 Breeding, with Sanderling (sleeping), Western (left), and Semipalmated Sandpipers, Aug., N.J.**

**9 With Piping Plover, Sept., N.J.**
• Atlantic Coast Piping is much paler; all have larger heads, thicker necks, less attenuated shape than Semipalmated
• What age is the Semipalmated?

**10 Breeding, Aug., N.J.**
• long wings; fairly prominent wing stripe; dark tail tip
• flight call is a strong, whistled *chu-EEp*

**11 Breeding, Aug., Calif.**

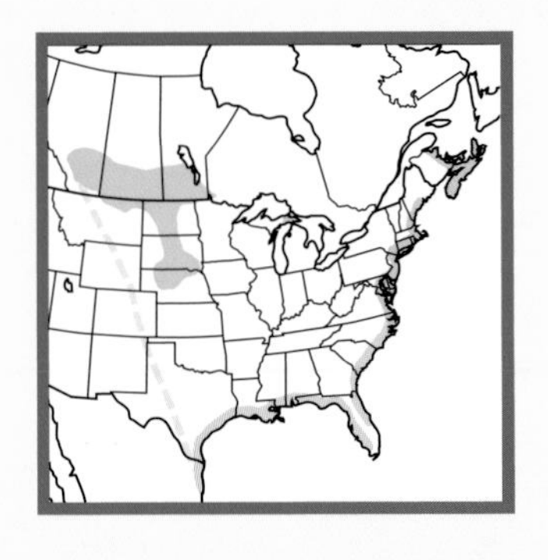

# Piping Plover

***Charadrius melodus*** **p. 335**

**Size:** **L 6¾–7¼ in. (17–18 cm); WS 18–18¾ in. (45–47 cm); WT 43–63 g**
**intermediate between "peep" and Sanderling**

**Structure:** **chunky with large rounded head, short thick neck; stubby bill**

**Behavior:** **run-stop-pluck feeding style; sandy beaches, often in high, dry sections away from water**

**Status:** **uncommon and local; globally threatened and endangered**

**1 Juvenile, Aug., N.J.**
- all plumages show orange-yellow legs, stubby bill, pale upperparts, and pale face with prominent dark eye
- juvenile and nonbreeding with mostly dark bill, partial collar, no black in plumage
- aged by uniformly fresh upperparts with subtle pale fringing (often very difficult to see)

**2 Nonbreeding and molting adult, Jan., Fla.**
• pale face with prominent dark eye visible at long range
• adult and first winter difficult to distinguish
• single bird showing partial black collar is an Atlantic Coast breeder *(C. m. melodus),* which molts and nests earlier than interior birds; note slightly paler upperparts, paler lores

**3 Breeding male on nest, June, N.J.**
• black forehead and breast-band, extensive orange bill-base indicate breeding plumage; extent of black collar variable
• male shows blacker markings than female
• predator-exclosure cages are erected around most nests of this closely monitored species

**4 Breeding male with chicks, July, N.J.**

**5 Breeding pair *(melodus),* June, N.J.**
• Atlantic Coast *melodus* paler overall with less contrasting gray cheeks, pale lores, more limited black; breast-band may be partial or complete
• male shows blacker markings than female

**6 Breeding and presumed first summer *C. m. circumcinctus,* Apr., Tex.**
• Interior *circumcinctus* darker overall with more contrasting gray cheeks and lores, more extensive black markings; often shows flecks of black above eye and at bill-base; breast-band often complete
• first summer lacks black in spring

**7 Nonbreeding with Semipalmated Plovers and Least Sandpiper (bottom left), Jan., Fla.**
• How can you rule out Snowy Plover?

**8 Nonbreeding (center) with Semipalmated Plovers, Jan., Fla.**
• flight call is a soft, whistled *peep* or *pee-werp*

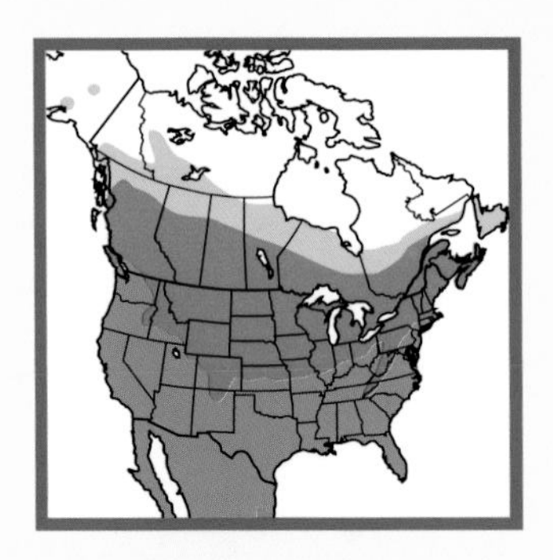

# Killdeer

***Charadrius vociferus*** **p. 338**

**Size:** L 9¼–10½ in. (23–26 cm); WS 23½–25¼ in. (59–63 cm); WT 65–90 g
slightly larger than American Robin; much larger than Semipalmated Plover

**Structure:** slim, long-tailed, small-headed, small-billed

**Behavior:** run-stop-pluck feeding style; lawns, fields, mudflats; nests in gravel or sparse vegetation, often close to human activity

**Status:** common and familiar virtually throughout the continent

**1 Young juvenile, July, N.J.**

- all plumages show dark brown upperparts, pale pinkish legs, slim black bill, and double breast-band (but note, downy young has *single* breast-band)
- aged by uniformly fresh plumage with thin pale fringing on upperparts and contrastingly pale wing coverts; much down remains on this very young individual; older juveniles often retain traces of down on tail tip

**2 Juvenile, Aug., N.J.**
• when prey is spotted, plovers quickly dash to pluck it from the surface

**3 Sept., N.J.**
• What are the four species? Look at size and structure, then color.

**4 Juvenile, Nov., N.M.**
• much variation in upperparts color; this individual is rather cold and dark
• aged by thin pale fringing above with wing coverts paler than scapulars

**5 First winter, Jan., N.J.**
• fresh nonbreeding scapulars and median coverts are dark brown with bold rufous edges; note the pale, whitish-fringed lesser coverts and darker, buff-fringed scapulars retained from juvenal plumage

**6 Breeding male at nest, June, N.Y.**
- a very rufous-toned individual
- males show more extensive black in face, brighter red orbital-ring than females
- Killdeers nest in short-grass fields or sparsely vegetated areas, often close to human habitation (note four-egg nest behind bird)

**7 Breeding female, June, N.Y.**
- females show less extensive black in face, duller orange orbital-ring
- in breeding season, adults often perform distraction displays during which they act injured to lure predators away from brood; such displays reveal bright rufous rump and base of tail

**8 With Semipalmated Plover, July, N.J.**
- distinctly larger than Semipalmated
- slim, long-tailed shape and double breast-band always distinctive

**9 Jan., N.J.**
• often found along roadsides
• a very hardy species

**10 Mar., Tex.**
• very long tail and gleaming white underparts with double breast-band
• flight call is a strident, rising *deeee;* the familiar *kill-deea* call is given when alarmed

**11 Nov., N.J.**
• long tail with rufous rump and long angular wings with bold wing stripe

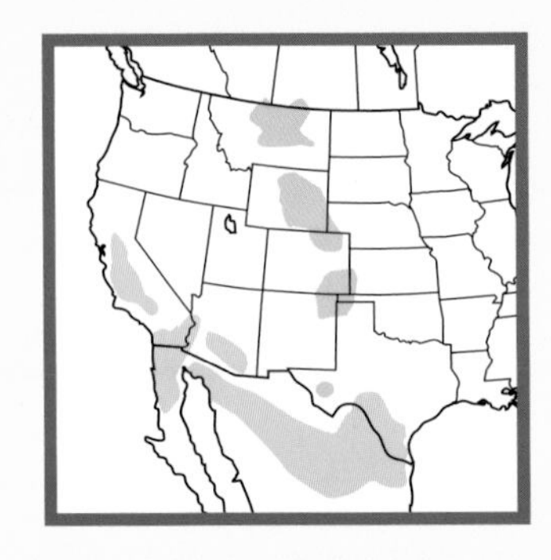

# Mountain Plover

***Charadrius montanus*** **p. 340**

**Size:** **L 8½–9½ in. (21–24 cm); WS 21½–24 in. (54–60 cm); WT 90–110 g**
**similar to Killdeer; smaller than American Golden-Plover**

**Structure:** **bigger-headed, thicker-necked, more compact than American Golden-Plover**

**Behavior:** **run-stop-pluck feeding style; arid, short-grass plains in summer, huge plowed fields in winter; often in large flocks**

**Status:** **uncommon and declining; rarely seen near coast**

**1 Juvenile, Jan., Calif.**

- all plumages very plain with sandy brown upperparts, white underparts, drab legs, slim black bill
- aged by uniformly fresh plumage with neat buff fringing on upperparts
- often holds juvenal plumage throughout first winter

**2 Nonbreeding, Feb., Calif.**
• aged by relatively plain upperparts with mix of old (plain) and new (rust-fringed) feathers

**3 Breeding male, late June, Colo.**
• black cap and lores indicate breeding plumage; female has slightly duller black markings

**4 Apr., Calif.**
• very plain with dark tail tip, pale stripe in primaries
• generally quiet; occasionally gives rough, grating *kirrp* flight call

**5 Feb., Calif.**
• bright white belly and underwing
• How do golden-plovers differ?

# American Oystercatcher

***Haematopus palliatus*** p. 343

**Size:** L 16–17½ in. (40–44 cm); WS 29¼–32½ in. (73–81 cm); WT 400–700 g
larger than godwits or curlews

**Structure:** bulky with long heavy bill, longish neck

**Behavior:** walks with clumsy gait; feeds by probing, stabbing, and hammering with bill; beaches, mudflats, shellbars

**Status:** fairly common but strictly tied to intertidal areas; northern populations migratory; rare in s. California.

**1 Juveniles with Laughing Gulls, Aug., N.J.**

- bold black, brown, and white pattern with orange bill in all plumages
- aged by dark eye and bill tip, dull orange-brown orbital-ring, and uniformly fresh plumage with subtle pale fringing on upperparts (difficult to see if not close)

**2 Adult, Mar., Fla.**
• bill and orbital-ring become clear, bright orange by fourth year; duller and variably smudged with dark in earlier stages
• eye becomes dull yellow by second year, bright yellow by third year

**3 Adult feeding chicks, June, N.J.**
• young are dependent on adults for at least two months after hatching and sometimes remain with adults throughout first winter

**4 With Marbled Godwit, Western Willet, Short-billed Dowitcher, Red Knot, Laughing Gull, and Common Tern, Sept., N.J.**
• Are all the oystercatchers adults?

**5 Adults in territorial/posturing display flight, July, N.J.**
• in spring and summer, noisy territorial/posturing display flights are performed by pairs and groups
• whistled flight call is a loud, easily imitated *wheeu,* mixed with piping calls during display

**6 Adult (left) and juveniles, Aug., N.J.**
• juveniles show dark bill tips, dark eyes, uniformly fresh wings, and subtly more pointed primaries than adults
• adults undergo primary molt in late summer; the new p4 is only half grown in this individual
• white wing stripe extends to inner primaries in nominate race; restricted to secondaries and greater coverts in *H. p. frazari* of Pacific Coast

**7 Adults and juveniles with other species, Sept., N.J.**
• What other species are visible in this photo?

# Black Oystercatcher

***Haematopus bachmani*** **p. 345**

**Size:** **L 16¾–18¾ in. (42–47 cm); WS 30¾–35¼ in. (77–86 cm); WT 500–700 g larger than godwits or curlews**

**Structure:** **bulky with long heavy bill, longish neck**

**Behavior:** **walks with clumsy gait; feeds by probing, stabbing, and chiseling with bill; rocky coastlines, adjacent beaches and headlands; often in pairs**

**Status:** **locally common in North, uncommon in South; strictly coastal**

**1 Juvenile, Aug., Calif.**
- solid blackish brown with orange bill in all plumages
- aged by dark eye and bill tip, dull orange-brown orbital-ring, and uniformly fresh plumage with subtle pale fringing on upperparts (difficult to see if not close)

**2 Adult, May, Alaska**
• oystercatcher bills are laterally compressed for "shucking" bivalves

**3 Adult, Dec., Calif.**
• adult shows bright orange-red bill and orbital-ring, bright yellow eyes; subadults have duller bills and eyes

**4 Adults with Glaucous-winged Gulls, May, Alaska**

**5 Adult on nest, May, Alaska**
• nests just above high-tide line on rocky beaches or islands

**6 (bottom left) Adults, July, Calif.**
• whistled flight call is a loud, easily imitated *wheeu,* mixed with piping calls during display

**7 (bottom right) Adult, May, Alaska**
• long broad wings enable short swift flights, usually low to the water

# Black-necked Stilt

***Himantopus mexicanus*** **p. 347**

**Size:** **L 14–15½ in. (35–39 cm); WS 29¼–32½ in. (73–81 cm); WT 136–220 g**
**same as or slightly larger than Greater Yellowlegs**

**Structure:** **extremely long legs; slim body; long slim neck; small head; long, needle-thin bill**

**Behavior:** **walks slowly, picking at surface of water; still, shallow ponds, lakeshores; active and often noisy**

**Status:** **locally common in West and South; scarce in Northeast; spring "overshoots" occur north of mapped range**

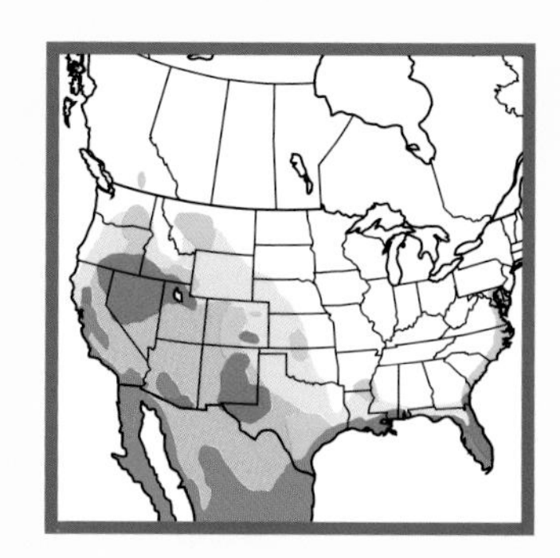

**1 Juvenile, Aug., Calif.**

- starkly contrasting pattern similar in all plumages
- aged by brownish upperparts, including crown, with pale-fringed feathering; dull legs; pink-based bill; dark eye; head pattern often less contrasting than on adult; wings show thin pale trailing edge in flight
- this is a very fresh individual; older juveniles show more subtle pale fringing above

**2 Adult male and female with American Avocet and Ruddy Duck, Feb., Calif.**
• male has blacker upperparts, brighter legs and, in spring, a blush of pink on the breast; female has browner back, usually duller legs
• often associates with American Avocet

**3 Adult male with chick, Apr., Calif.**
• all adults have red eyes
• often rests on heels
• adults shield chicks under their wing for protection against danger or extreme temperatures

**4 Mar., Cayman Brac, Cayman Is.**
• active, gregarious, and noisy birds
• often difficult to age after late fall, although first-year birds typically have less distinct head patterns than adults through spring

**5 Adult female with Lesser Yellowlegs (right) and two dowitchers, mid-Apr., Tex.**
• What species are the dowitchers?

**6 Adult female, Feb., Calif.**

**7 With American Avocets, Feb., Calif.**

**8 Adult male, Apr., Tex.**
• white rump and pale gray tail visible in flight
• flight call is a loud, yapping *yip, yip, yip, yip,* often given in a long series; very noisy

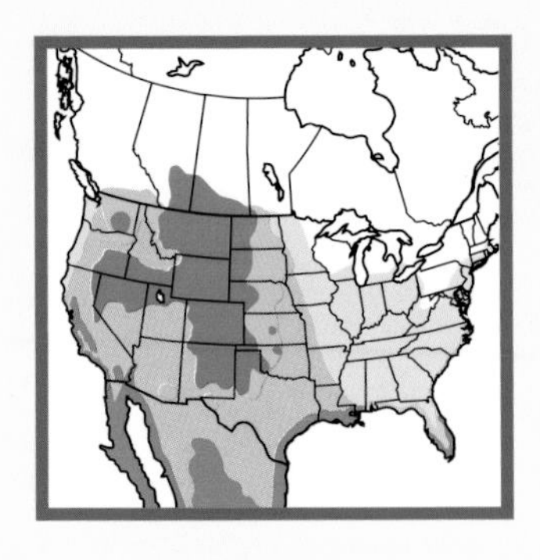

# American Avocet

***Recurvirostra americana*** **p. 348**

**Size:** **L 17¼–18¾ in. (43–47 cm); WS 29½–32½ in. (74–81 cm); WT 275–350 g**
**distinctly larger than Black-necked Stilt; about the same as Western Willet**

**Structure:** **long legs and neck; chunky body; small head; long upturned bill**

**Behavior:** **walks steadily, swishing bill from side to side; shallow lakes, impoundments, protected coastal waters; sometimes belly-deep**

**Status:** **locally common through much of West and South; scarce in the Northeast**

**1 Juvenile (right) and nonbreeding, Aug., Del.**
• bold wing and body pattern similar in all plumages
• aged with close view by uniformly fresh plumage with relatively small, subtly pale-fringed scapulars and coverts; dark areas are duller, more brownish than on adults; on fresh birds, gray-washed face blends to cinnamon-washed nape (not visible on this individual)

**2 Nonbreeding, Feb., Calif.**
- from juvenile by less neatly arranged upperparts with larger, blacker wing coverts and upper scapulars

**3 Breeding, Apr., Tex.**
- cinnamon-washed head, neck, and breast

**4 Breeding and nonbreeding with dowitchers, Dunlin, and Black-necked Stilt, Apr., Tex.**
- when feeding, walks actively, swishing bill from side to side
- Can you identify any of the dowitchers?

**5 Breeding, with Marbled Godwit, Apr., Tex.**
• often gathers in dense flocks at favored wintering or stop-over sites

**6 Breeding male and female copulating, Apr., Calif.**
• female has slightly more upturned bill than male

**7 Nonbreeding, Aug., Del.**
• flight feather molt takes place mostly at stop-over sites
• primary call is a loud, whistled *bleet,* given as much while foraging as in flight

# Greater Yellowlegs

***Tringa melanoleuca*** **p. 351**

**Size:** **L 11½–13¼ in. (29–33 cm); WS 28–29½ in. (70–74 cm); WT 111–235 g**
**close to Black-necked Stilt; 2x the weight of Lesser Yellowlegs**

**Structure:** **rangy with long legs, neck, and bill; small head; sculpted body and neck, often with prominent "Adam's apple"**

**Behavior:** **active; walks steadily, picking at surface; often runs frantically to chase fish; flooded fields, marshes, tidal creeks; nests in open boreal-forest bogs**

**Status:** **common at interior and coastal sites**

**1 Juvenile, Sept., N.J.**
• in all plumages, bill sturdy, pale-based, slightly upturned; often distinctly longer than depth of head but not always easy to judge; legs bright yellow; sculpted body contour with angular chest or neck, often with prominent "Adam's apple"
• aged by brownish upperparts with relatively large, sharply defined white spots

**2 Juvenile molting to first winter, Nov., N.J.**
• retained juvenal wing coverts and tertials contrast with fresh nonbreeding scapulars; nonbreeding feathers are paler and grayer with alternating dark and pale marginal spots

**3 Juvenile (left) and nonbreeding with juvenile Lesser Yellowlegs (center right), Aug., N.Y.**
• compared to Lesser, note Greater's larger size, heavier chest, more distinct eye-ring, more extensive and distinct breast markings
• compared to adult Greater, note juvenile's darker, browner upperparts with larger, more sharply defined spots
• Can you find a second Lesser Yellowlegs and a second juvenile Greater?

**4 Nonbreeding (right) with nonbreeding Lesser Yellowlegs, Eastern Willet, and Black-necked Stilt, Mar., Cayman Brac, Cayman Is.**
• similar body size to Eastern Willet and Black-necked Stilt; much larger than Lesser Yellowlegs
• note Lesser's shorter, finer bill, smoother body contour, less streaked appearance to head and neck, cleaner flanks

**5 Nonbreeding, Nov., Tex.**
• much paler and grayer than juvenile, with distinct dark and light marginal spots on upperparts
• stands very tall when neck is extended
• note rudimentary hind toes

**6 Breeding, Apr., N.J.**
• breeding birds attain extensive barring on underparts, often extending across belly, and variable number of black feathers in upperparts
• bulging base to the neck may give "Adam's apple" look at times

**7 Breeding, with Lesser Yellowlegs and Snowy Egret, July, N.J.**
• heavier flank barring than breeding Lesser Yellowlegs
• often chases fish with Snowy Egrets

**9 Molting adult with molting adult Stilt Sandpiper, early Aug., N.Y.**
• mix of worn, dark breeding feathers and fresh, gray nonbreeding feathers is typical of late-summer migrants

**10 Juvenile, Aug., Calif.**
• heavy body with angular chest, long sturdy bill, long yellow legs, and white rump are often evident in flight
• fine spotting on secondaries and inner primaries is lacking in Lesser Yellowlegs, but hard to see in the field
• flight call is a loud, strident *dee-dee-deer,* last note slightly lower

**11 Molting adults (the 5 big birds with bright yellow legs) with Semipalmated Sandpipers and Short-billed Dowitchers, Aug., N.Y.**
• molts flight feathers at stop-over sites during migration (Lesser Yellowlegs does so on wintering grounds)

# Lesser Yellowlegs

*Tringa flavipes* p. 353

**Size:** L $9^{1}/_{4}$–10 in. (23–25 cm); WS $23^{1}/_{2}$–$25^{1}/_{2}$ in. (59–64 cm); WT 48–114 g
slightly smaller than dowitchers

**Structure:** delicate with slim chest and smooth body contour; small head; long legs and neck; slim straight bill

**Behavior:** walks rapidly and methodically, picking at surface; seldom runs; flooded fields, shallow ponds, mudflats; nests in open boreal forest and edge of tundra

**Status:** common at interior and coastal sites in fall; mostly midcontinent in spring

**1 Juvenile, Aug., N.J.**
• in all plumages, bill slim, straight, and mostly dark; often just barely longer than depth of head but not always easy to judge; legs bright yellow
• aged by uniformly fresh, brownish upperparts with sharply defined white spots

**2 Juvenile (left) with juvenile Greater Yellowlegs and Stilt Sandpiper, Sept., N.Y.**
- compared to Greater, note smooth contour to body, slimmer chest and neck, smaller head and bill; also note more capped appearance, less distinct eye-ring, more smoothly patterned breast, finer spotting above
- Stilt Sandpiper has a more clumsy shape with larger head; longer, heavier, drooped bill; shorter legs

**3 Adults, Aug., N.Y.**
- often fights with other Lessers over feeding territory in migration

**4 Juvenile with juvenile Greater Yellowlegs (1), Stilt Sandpipers (4), and Short-billed Dowitchers (3), Sept., N.Y.**
- Can you find the Lesser Yellowlegs?

**5 Nonbreeding with nonbreeding Greater Yellowlegs, Nov., Tex.**
• nonbreeding feathers are paler, grayer than juvenile with alternating dark and pale marginal spots
• Greater is nearly twice the bulk of Lesser with more sculpted body and neck and longer, heavier, paler-based bill; also note Greater's more streaked head and neck, more distinct eye-ring, less distinct cap, and more prominently spotted upperparts

**6 Breeding, June, Manitoba**
• breeding birds attain variable black markings on upperparts; fine breast and flank barring contrasting with clean white belly (compare with Greater)
• frequently perches atop spruce trees on breeding grounds

**7 Molting adult, Aug., N.Y.**
• mix of worn, dark breeding feathers and fresh, gray nonbreeding feathers is typical of southbound migrants in late summer

**8 Molting adult with Short-billed Dowitcher and Semipalmated Sandpipers, late July, N.Y.**
• Which is the largest bird? When judging size in the field, it is most instructive to evaluate body bulk and leg length separately.

**9 Juvenile, Aug., N.J.**
• white rump and long legs obvious in flight
• note unmarked primaries and secondaries (finely spotted in Greater)
• flight call is a low, whistled *tu,* often doubled; alarmed birds regularly give a long, measured series of more emphatic *tu* or *kewp* notes

**10 With Greater Yellowlegs (fourth from left) and Stilt Sandpipers (right, left, center), Sept., N.J.**
• smaller size, delicate bill, slim neck and chest are usually best distinctions from Greater in flight; this Greater appears only subtly larger but note broader wings, larger head, and longer, paler-based bill
• Stilt Sandpipers appear more front-heavy with longer bill, thicker neck; wings are narrower and more swept back with whiter underwing, quicker wingbeats

# Solitary Sandpiper

***Tringa solitaria*** **p. 361**

**Size:** **L 7½–9¼ in. (19–23 cm); WS 22–23½ in. (55–59 cm); WT 31–65 g**
**midway between "peep" and Lesser Yellowlegs**

**Structure:** **more compact than Lesser Yellowlegs with shorter wings, legs, and neck; slightly larger head**

**Behavior:** **moves slowly and nervously, picking at surface; often by itself at sheltered ponds, creeks, or rain pools in forested areas; seldom associates with other shorebirds; nests in boreal-forest bogs**

**Status:** **fairly common throughout but never in large groups**

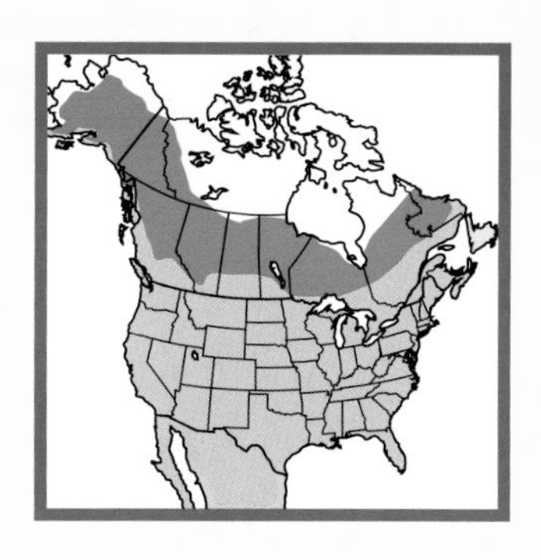

**1 Juvenile, Aug., N.J.**
• all plumages show dark upperparts with fine spots; blackish bend of the wing; darkish breast contrasting with clean white belly; bold eye-ring, greenish legs, and greenish-based bill
• aged by uniformly fresh, neatly arranged plumage with smallish scapulars, softly patterned breast, relatively large spots on upperparts, and buff-tipped primaries

**2 Juvenile (center) with Spotted Sandpiper (left) and Lesser Yellowlegs, Sept., N.J. (inset, juvenile *T. s. cinnamomea,* Aug., Calif.)**
• smaller and more compact than Lesser Yellowlegs with shorter, greenish legs; darker upperparts with blackish bend of the wing; darker, more contrasting breast; darker loral stripe; bolder eye-ring
• slightly larger and more elegantly proportioned than Spotted with longer legs and bill; blackish bend of the wing; no white wedge at breast-sides; spotted upperparts; different face-pattern
• northern and western *cinnamomea* differs subtly from nominate in broader tail barring, weaker loral stripe, less contrasting bend of wing, buffier spots on juvenile upperparts; much variation and overlap

**3 Nonbreeding, Jan., Tex.**
• differentiate from juvenile by larger scapulars with smaller spots and blackish markings in breast; difficult to distinguish from first winter

**4 Breeding, May, Conn.**
• barred breast and extensive black markings in upperparts indicate breeding plumage
• on breeding grounds, often perches on treetops; only New World shorebird to regularly nest in trees, where it uses abandoned songbird nests

**5 Worn breeding, July, N.J.**
• when worn, upperparts and breast-sides quite dark, with most white markings worn off
• most adults in fall migration are in worn breeding plumage with very few feathers replaced; the molt to nonbreeding takes place largely on the wintering grounds

**6 Aug., N.J.**
• compact and short-tailed with sharply angled wings, short toe projection; dark underwing and breast contrast strongly with white belly
• quick, flickering wingbeats on takeoff; swooping flight and dark underwing often give swallowlike appearance when high in the sky (size about the same as Purple Martin)

**7 Juvenile, Sept., N.J.**
• dark rump and central tail; boldly barred sides of tail
• flight call a shrill, whistled *psEEt-weet-weet;* higher and more emphatic than Spotted Sandpiper

# Eastern Willet

***Tringa semipalmata semipalmata*** **p. 363**

**Size:** **L 12½–14 in. (31–35 cm); WS 21½–24½ in. (54–61 cm); WT 199–263 g**
**same or slightly larger than Greater Yellowlegs**

**Structure:** **stockier and more compact than Greater Yellowlegs with larger head, thicker legs, neck, and bill; see Willet subspecies comparison (p. 93)**

**Behavior:** **walks steadily, picking, swishing, or shallowly probing; noisy in summer; salt marshes, tidal creeks, tidal flats, mangroves**

**Status:** **common and conspicuous breeder in coastal salt marshes; vacates continent in winter**

**1 Juveniles, July, N.J.**

- at all ages, darkish overall with heavy, pale-based bill; grayish legs may be tinged with blue, green, or pink
- aged by uniformly fresh plumage with pale marginal spots and dark subterminal markings on upperparts, smooth patterning on breast
- juveniles show relatively dark scapulars contrasting with pale wing coverts and prominent buff marginal spots

**2 Mostly nonbreeding, early Mar., Tex.**
• nonbreeding birds show plain grayish plumage with scapulars slightly darker than wing coverts; a few dark bars on upperparts and flanks are first incoming breeding feathers
• rarely if ever seen in full nonbreeding in North America; absent from North America in winter (Nov.–Feb.)

**3 Breeding, with 2 Greater and 2 Lesser Yellowlegs, Apr., Tex.**
• Which is the Eastern Willet? Look for heavy bill, grayish legs, barred upperparts, more uniform overall tone; body size is about the same as Greater Yellowlegs

**4 Breeding, June, N.J.**
• dark barring on wings and body indicate breeding plumage
• often shows pinkish cast to bill and legs
• perches on poles, wires, and other tall structures near salt-marsh breeding grounds; Western Willet does so only on the prairies where they breed

**5 Worn breeding (2 dark birds) with worn breeding Western Willets, July, N.J.**
• Western Willets average larger and paler in all corresponding plumages

**6 Breeding, Apr., Tex.**
• flashy wing pattern in all plumages; wing stripe averages narrower than Western's but much overlap
• flight call is a loud, strident *klaay-dr* or *klaay-dr-dr;* rapidly repeated *pidl-will-willet* song given mostly on breeding grounds

**7 Breeding (upper left) with molting adult Western Willets, late July, N.J.**
• more compact appearance with shorter body and legs than Western (toe projection similar)
• Westerns molt flight feathers at coastal stop-over and wintering sites; Easterns molt flight feathers after departing North America

**8 Eastern Willet, June, N.J.**

# Willet Subspecies Comparison

**1 Juvenile Western (left) and Eastern Willets, late July, N.C.**
• most Westerns are obviously larger than Easterns, but some overlap (Westerns are highly variable in size)
• structural differences often obvious, but sometimes subtle: Eastern is shorter legged with usually shorter, heavier, thicker-tipped, often subtly drooped bill; body is more compact and often subtly more angular; Western is more godwit-like with longer legs, thinner bill, and longer, more gracefully shaped body
• in all plumages, Eastern is darker and browner than corresponding plumage of Western

**2 Juvenile Western, Aug., N.J.**
• pale and buffy gray overall with low-contrast patterning; little contrast between scapulars and coverts; pale head contrasts with often darkish bill
• more elongated body and graceful curve to back; longer primary projection (juveniles only)

**3 Juvenile Eastern, Aug., Conn.**
• dark and brownish overall with contrasty patterning; dark scapulars contrast with pale coverts and have prominent buff spots; dark cap and lores contrast with thin, pale supraloral stripe
• angular shape with flat back; short primary projection (juveniles only)

**4 Nonbreeding Western, Nov., Fla.**
• typically very pale and grayish overall
• relatively long, tapered, fine-tipped bill with blue-gray base; next to pale head, bill often looks contrastingly dark
• birds in this plumage are common in winter on southern coasts

**5 Nonbreeding Eastern, early Mar., Cayman Brac, Cayman Is.**
• slightly darker and browner overall with subtle contrast between darker scapulars and paler coverts
• bill is relatively heavy all the way to tip and has pinkish cast to base
• this plumage of Eastern is undocumented in North America

**6 Breeding Western (left) and Eastern Willets with Long-billed Dowitcher, Apr., Tex.**
• many spring Westerns (particularly females) are strikingly large, pale, and lightly marked compared to Easterns

**7 Breeding Western, June, Calif.**
• some breeding Westerns (particularly males) can be rather small and heavily marked, but note grayish cast to crown and upperparts; relatively thin, fine-tipped, mostly dark bill; rounded body contour

**8 Breeding Eastern, Apr., Tex.**
• some Gulf Coast Easterns in spring can be rather large and lightly marked and often have narrower-based bills, but note brownish overall color; heavy-tipped, pinkish-tinged bill; flat back

**9 Worn breeding Western, late July, N.J.**
• worn breeders may become quite dark on upperparts and more closely resemble Eastern; note fine-tipped bill, attenuated shape with rounded contour to back, and overall gray cast to plumage

**10 Worn breeding Eastern, July, N.J.**
• worn breeders may become quite dark on upperparts; note relatively thick, blunt-tipped bill with extensively pinkish base, flatter back, and shorter body and legs

# Western Willet

***Tringa semipalmata inornata*** **p. 363**

**Size:** **L 13½–16½ in. (34–41 cm); WS 23½–28½ in. (59–71 cm); WT 203–339 g**
**usually larger than Greater Yellowlegs; overlap with Marbled Godwit**

**Structure:** **lanky and long-legged; bigger head, thicker neck, shorter bill than godwits; see Willet subspecies comparison (p. 93)**

**Behavior:** **walks steadily with long, stiff-legged gait; picks, swishes, or shallowly probes; prairie potholes in summer, beaches, tidal flats in winter; often with Marbled Godwits**

**Status:** **common in many coastal areas during migration and winter; more sparsely distributed breeder and migrant inland**

**1 Juvenile, late July, N.C.**
• all plumages pale overall with slim, straight, fine-tipped bill; grayish legs may be tinged with blue or green
• aged by uniformly fresh plumage with pale buff marginal spots and dark subterminal markings on upperparts; smooth patterning on breast
• juveniles show low-contrasting patterning on upperparts

**2 Nonbreeding with Marbled Godwits and juvenile Eastern Willet (center right), Sept., N.J.**
• plain gray plumage indicates nonbreeding; most are slightly smaller than Marbled Godwit but some overlap
• Eastern looks smaller, darker, less attenuated (see Willet subspecies comparison, p. 93)

**3 Nonbreeding with Marbled Godwits, early Dec., N.J.**
• very difficult to distinguish from first winter

**4 Breeding, mid-Apr., Tex.**
• dark barring on wings and body indicate breeding plumage
• many spring birds (particularly females) are distinctively large, pale, and lightly marked
• pale coloration, rounded body contour, and evenly tapered, fine-tipped bill distinctive
• bill is often all-dark on breeding birds

**5 Worn breeding, late July, Fla.**
• when worn, more heavily marked individuals may look as dark as Eastern Willet
• slim, fine-tipped bill, rounded body contour, long legs, and clumsy, stiff-legged gait distinctive

**6 First summer, July, Fla.**
• first-summer birds may look like nonbreeding or attain partial breeding; many spend their first summer at wintering grounds, where they molt earlier than adults; this bird is midway through primary molt—note fresh (blacker) p1–5, half-grown p6 (visible on underwing), worn juvenal p7–10; note also that juvenal primaries are narrower and more pointed than adult feathers

**7 Nonbreeding with Marbled Godwits, Dec., N.J.**
• flashy wing-pattern in all plumages
• flight call is a loud, strident *klaay-dr* or *klaay-dr-dr,* similar to Eastern but averages lower pitched and hoarser (variable)
• *p'd-weeel-will-wit* song given mostly on breeding grounds; lower pitched, slower than Eastern's with second note more drawn out

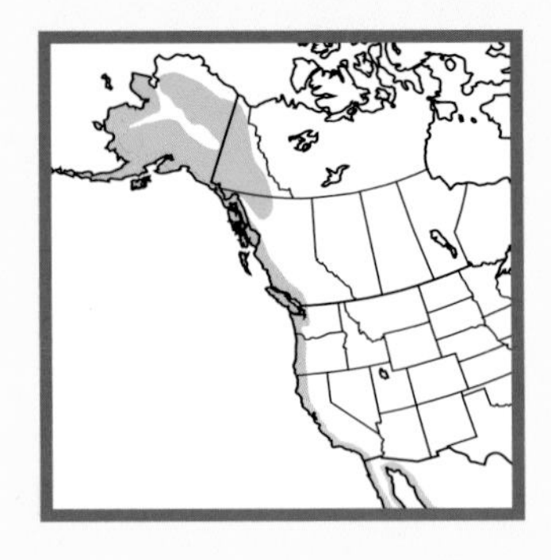

# Wandering Tattler

***Tringa incanus*** **p. 366**

**Size:** L $10^1/_2$–12 in. (26–30 cm); WS 20–22 in. (50–55 cm); WT 100–140 g
close to Red Knot

**Structure:** slim and attenuated with short legs; medium-length bill; horizontal stance

**Behavior:** active; walks quickly over rocks with teetering motion like Spotted Sandpiper; picks or probes at surface; rocky coastlines in winter, gravel alpine streams in summer; usually solitary

**Status:** fairly common but strictly coastal and never in large groups; very rare inland

**1 Juvenile, Nov., Calif.**
- all plumages dark gray overall, including breast and flanks; yellowish legs; dark lores; pale supraloral
- aged at close range by uniformly fresh plumage, smallish scapulars and coverts, subtly pale-fringed upperparts, sharply pointed primaries

**2 Nonbreeding, Dec., Calif.**
• differs from juvenile at close range by broader scapulars and coverts, lack of pale fringing on upperparts

**3 Breeding, Apr., Calif.**
• barred underparts indicate breeding plumage

**4 Breeding (right), with breeding Surfbird, July, Calif.**
• seldom associates with other shorebirds

**5 Breeding, Aug., Calif.**
• all dark above
• flight call is a ringing, trilled whistle, *didididididi,* all on one pitch

**6 Breeding, May, Alaska**
• walks quickly over rocks with teetering motion like Spotted Sandpiper

# Spotted Sandpiper

***Actitis macularia*** p. 370

**Size:** L 7¼–8 in. (18–20 cm); WS 14¾–16 in. (37–40 cm); WT 43–50 g
slightly larger than "peep"; same as White-rumped Sandpiper

**Structure:** short legs and bill like peep but longer neck and tail; horizontal stance

**Behavior:** active; walks quickly, picking at surface and bobbing tail almost constantly; sandy or rocky shorelines, streams; usually solitary or in small groups

**Status:** common throughout interior and along coast

**1 Juvenile, July, N.J.**
- all plumages show brown upperparts with dark eye-line and pale supercilium; white of underparts extends in a spur at breast-sides; yellowish legs and pinkish-based bill; tail projects past wings
- aged by uniformly fresh plumage with boldly barred coverts; subtle pale fringes and thin dark highlights to scapulars and tertial tips

**2 Juvenile with juvenile Semipalmated Plover, Aug., N.Y.**

**3 Worn juvenile, Oct., N.J.**
• aged with close view by uniformly worn plumage, boldly barred coverts, relatively small scapulars with narrow pale fringing

**4 Nonbreeding, Jan., Andros, Bahamas**
• differentiate from juvenile with close view by uniform upperparts without pale fringing, relatively large scapulars; adults typically retain fine spotting on rear flanks

**5 Juvenile with Solitary Sandpiper, Sept., N.J.**
• at long range, Solitary looks slightly larger and more elegantly proportioned with longer legs and bill, dark bend of the wing, all-dark breast-sides, bolder eye-ring

**6 Breeding, early May, Fla.**
- extensive dark spotting on underparts and upperparts indicates breeding plumage

**7 Molting adult, July, N.J.**
- during fall migration, adults may begin to lose some breeding-plumage spots, but only limited molt takes place before the wintering grounds are reached
- Can you identify the bird in the background?

**8 Breeding, Sept., N.J.**
- flies low to the water on shallow, vibrant wingbeats mixed with stiff-winged glides; longer, higher-altitude flights are on more "typical shorebird" wingbeats

**9 Breeding, Sept., N.J.**
- flight call is a clear, whistled, two- to three-note *peet-weet-weet;* lower pitched, slower, and more wavering than Solitary Sandpiper's call

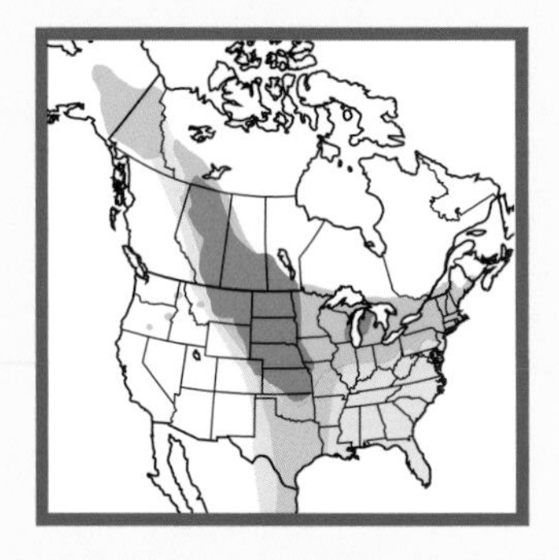

# Upland Sandpiper

***Bartramia longicauda*** **p. 373**

**Size:** **L 11¼–12¾ in. (28–32 cm); WS 25½–27¾ in. (64–68 cm); WT 97–226 g**
**larger than Killdeer; slightly larger than Mourning Dove**

**Structure:** **small head and short bill; long neck and tail; upright stance**

**Behavior:** **walks steadily with head-bobbing, picking occasionally at ground; prairies in summer, dry pastures, sod farms during migration; does not usually associate with other shorebirds**

**Status:** **uncommon and declining in most of range; most numerous in Great Plains**

**1 Juvenile, Oct., England**
- unique among shorebirds with long neck, small dovelike head, short bill, and long tail projecting well past wingtips
- in all plumages, buff brown overall with extensive barring; yellowish legs and bill; large eye conspicuous on plain face
- aged by uniformly fresh, neatly arranged plumage with crisp pale fringes and dark subterminal lines on scapulars, coverts, and breast-sides

**2 Adult, late Apr., Fla.**
• from juvenile by less neatly arranged plumage without dark subterminal lines and crisp pale fringes to scapulars, coverts, and breast-sides; breeding and nonbreeding similar

**3 Adult, May, Neb.**
• frequently perches on fenceposts or rocks on breeding grounds
• often holds wings raised momentarily after landing

**4 Adult, June, N.J.**
• typical upright stance; at long range, may resemble Mourning Dove but longer necked
• found in grassland habitats at all seasons, seldom with other shorebirds

**5 Juvenile, Oct., England**
• Upland Sandpiper is a long-distance migrant and a rare stray to w. Europe. After an exhausting ocean crossing, this starving vagrant resorted to taking mealworms from the hand. Note the birder carefully taking notes, the best way to learn about birds.

**6 (below, left) Adult, May, Neb.**

**7 (above, right) Adult, May, Neb.**
• barred underwing
• flight call is a mellow, whistled *qui-pi-pi-pi;* commonly heard from migrants high overhead

**8 Adult, May, Neb.**
• plain upperparts with dark rump and primary coverts, strikingly long rounded tail
• flies with stiff, shallow wingbeats

# Whimbrel

***Numenius phaeopus*** **p. 376**

p. 376

**Size:** **L 16–16¾ in. (40–42 cm); WS 30½–35½ in. (76–89 cm); WT 312–493 g**
**same as or larger than Western Willet; noticeably smaller than Long-billed Curlew**

**Structure:** **bulky but attenuated; small head; long neck; long decurved bill**

**Behavior:** **walks steadily, picking and probing in mud; marshes, mudflats, rocky coasts; breeds on tundra**

**Status:** **locally common along coast; generally rare inland**

**1 Juvenile, Sept., Calif.**
• in all plumages, brown overall with extensive barring; bold head stripes; gray legs • aged by uniformly fresh plumage with dark centers and crisply notched fringes to scapulars, tertials, and coverts; coverts often contrastingly paler than scapulars; bill often shorter than adult's

**2 Adult (left) with Western Willet, Jan., Fla.**
• differs from juvenile by more softly patterned upperparts with less sharply defined pale markings; often difficult to distinguish from first winter

**3 Adult, Jan., Fla.**
• bill usually pink-based in nonbreeding season

**4 Adult, June, Manitoba**
• bill usually all dark in breeding season; plumage similar to nonbreeding

**5 Adult with Marbled Godwit, Feb., Calif.**
• Aside from bill shape, what differences are apparent from Marbled Godwit?

**6 With Red Knots (2, center) and Short-billed Dowitchers, July, N.J.**
• during migration, often shares beachfront habitat with other shorebirds

**7 Adults, Apr., N.J.**
• flight call is a far-carrying staccato whistle, *pi-pi-pi-pi-pi-pi-pi,* all on one pitch

**8 Adults, Apr., N.J.**
• brown overall with heavy chest, long pointed wings, and long decurved bill

## Whimbrel Subspecies Comparison

**9 *N. p. variegatus,* Apr., Hong Kong**
- white rump; usually dark uppertail coverts and tail

**10 *N. p. phaeopus,* Aug., England**
- white rump; usually whiter uppertail coverts, tail, and underwing than *variegatus*

**11 *N. p. variegatus,* Apr., Hong Kong**
- whitish underwing like *phaeopus* but often more heavily marked

**12 Adult *N. p. hudsonicus,* Jan., Fla.**
- brown underwing, rump, and tail
- warmer tone and less contrasting neck and breast markings than Eurasian subspecies

**13 Adult *N. p. variegatus,* Apr., Australia**
- colder overall tone than *hudsonicus* with more contrasting streaks on neck and breast; often bolder eye-ring, less distinct post-ocular line

**14 Adult *N. p. phaeopus,* Aug., England**
- very much like *variegatus* but typically more lightly marked flanks and undertail coverts

# Long-billed Curlew

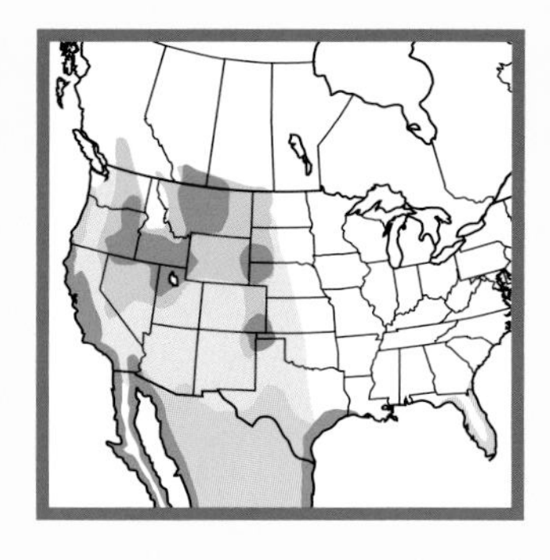

***Numenius americanus*** **p. 383**

**Size:** **L 20–26 in. (50–65 cm); WS 30½–39½ in. (76–99 cm); WT 445–792 g**
**distinctly larger than Marbled Godwit or Whimbrel**

**Structure:** **heavy rounded body; long neck; extremely long decurved bill; less attenuated, longer billed than Whimbrel**

**Behavior:** **walks steadily, picking and probing in mud or grass; pastures, mudflats; breeds in prairies**

**Status:** **locally common, especially near coast but also at interior sites**

**1 Juvenile, Sept., Calif.**
• in all plumages, buff brown overall with barred upperparts; unpatterned head, gray legs
• aged by uniform plumage condition with broad dark centers and narrow dark bars on scapulars, tertials, and coverts; coverts usually contrastingly paler than scapulars; breast and neck mostly unmarked; often grayish white cast to face and foreneck; often difficult to age

**2 Adult, Jan., Fla.**
• differs from juvenile by narrower dark centers, broader dark bars on upperparts; fine barring and streaking on breast; difficult to distinguish from first winter

**3 With Marbled Godwit and Short-billed Dowitchers, Feb., Calif.**
• note paler legs than Marbled Godwit

**4 Adult, Jan., Fla.**
• extraordinarily long decurved bill is adapted to extract Fiddler Crabs and other crustaceans from deep burrows

**5 Adults with White-faced Ibis, Feb., Calif.**
• very few shorebirds are as big as an ibis

**6 Adult with two yellowlegs, Jan., Fla.**
• What species are the yellowlegs?

**7 Adult, Jan., Fla.**
• prominent cinnamon wash in wings similar to Marbled Godwit; note barred flight feathers

**8 Adult, Feb., Calif.**
• flight call is a clear, whistled *cur-IEE*

# Hudsonian Godwit

***Limosa haemastica*** **p. 386**

**Size:** **L 14½–16¾ in. (36–42 cm); WS 26¾–31½ in. (67–79 cm); WT 196–358 g**
**same as Western Willet**

**Structure:** **"athletic" look with heavy chest, sleek body, and long pointed wings; long legs; long upturned bill**

**Behavior:** **walks steadily, probing deeply in mud; flooded fields, impoundments, mudflats; breeds on tundra**

**Status:** **uncommon; migrates through Great Plains in spring (sometimes large flocks), along Northeast coast in fall (mostly small groups); most fall migrants pass offshore**

**1 Juvenile, Oct., N.J.**
• in all plumages, bicolored upturned bill, dark lores, pale supraloral, black legs • juvenile and non-breeding mostly grayish overall with unpatterned breast • aged by uniformly fresh plumage with barred tips to scapulars, tertials, and greater coverts; buff-tinged breast when fresh

**2 Mostly nonbreeding, Sept., N.J.**
• differs from juvenile by plain gray scapulars
• wing coverts and tertials are retained from breeding plumage and will not be molted until the wintering grounds are reached; full nonbreeding plumage is rarely seen in North America

**3 Breeding female, June, Manitoba**
• rufous belly and streaked gray-brown neck distinctive
• slightly larger and longer billed than male; duller plumage with fewer "breeding" wing coverts and tertials

**4 Breeding male, June, Manitoba**
• slightly smaller, shorter billed, and more richly colored than female

**6 Molting adults, Aug., N.Y.**
• molting birds in fall typically retain scattered rufous feathers on belly, but much of head, underparts, and scapulars have been replaced; breeding wing coverts and tertials are retained throughout fall migration

**7 Molting adults with Lesser Yellowlegs, Sept., N.J.**
• Hudsonian is the smallest godwit and often looks surprisingly small; they can easily blend in with groups of yellowlegs or Willets

**8 Molting adults, Sept., N.J.**
• very long pointed wings show striking black wing-linings and narrow white stripe

**9 Molting adult, Sept., N.J.**
• white rump and black tail
• flight call is an emphatic, piping *pEEd-wid*

# Marbled Godwit

***Limosa fedoa*** **p. 390**

**Size:** **L $16^{3}/_{4}$–$19^{1}/_{4}$ in. (42–48 cm); WS 28–32 in. (70–80 cm); WT 285–454 g**
**same or larger than Western Willet**
**Structure:** **bulky body; long legs; very long upturned bill**
**Behavior:** **walks steadily, probing deeply in mud; prairies in summer; mudflats, beaches in winter**

**Status:** **locally common in West and South; uncommon in East**

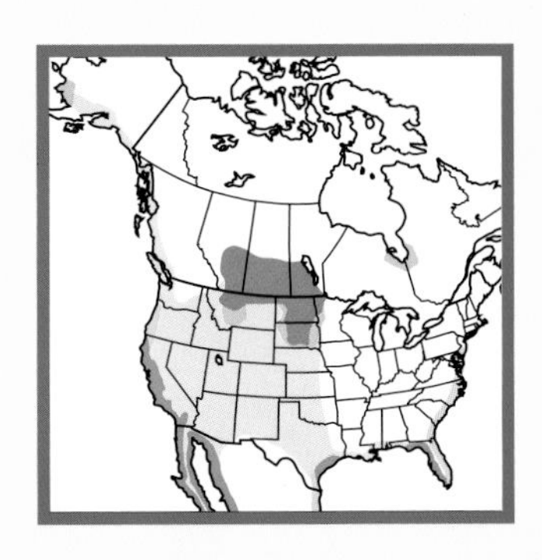

**1 Juvenile, Aug., N.Y.**
• in all plumages, buffy overall with barred upperparts, long bicolored bill, blackish legs
• juvenile and non-breeding show mostly unmarked breast
• aged in close view by uniformly fresh plumage with broad dark centers to scapulars, greater coverts, and tertials; contrastingly pale, buff-fringed coverts; smudgy grayish crown and nape
• molts out of juvenile plumage early and quickly becomes very difficult to distinguish from nonbreeding

**2 Nonbreeding with Long-billed Curlew, Feb., Calif.**
• differs from juvenile by narrower dark centers to scapulars and tertials, buffy crown and nape with crisp streaking; usually indistinguishable from first winter in the field • leg color and overall bulk are often clearest distinctions from sleeping Long-billed Curlew; also note longer primary projection

**3 Breeding male, Apr., Tex.**
• differentiate from nonbreeding by barred underparts

**4 Breeding male (center) and female with chick, July, Mont.**
• female slightly larger and longer billed, often with darker culmen

**6 With American Oystercatcher, Western Willet, and Short-billed Dowitcher, Sept., N.J.**

**7 Jan., Calif.**
- How does this silhouette differ from Hudsonian?

**8 Nonbreeding with Western Willet, Dec., N.J.**
- note rich buff underwing coverts

**9 Sept., N.J.**
- dark outer primaries contrast with buff inner primaries
- flight call is a strong, nasal *kwEH* or *kwEH-wed*

# Ruddy Turnstone

***Arenaria interpres*** **p. 392**

**Size:** **L 8½–10½ in. (21–26 cm); WS 20–22¾ in. (50–57 cm); WT 84–190 g slightly larger than Sanderling**

**Structure:** **chunky body with short legs and neck; short, stout chisel-shaped bill; low, crouched stance**

**Behavior:** **active; walks quickly, flipping over debris; often digs in sand; sandy or rocky shorelines; nests on tundra**

**Status:** **common along coast; uncommon to rare inland**

**1 Juvenile, Sept., N.J.**

- all plumages show white underparts, dark breast-band, orange legs
- juvenile and non-breeding show brownish upperparts and head
- aged by uniformly fresh plumage with crisp pale fringing on upperparts; juvenal plumage highly variable; this individual is rather pale

**2 Molting juvenile, Oct., N.J.**
• a darker but more colorful individual than the bird in photo #1; much variation
• fresh nonbreeding scapulars are more rounded with diffuse pale fringes

**3 Juvenile and adult *A. i. interpres,* Aug., England**
• mostly Eurasian *interpres* averages darker than widespread American *morinella,* but much overlap; populations of *interpres* in Alaska and Pacific Coast are somewhat intermediate, rendering subspecific identification in the field nearly impossible

**4 Nonbreeding, Feb., N.J.**
• chisel-like bill well-suited for prying open mollusks

**5 Nonbreeding, Mar., N.J.**
• differs from juvenile by less neatly arranged upperparts with diffuse pale fringing on wing coverts

**6 Nonbreeding with Black Turnstones, Feb., Calif.**
• Black Turnstone is darker overall with duller legs; no pale patch at breast-sides

**7 Breeding male with Eastern Willet and Pectoral Sandpiper, Apr., Tex.**
• Could you identify these birds in silhouette?

**8 Breeding male and female *morinella* with Red Knot, Semipalmated Sandpiper, and Laughing Gull, May, N.J.**
• males are brighter than females, with whiter head markings and cleaner rufous wing coverts and upper scapulars; *morinella* average less black on upperparts than *interpres*

**9 Breeding, with Red Knot, Semipalmated Sandpiper, and Laughing Gull, May, N.J.**
• large flocks gather in spring along Delaware Bayshore to feed on Horseshoe Crab eggs

**10 Breeding male *interpres* at nest, June, Alaska**
• breeding male *interpres* averages more black on rear scapulars, tertials, and greater coverts than *morinella,* but much overlap

**11 Breeding female *interpres* at nest, June, Alaska**
• like males, more black on upperparts than corresponding plumage of *morinella,* but much overlap

12 **Molting adult, Sept., N.J.**

13 **Nonbreeding, Feb., N.J.**
• bright white underwing

14 **Breeding male, May, Del.**
• pied pattern on upperparts similar in all plumages

15 **Breeding, with Red Knots and Sanderling (lower left), June, N.J.**
• flight call is a low, rapid *cut-a-cut,* sometimes extended into a long chatter

# Black Turnstone

***Arenaria melanocephala*** **p. 394**

**Size:** **L 8¾–10 in. (22–25 cm); WS 20–22¾ in. (50–57 cm); WT 100–170 g same as Ruddy Turnstone**

**Structure:** **slightly chunkier than Ruddy; short legs and neck; short, stout chisel-shaped bill; low, crouched stance**

**Behavior:** **active; walks quickly, flipping over debris; often digs in sand; sandy or rocky shorelines**

**Status:** **common along coast; accidental inland**

**1 Juvenile, Sept., Calif.**
- all plumages dark overall with white belly, dull orange-brown legs
- juvenile and nonbreeding lack white markings on face
- aged by relatively small round-tipped scapulars and coverts with thin, even-width pale fringing; note moderately pointed, pale-fringed primaries; a few freshly molted, nonbreeding scapulars and median coverts appear larger and more tapered and are darker with prominent white edges

**2 Nonbreeding, Nov., Calif.**
• differentiate from juvenile by larger, more tapered, and darker scapulars and coverts with prominent white edges; also note more blunt-tipped primaries; adults generally appear darker and less brownish overall than juveniles

**3 Nonbreeding with Ruddy Turnstones, Feb., Calif.**
• Which are the two Ruddy Turnstones? Look for paler overall color, brighter orange legs, whitish patch at breast-sides.

**4 Nonbreeding, Feb., Calif.**
• chisel-shaped bill unique to turnstones

**5 Nonbreeding with Rock Sandpiper and Black-bellied Plover, Mar., Calif.**

**6 Breeding, Apr., Calif.**
• differentiate from nonbreeding by white markings in face and breast

**7 Breeding, Apr., Calif.**
• pied upperparts similar in all plumages
• flight call is a shrill, rattled *breerp,* often extended into a chatter; much higher pitched than Ruddy

# Surfbird

***Aphriza virgata*** **p. 395**

**Size:** **L 9½–10½ in. (24–26 cm); WS 24¾–27¼ in. (62–68 cm); WT 133–230 g**
**slightly larger than turnstones**

**Structure:** **chunky with short legs and neck, short stout bill**

**Behavior:** **active; walks quickly over rocks, stopping to peck and tug at prey; rocky, wave-pounded coastlines; nests on tundra**

**Status:** **locally common along coast; accidental inland**

**1 Juvenile, Aug., Calif.**
- yellow legs and orange bill-base in all plumages
- aged by uniformly fresh plumage with subtle pale fringes and dark subterminal lines on scapulars, coverts, and tertials

**2 Nonbreeding, Nov., Calif.**
- differs from juvenile by less neatly arranged, plain gray upperparts
- stout, ploverlike bill distinctive

**3 Breeding (6), with two other species, Mar., Calif.**
- The Surfbirds are the largest in this roost of rock-dwelling shorebirds. What are the other two species?

**4 Breeding, late Mar., Calif.**
- breeding birds show a variable number of orange-and-black scapulars and heavily marked underparts

**5 First summer, Apr., Calif.**
- extremely worn tertials and coverts in spring indicate first summer

**6 Breeding, late Mar., Calif.**
• extent of rufous in scapulars highly variable, perhaps relating to health or hormonal levels; most breeding birds retain many "nonbreeding" wing coverts and tertials from midwinter molt

**7 Breeding, July, Calif.**
• flight call is a soft, mellow *whiff-if-if;* generally quiet

# Red Knot

***Calidris canutus*** **p. 398**

**Size: L 9¼–10 in. (23–25 cm); WS 22¾–24½ in. (57–61 cm); WT 93–215 g**
**about equal to dowitchers**

**Structure: chunky but long winged; short legs and medium-length bill; horizontal stance**

**Behavior: walks slowly, picking and shallowly probing; forages on sandy beaches; high-tide roost in marshes; breeds on tundra**

**Status: locally common but declining at scattered stop-over and wintering sites along coast; mostly rare inland**

**1 Juvenile, Aug., N.J.**

- juvenile and non-breeding mostly gray with barred flanks, dull yellowish legs
- aged by uniformly fresh plumage with delicate pale fringing and dark subterminal lines on scapulars, coverts, and tertials; breast washed with pinkish buff when fresh

**2 Juvenile (right) and nonbreeding with Black-bellied Plover, late Sept., N.J.**
• nonbreeding shows larger, less neatly arranged scapulars and less distinct pale fringing on upperparts without dark subterminal lines

**3 Nonbreeding with Dunlin, Western Sandpiper, and Black-bellied Plover, Nov., N.J.**
• Dunlin smaller and darker with whiter flanks, darker breast
• Which one is the Western Sandpiper?

**4 Breeding (right), with Semipalmated Plover and Dunlin, May, N.J.**
• Would you recognize these birds in silhouette?

**5 Breeding, May, Del.**
• with breeding Sanderlings and Horseshoe Crab
• colored leg-bands enable biologists to track the movements of this disappearing species

**6 Breeding, May, Fla.**
• wide variation in breeding plumage between subspecies and individuals; subspecies identification seldom possible in the field

**7 Breeding, with Ruddy Turnstones, Sanderlings, and Laughing and Herring Gulls, late May, N.J.**
• Large flocks gather in mid-late May along Delaware Bayshore to feed on Horseshoe Crab eggs. At this critical stop-over site, Red Knots must double their weight in order to fly nonstop to the Arctic and begin nesting in a landscape, at that early season, still dominated by snow and ice.

**8 Worn breeding, Aug., N.J.**
• because pale feather markings wear more quickly than dark markings, worn birds in late summer may become quite dark above

**9 May, N.J.**
• stocky shape and broad-based, pointed wings distinctive in flight

**10 Nonbreeding, Oct., N.J.**
• flight is fast and direct
• flight call, given infrequently, is a godwit-like *kawit* or *kwit-wit-wit*

**11 Nonbreeding with Black-bellied Plover and Dunlin, Dec., N.J.**
• gray rump unique among similar species

# Sanderling

***Calidris alba*** p. 400

**Size:** **L 7½–8 in. (18–20 cm); WS 16–18 in. (40–45 cm); WT 40–100 g**
**much larger than "peep"; same as Dunlin**

**Structure:** **chunky with rounded body, large head, thick neck; stout, medium-length bill; lacks hind toe**

**Behavior:** **walks steadily, picking at surface; sometimes digs or shallowly probes; "chased by waves" on sandy beaches; breeds on tundra**

**Status:** **common along coast; mostly rare inland except at a few favored stop-over sites**

**1 Juvenile, early Sept., N.J.**
• juvenile and non-breeding mostly white on face and underparts
• aged by uniformly fresh plumage with distinct black checkering in upperparts, black central crown stripe; breast washed with buff when fresh

**3 Juvenile, Sept., N.J.**
• Sanderlings look chunky from all angles
• pinkish cast is due to low-angle lighting

**4 Juvenile (center left) and nonbreeding, Sept., N.J.**
• nonbreeding shows plain, very pale gray upperparts; bland face with prominent dark eye

**5 Molting juveniles, Oct., N.J.**
• all plumages may show blackish bend of the wing, though this is sometimes concealed

**6 Nonbreeding with Western Sandpiper (center left) and Dunlin (center right and back), Dec., N.J.**
• What differences do you see in size, structure, and color?

**7 Nonbreeding with Dunlin, Nov., Fla.**
• both species feed on marine invertebrates washed up by waves

**8 Breeding, with Red Knot, Ruddy Turnstone, and Laughing Gulls, May, N.J.**
• general appearance like peep, but larger and with blander face-pattern

**9 Breeding, May, N.J.**
• breeding plumage highly variable; some show rich chestnut head and upperparts, while others remain relatively colorless; all show bland face-pattern
• very active and often seen running or "chased by waves"

**10 Breeding, May, Fla.**
• aggression display toward other Sanderlings for feeding space includes lowering of head and raising of back feathers
• lacks hind toe, unlike peep

**11 Nov., N.J.**
• rounded body distinctive

**12 Molting adults with Semipalmated Sandpipers, late July, N.J.**
• molting adults in late summer often show a reddish wash to throat and a contrasty mix of breeding and nonbreeding feathers above
• Semipalmated Sandpipers are smaller with less contrasty upperparts, bolder supercilium

**13 Molting adult with Semipalmated Sandpipers, Aug., N.J.**
• much larger, chunkier, and larger-headed than Semipalmated; incoming nonbreeding scapulars are very pale, so upperparts more contrasty than Semipalmated's

**14 Oct., N.J.**
• very white head, body, and underwing with bold wing stripe
• many show bold blackish leading edge to the wing (the blackish "shoulder" mark sometimes visible on standing birds)
• flight call is a soft, squeaky *pweet*

**15 Breeding, with four other species, May, N.J.**
• What are the other species?

# Semipalmated Sandpiper

***Calidris pusilla*** p. 402

**Size:** L 5¼–6 in. (13–15 cm); WS 13½–14¾ in. (34–37 cm); WT 14–41 g
slightly larger than Least Sandpiper

**Structure:** plump body with more "balanced" (less front-heavy) proportions than Western; straight, blunt-tipped bill; partial webbing between middle and outer toes

**Behavior:** walks steadily, picking nervously at surface; frequent aggressive encounters with other birds; found mostly on broad mudflats, often in large flocks; nests on tundra

**Status:** most abundant "peep" at many Atlantic Coast sites; uncommon inland; rare in West; a few winter in s. Florida

**1 Juvenile, Sept., N.J.**
• all plumages show dark olive legs, short primary projection
• aged by uniformly fresh plumage with crisply pale-fringed upperparts imparting scaly look
• juveniles look grayish overall with blurry streaking at breast-sides, variably darkish ear coverts; very fresh birds show a buffy wash on breast, crown, and upperparts

**2 Juveniles with juvenile Western (left) and Least (right) Sandpipers, Sept., N.J.**
• worn juveniles show darker cap, ear coverts, and lores than fresh birds
• juvenile Least looks smaller, darker, and browner with smaller, darker head, yellowish legs
• juvenile Western looks longer billed and whiter faced with reddish-fringed upper scapulars and grayish cap; compared to the long-billed Semipalmated next to it, note the Western's pale lores, ear coverts, and cap and narrower dark smudge at breast-side

**3 Juvenile, Aug., N.Y.**
• bill straight and blunt-tipped with a subtle expansion near tip visible from above
• on this fresh juvenile, note the evenly buffy-washed cap, similar in color to the scapulars

**4 Juvenile molting to first winter, Sept., N.J.**
• much juvenal plumage remains, as is typical on late fall migrants; molt to first winter is completed on the wintering grounds, mostly outside the United States

**5 Worn juvenile (left) with worn juvenile and nonbreeding Western Sandpipers, Sept., N.J.**
• Westerns typically look longer billed and larger-headed than Semipalmated
• juvenile Westerns begin the molt to first winter much earlier than Semipalmateds and typically show a pale gray cap from early in the season; the rufous-fringed upper scapulars, often clearly brighter than on juvenile Semipalmated, have become quite dull on this individual

**6 Nonbreeding, Mar., Bahamas**
• compared to Western, note smallish head, stubby bill, sharply defined cap, smudgy breast-sides, smooth gray-brown upperparts
• both species show partially webbed toes

**7 Nonbreeding Western Sandpiper, Sept., N.J.**
• compared to Semipalmated, note larger head, thicker neck, and longer, finer-tipped bill; slightly paler and grayer with less contrasting cap, more prominent streaks at breast-sides

**8 Breeding, with Sanderling, May, N.J.**
• Sanderling is much larger with more rotund body, shorter legs, and blander face-pattern; bright Sanderlings often show rich rufous color and dark hood
• note long, fine-tipped bills on some of these eastern birds

**9 Breeding, May, N.J.**
• bill size variable; females longer billed than males; western birds have shorter, more blunt-tipped bills than eastern birds

**10 Breeding, with chick, June, Alaska**
- limited streaking at breast-sides
- Do you notice anything distinctive about the chick?

**11 Molting adult, late July, N.Y.**
- typical appearance during fall migration with a mix of blackish (worn breeding) and gray (fresh nonbreeding) feathers on upperparts; adults retain at least a few breeding scapulars and tertials throughout fall migration; molt is not completed until wintering grounds are reached

**13 Molting adult (left) with Western Sandpiper, Western Willet, and Short-billed Dowitcher, late July, N.J.**
• note Western Sandpiper's more front-heavy proportions with larger head, broader chest, longer bill; Semipalmated's proportions are relatively more "balanced"

**14 Molting adults, Aug., N.Y.**
• Semipalmateds are more aggressive than other peeps while feeding, frequently engaging in squabbles with nearby birds

**15 Breeding, June, N.J.**
• grayish brown above with thin wing stripe; mostly white below with limited dark at breast-sides
• flight call is a rough *chrrk* or higher, sharper *chit*

# Western Sandpiper

***Calidris mauri*** p. 404

**Size:** **L 5½–6¾ in. (14–17 cm); WS 14–15 in. (35–37 cm); WT 22–35 g**
**slightly larger than Semipalmated Sandpiper**

**Structure:** **more front-heavy than Semipalmated with larger head, thicker neck; longer, finer-tipped bill, often with slight droop; roosting birds stand more upright than Semipalmated; partial webbing between middle and outer toes (as on Semipalmated)**

**Behavior:** **walks steadily, picking nervously at surface; found mostly on broad mudflats, often in large flocks; inland wetlands on migration; nests on tundra**

**Status:** **most abundant "peep" at many Pacific Coast sites; less common in Northeast; uncommon to rare inland**

**1 Juvenile, Sept., N.J.**
• all plumages show dark legs, short primary projection
• aged by uniformly fresh plumage and crisply pale-fringed upperparts imparting scaly look
• juveniles look grayish overall with pale heads, contrasting rufous-edged upper scapulars, finely streaked breast-sides

**2 Juvenile with juvenile Semipalmated (center) and breeding Least (right) Sandpipers, Aug., N.J.**
• paler head and more contrasting rufous-edged upper scapulars than Semipalmated; Western's scapulars and coverts are subtly narrower with thinner fringes near tip; overall impression of more uniformly scaled appearance on Semipalmated

**3 Molting juvenile, Sept., N.J.**
• many juveniles become very pale-headed and begin to replace scapulars soon after arrival at stop-over sites; Semipalmateds hold full juvenal plumage longer
• very long bill indicates female

**4 First winter, Oct., N.J.**
• aged by contrast between fresh, plain gray scapulars with more worn (brownish-tinged) and relatively narrower retained juvenal coverts and tertials

**5 Nonbreeding, Jan., Fla.**
• pale gray upperparts with indistinct cap, fine dark shaft streaks; best identified by structure

**6 Nonbreeding, Jan., Fla.**
• from first winter by lack of contrast between scapulars and coverts

**7 Nonbreeding, with Sanderling and Dunlin, Jan., Fla.**
• superficially similar to Dunlin but much smaller and whiter-chested
• relatively short bill suggests male; males winter farther north than females and return to the breeding grounds earlier

**8 First summers with Sanderlings, June, N.J.**
• One-year-old birds may attain a plumage like breeding, nonbreeding, or somewhere in between. Those that do not attain full plumage remain at wintering or stop-over sites for the summer and begin the complete molt to nonbreeding plumage as early as May. The stretching bird is already beginning its first flight feather molt; note the partially grown (blackish) inner primaries and outer secondaries.

**9 Breeding (right), with breeding Semipalmated Sandpiper, Apr., Tex.**
• early spring Westerns may show very little color, but note more front-heavy structure and more extensive breast-streaking than Semipalmated

**10 Breeding, with Dunlin (left), Apr., Tex.**
• extensive rufous in crown and scapulars; breast-streaking usually extends down to legs

**11 Breeding, May, Alaska**
• huge flocks gather at favored spring stop-over sites, such as this one in Cordova, Alaska, where 3 to 4 million Westerns stage en route to the breeding grounds

**12 Breeding, at nest, June, Alaska**
• Due to short summers, Arctic-breeding shorebirds must begin nesting under very harsh conditions. Their survival during these lean times depends on fat reserves accumulated at spring stop-over sites.
• Can you see enough to identify this bird?

**13 Breeding, June, Alaska**
• extensive triangular spots on underparts

**14 Breeding, May, Alaska**
• body often looks quite rounded, especially after feeding continuously for a week at staging area

**15 Molting adults (4) with molting adult Semipalmated Sandpipers (4), Aug., N.J.**
- adults molt rapidly upon arrival at coastal stop-over sites
- fresh nonbreeding scapulars are neutral gray on Western, darker and more brownish gray on Semipalmated; retained rufous-based scapulars and spotted underparts distinctive

**16 Molting adults with Semipalmated Sandpiper (center back), Aug., N.J.**
- by late summer, most Westerns have replaced virtually all upperparts including tertials and wing coverts, but retain a few telltale spots on belly; Semipalmateds always retain coverts, tertials, and some scapulars during fall migration
- note Semipalmated's smaller head, slimmer chest, shorter bill, and less upright stance

**17 Molting adults with Semipalmated Sandpipers, Aug., N.J.**
- note flight feather and wing-covert molt on Westerns but not on Semipalmateds; Atlantic Coast Semipalmateds wait until the wintering grounds are reached to molt flight feathers
- flight call is a husky *jrrk* or higher, sharper *jeek;* usually higher pitched than Semipalmated (almost recalling White-rumped but not nearly as high pitched)

# Least Sandpiper

***Calidris minutilla*** p. 411

**Size:** **L 4½–4¾ in. (11–12 cm); WS 13¼–14 in. (33–35 cm); WT 9–27 g**
**smallest shorebird; slightly smaller than Semipalmated**

**Structure:** **chunky, short-tailed, short-winged; smaller-headed, less attenuated than other "peeps" with finer, slightly drooped bill; crouched posture**

**Behavior:** **walks slowly, often sweeping side-to-side as it picks rapidly at surface; muddy areas near grass or debris; nests in boggy areas near treeline**

**Status:** **common and widespread; most numerous peep at interior sites, but outnumbered by Semipalmated or Western along coast**

**1 Juvenile, Aug., N.Y.**

- all plumages browner than other peeps with limited white in face or breast, usually dark cheek-patch, yellowish legs, very short primary projection; tail projects slightly beyond primaries
- aged by uniform fresh plumage with crisply pale-fringed upperparts imparting scaly look
- juveniles look reddish brown overall with extensive streaking at breast-sides often forming complete breast-band

**2 Juveniles with juvenile Semipalmated Sandpipers (center and 2 at right), Aug., N.Y.**
• juvenile Semipalmated looks larger and grayer with whiter throat and supercilium, larger head, heavier bill, darker legs

**3 Juvenile, Sept., N.J.**
• rich rufous upperparts with thin white mantle "V" and white outer tips to scapulars

**4 Juvenile, Aug., N.J.**
• even pale individuals lack extensive clean white throat of Semipalmated and Western

**5 Juveniles and molting to first winter, Sept., N.J.**
• Molt timing varies from one individual to the next. While three birds are in full juvenal plumage, second bird from left has replaced most of head, breast, and scapulars but retains juvenal wing coverts. In general, longer-distance migrants undergo more of their molt on the wintering grounds, saving energy for migration. The heavily molted bird may be one that will winter nearby.

**6 Nonbreeding, Jan., Fla.**
• browner than Semipalmated or Western with clean white limited to belly and chin
• note smudgy dark centers to scapulars and virtually no primary projection
• typical crouched posture

**7 Breeding, Apr., Tex.**
• breeding birds in fresh spring plumage show blackish scapulars with frosty edges

**8 Breeding, with breeding Semipalmated Sandpiper (center), May, N.J.**
• at long range, Least looks smaller, darker, and browner than Semipalmated, with darker head and breast; the head looks smaller and the body "stubbier" (less attenuated)
• at closer range, also note Least's finer-tipped, subtly decurved bill and yellowish legs

**9 Worn breeding with molting adult Western Sandpipers (foreground), July, N.J.**
- much smaller, darker, and browner than Western with smaller, darker head; note tail projecting past primaries
- What is the fourth bird in the background?

**10 Worn breeding, Aug., N.J.**
- when worn, upperparts may become very dark

**11 Worn breeding, July, N.J.**
- complete breast-band of streaks
- often prefers to forage near mats of vegetation rather than on open mudflat

**12 Worn breeding and fresh nonbreeding (center), July, N.J.**
• Molt timing varies between populations. Birds migrating to South America show little or no molt while at stop-over sites, whereas birds wintering in North America may begin to molt by early July and attain nearly full nonbreeding plumage by mid-August. Additionally, first-summer birds attain this plumage up to a month earlier than adults. The bird at center may be either an adult or a first summer.

**13 Mixed flock, Aug., N.J.**
• Which one is the Least, and what plumage is it in?
• What other species are present?

**14 Sept., N.J.**
• small, brownish head; short wings with darker underwing than Semipalmated or Western
• flight call is a variable, drawn-out *kreeet* with a trilled musical quality

**15 Juveniles with juvenile Semipalmateds (right and top), Aug., N.Y.**
• browner than Semipalmated with smaller, darker head, shorter wings and tail, broader dark midline down rump

# White-rumped Sandpiper

*Calidris fuscicollis* p. 412

**Size:** L 6–6¾ in. (15–17 cm); WS 16–18 in. (40–45 cm); WT 25–51 g
slightly larger than "peep"

**Structure:** "athletic" look with bulky chest, sleek body, long wings; primaries project well past tertials and tail tip; short legs; medium-length, fine-tipped bill; horizontal stance

**Behavior:** walks steadily, picking and shallowly probing; grassy borders of mudflats, shallow water; nests on tundra

**Status:** fairly common but easily overlooked among large flocks of shorebirds; most migrate through Great Plains in spring, Atlantic Coast in fall

**1 Juvenile, Aug., N.Y.**
- all plumages show thin, fairly prominent supercilium on otherwise darkish head and breast; extremely long primaries project well past tertials and tail tip (similar only to Baird's Sandpiper)
- aged by uniformly fresh plumage with crisply pale-fringed upperparts imparting scaly look
- juveniles look grayish overall with contrasting reddish-fringed upper scapulars

**2 Juvenile with nonbreeding Dunlin, Oct., N.J.**
- often holds juvenal plumage late into the fall
- juveniles migrate relatively late in fall, not appearing until October in most areas

**3 Nonbreeding, Oct., N.Y.**
- extensive smudgy streaking on face, breast, and flanks; often looks hooded
- most shorebirds have extremely sensitive, prehensile bill-tip for feeling and gripping prey beneath mud

**4 Breeding, May, Neb.**
- extensive fine streaking on breast and flanks; birds in early spring may look quite pale, with broad pale fringes to upperparts
- all plumages may show unique reddish or brownish base to lower mandible (particularly obvious here)

**5 Breeding (center front), with three other species, early June, N.J.**
- What are the other species? Focus on size, structure, and overall color pattern.

**6 Breeding, with Semipalmated Sandpiper, early June, N.J.**
• note extensive fine breast-streaking and long primaries projecting well past tertials and tail tip
• even though way out of focus, the Semipalmated behind looks slightly smaller with much whiter flanks and throat, less attenuated silhouette

**7 First summer (2nd from right) with breeding Semipalmated Sandpipers, June, N.J.**
• One-year-old birds may attain a plumage like breeding, nonbreeding, or somewhere in between. Those that do not attain full breeding plumage remain at wintering or stop-over sites for the summer. Note size, attenuated shape, and "dirty" flanks; primaries project well past tertials and tail tip.
• Are all the other birds Semipalmateds? Consider size, structure, and patterns of face, breast, and scapulars

**8 Nonbreeding (left), with molting adult Semipalmated Sandpiper, late Aug., N.J.**
• White-rumped is larger, much more attenuated, and has a dark-hooded look with isolated whitish supercilium; Semipalmated shows less attenuated shape, whiter flanks and throat, more isolated dark patch at breast-side

**9 Molting adult, Aug., N.Y.**
- fall migrants may show extensive nonbreeding plumage but wing coverts, tertials, and some scapulars are often retained until the wintering grounds are reached
- primaries extend well past tertials and tail tip

**10 Breeding, with single Semipalmated Sandpiper (lower left), May, Neb.**
- very long wings and complete white rump-band set off by dark tail
- flight call is a very high, insectlike *tzeet;* much higher-pitched than calls of other shorebirds

# Baird's Sandpiper

*Calidris bairdii* p. 414

**Size:** L 5¾–7¼ in. (14–18 cm); WS 16–18½ in. (40–46 cm); WT 28–58 g
slightly larger than "peep"

**Structure:** much like White-rumped but with steeper forehead and straighter, often finer-tipped bill; much longer wings than peep

**Behavior:** walks steadily, picking at surface; dry edges of mudflats, short-grass pastures; breeds on tundra

**Status:** uncommon to fairly common migrant through Great Plains and Mountain West, where it is sometimes the most numerous small shorebird; scarce elsewhere in fall

**1 Juvenile, Aug., Conn.**
- all plumages with pale but complete breast-band, clean white flanks, and plain face with dark lores invaded by pale supraloral spot; extremely long primaries project well past tertials and tail tip
- aged by uniformly fresh plumage with crisply pale-fringed upperparts imparting scaly look
- juveniles look buffy overall, brightest on head; full juvenal plumage usually held throughout fall migration

**2 Juvenile (left) with juvenile Least Sandpipers, Sept., N.J.**
• larger, longer winged, and buffier than Least, with plainer face-pattern, dark legs

**3 Juvenile (right) with juvenile Semipalmated and Stilt Sandpipers and Short-billed Dowitcher, Sept., N.Y.**
• juvenile Semipalmated shows similar scaly upperparts but is smaller with much shorter primary projection, whiter breast, more contrasty head-pattern, less buffy overall

**4 Nonbreeding, Dec., Chile**
• plain buffy head and breast with bland face-pattern
• bill straight and fine-tipped

**5 Nonbreeding (left), with nonbreeding White-rumped Sandpiper, Nov., Argentina**
• buffier than White-rumped, with plainer head-pattern and straighter bill

**6 Molting adult, May, Neb.**
- spring migrants typically look much like nonbreeding but with a scattering of dark scapulars

**7 Molting adult, with Pectoral Sandpiper, Aug., Kans.**
- small female Pectoral may approach Baird's in size, but note Pectoral's strongly marked breast-band, yellowish legs and bill-base, heavier chest, shorter wing point, and smaller head

**8 Breeding, June, Alaska**
- in full breeding plumage, breast and head are more heavily streaked, and scapulars are blackish with silvery edges
- note long wings and relatively plain face with dark lores invaded by pale supraloral spot

**9 Molting adult, with Least Sandpipers and Wilson's Phalaropes, July, Ariz.**
• Which is the Baird's, and what are the most noticeable differences from Least?

**10 Juvenile, Sept., N.J.**
• long pointed wings and relatively long tail; rather plain, pale, and uniform buffy above
• flight swift and steady with level glides
• flight call is a low, trilled *preep,* similar to Pectoral Sandpiper but higher pitched

**11 Juvenile (left) with Least Sandpipers, Sept., N.J.**
• larger, longer winged, and longer tailed than Least, with more uniformly buffy head and breast

# Pectoral Sandpiper

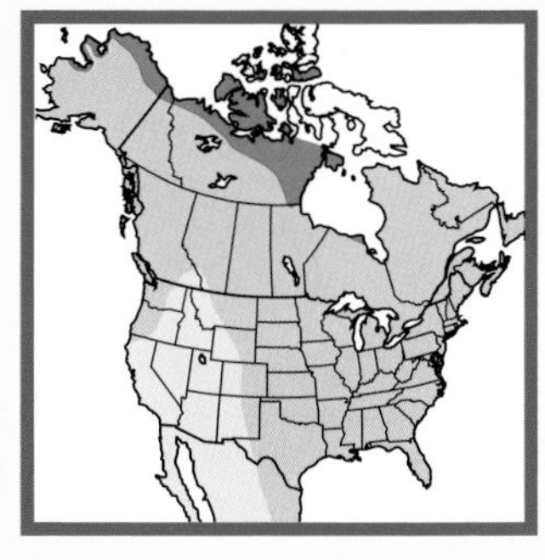

***Calidris melanotos*** p. 416

**Size:** **L 7¾–9¼ in. (19–23 cm); WS 16¾–19½ in. (42–49 cm); WT 50–117g**
**much larger than "peep"; usually larger than Dunlin; males 25–30 percent larger than females**

**Structure:** **bulky and broad-chested but attenuated; smaller head, longer neck and legs than Baird's; primaries equal with tail tip**

**Behavior:** **walks steadily, picking at surface; grassy borders of mudflats, wet pastures; breeds on tundra**

**Status:** **fairly common migrant through Great Plains in spring and fall, Atlantic Coast in fall; less common elsewhere**

**1 Juvenile, Aug., N.J.**
• all plumages show dark bib of streaks clean-cut from white belly; yellowish legs; long primaries project well beyond tertials but equal with tip of long tail; bill may be all dark or with dull orange-yellow base
• aged by uniformly fresh plumage and relatively small scapulars and coverts with crisp rufous fringes and white outer tips

**2 Juvenile, early Sept., N.Y.**
• rain can make birds look duller, darker, and messier than normal
• aged by fresh, round-tipped, relatively narrow coverts with whitish outer tips

**3 Breeding male, June, Alaska**
• differs from juvenile by larger, less neatly arranged scapulars and coverts with duller fringes
• males are larger with heavier, more extensive breast-markings than females

**4 Breeding male, late June, Alaska**
• On the breeding grounds, male has fatty inflatable throat sac that hangs down like a dewlap; the sac is used to produce hooting calls as male performs courtship display flight. Male is promiscuous, displaying and gathering a harem of females in his territory, but takes no part in nesting or brood rearing duties.

**5 Breeding female, June, Alaska**
• females are smaller (some half the weight), duller, and often darker billed than males; breast-markings are sparser though still clean-cut from white belly

**6 Worn breeding male with Short-billed Dowitcher, Aug., Del.**
• largest males are dowitcher-sized

**7 With juvenile Semipalmated Sandpiper, Aug., N.J.**
• What age and sex is this Pectoral?

**8 Worn breeding female (center right) with White-rumped (left) and Semipalmated Sandpipers and Greater Yellowlegs (right), Aug., N.Y.**
• smallest females may nearly overlap with White-rumped Sandpiper; breast pattern always distinctive
• size, leg color, and breast pattern quickly distinguish the three smaller birds

**9 Molting adult, Aug., N.J.**
• molt to nonbreeding takes place largely on the wintering grounds, so nonbreeding plumage is rarely seen in North America; the soft gray-brown tone of that plumage is visible here in a few newly grown scapulars

**10 Sept., N.J.**
• chunky and thick-chested with relatively small head; dark breast-band always prominent

**11 Juvenile, Sept., Ontario**
• long, broad-based, pointed wings and relatively long tail with broad dark divide; very faint wing stripe
• juveniles show reddish upperparts with white mantle lines
• flight call is a low, trilled *chrrk,* similar to Baird's Sandpiper but lower pitched

# Purple Sandpiper

***Calidris maritima*** p. 419

**Size:** **L 8–8¾ in. (20–22 cm); WS 16¾–18½ in. (42–46 cm); WT 52–105 g**
**equal to or slightly larger than Dunlin**

**Structure:** **chunky and rounded with short thick neck, short legs; medium to longish drooped bill; low, crouched posture**

**Behavior:** **walks steadily, rapidly picking at surface; hops and flutters from one rock to another; rocky, wave-washed coastlines, jetties; nests on barren tundra**

**Status:** **fairly common but highly localized along seacoast; rare but regular visitor to Great Lakes and Gulf Coast**

**1 Juvenile, Aug., England**
- all plumages very dark overall with dark head and chest, heavily streaked flanks, dull orange legs and bill-base
- aged by uniformly fresh plumage with crisply white-fringed coverts and lower scapulars, buff-fringed upper scapulars and mantle
- this plumage is rarely seen away from the breeding grounds

**2 First winter, Nov., N.J.**
• aged by contrast between crisply white-fringed coverts and tertials with plain gray scapulars; distinct purple sheen visible in good light

**3 Nonbreeding, Nov., N.J.**
• differs from first winter by broader, less neatly arranged coverts and tertials with diffuse pale fringes

**4 Nonbreeding with Sanderlings, Mar., N.J.**
• Purples are the darkest and Sanderlings the palest small shorebirds; both regularly roost on rocky coastlines

**5 Nonbreeding, Feb., N.J.**
• smoother, more solid gray head and breast and brighter orange legs and bill-base than Rock Sandpiper (no overlap in range)

**6 Nonbreeding (lower left) with Dunlin and Ruddy Turnstones, Feb., N.J.**
• a typical winter assemblage along rocky coastlines
• What differences do you see in size, structure, and color pattern?

**7 Breeding, May, N.J.**
• wintering birds molt to breeding plumage in April and May before departing to the Arctic

**8 Nonbreeding with Ruddy Turnstones, Feb., N.J.**
• very dark with thin wing stripe

**9 Nonbreeding, Dec., N.J.**
• flight call is a scratchy *kweesh*

# Rock Sandpiper

***Calidris ptilocnemis*** **p. 420**

**Size:** L 7¼–9½ in. (18–24 cm); WS 13½–18½ in. (34–46 cm); WT 71–114 g
equal to or slightly larger than Dunlin

**Structure:** chunky and rounded with short thick neck, short legs; medium to longish drooped bill; low, crouched posture

**Behavior:** walks steadily, rapidly picking at surface; hops and flutters from one rock to another; rocky, wave-washed coastlines, jetties; nests on tundra

**Status:** fairly common but highly localized along seacoast; less common in southern part of winter range; three subspecies in North America: *C. p. tschucktschorum* breeds on mainland Alaska, winters south to California; *C. p. ptilocnemis* breeds on Pribilof Is., winters to s. Alaska, rarely Washington; *C. p. couesi* breeds Aleutian Is. and is largely resident but may winter to Washington.

**1 Juvenile *tschuktschorum*, Aug., Alaska**
- all plumages with yellow to dark olive legs and bill-base
- aged by uniformly fresh plumage with crisply pale-fringed upperparts; this plumage rarely seen away from the breeding grounds
- juvenile *tschucktschorum* is relatively small, dark, and short-billed with heavily streaked flanks; juvenile *couesi* is similar

**2 Juvenile *ptilocnemis,* July, Pribilof Is., Alaska**
• larger and distinctly paler than other races, with lighter flank streaking

**3 Nonbreeding *tschuktschorum* or *couesi,* Feb., Calif.**
• from Purple Sandpiper by range, duller, more yellowish legs and bill-base, more speckled neck and breast, lack of purple sheen above

**4 Nonbreeding *ptilocnemis* (left), with *tschuktschorum* or *couesi*, Feb., Alaska**
• *ptilocnemis* distinctly paler than other races
• smallest, darkest bird (center) is likely *couesi* and others *tschuktschorum,* but variation precludes certain subspecific identification

**5 Nonbreeding *tschuktschorum* or *couesi* (center), with Black Turnstones and Surfbird, Feb., Calif.**

**6 Breeding *ptilocnemis*, June, Pribilof Is., Alaska**
• breeding birds show rufous upperparts and variable black belly-patch recalling Dunlin
• *ptilocnemis* is large and pale-headed with bright but pale rufous upperparts and well-defined cheek and belly patches; bright white underwing and very broad wing stripe distinctive

**7 Breeding *C. p. quarta*, June, Russia**
• relatively dull upperparts, darkish head, and heavily streaked flanks; poorly defined belly- and cheek-patches; *couesi* is similar
• *quarta* is largely resident on islands off Russia's Kamchatka Pen. but has occurred in w. Aleutian Is.

**8 Breeding *tschuktschorum*, June, Alaska**
• *tschuktschorum* similar to *ptilocnemis* but slightly smaller and shorter billed with slightly darker rufous upperparts and weaker belly-patch; best distinguished by darker underwing pattern and narrower wing stripe
• some birds are not safely identifiable to subspecies

**9 Breeding *ptilocnemis*, June, Pribilof Is., Alaska**
- wing-flash display is performed by males on the breeding grounds
- *ptilocnemis* has cleaner white underwing and broader wing stripe than other subspecies

**10 Molting adult *tschuktschorum*, Aug., Alaska**
- the molt to nonbreeding takes place largely at postbreeding staging areas

**11 Molting adults with Black Turnstones (the black-headed birds) and Surfbird, July, Alaska**
- wing stripe broader and underwing cleaner white in *ptilocnemis* than in other races but often difficult to judge on molting birds; all races with broader wing stripes than Purple Sandpiper
- flight call is a scratchy *kwit;* calls of *ptilocnemis* are lower pitched than those of other races
- Can you find the Surfbird?

# Dunlin

*Calidris alpina* p. 422

**Size:** L 7½–8¾ in. (18–22 cm); WS 15–17½ in. (38–44 cm); WT 48–64 g about equal to Sanderling

**Structure:** chunky with short thick neck, short legs, long drooped bill

**Behavior:** walks steadily, rapidly picking and probing; mudflats, beaches; often in large flocks; nests on tundra

**Status:** most abundant wintering shorebird in many coastal areas; scarce migrant at most inland sites

**1 Molting juvenile, Oct., N.J.**
- all plumages with short black legs and long, drooped black bill
- aged by crisply buff-fringed coverts and lower scapulars, rufous-fringed upper scapulars
- many head, mantle, and scapular feathers have already been replaced by plain gray nonbreeding feathers; most birds begin molting out of juvenal plumage before departing the Arctic; fresher juveniles are brighter above with more heavily streaked breast and belly

**2 First winter, Oct., N.Y.**
• most of upperparts have been replaced by plain gray feathers; most juvenal coverts and tertials are retained through the winter; the few remaining juvenal scapulars may be replaced later in the season

**3 Nonbreeding, Jan., N.J.**
• from first winter by broader, less neatly arranged coverts and tertials with diffuse pale edges
• nonbreeding birds have a gray-hooded appearance

**4 Nonbreeding with Black-bellied Plovers, Western Sandpipers, and Herring Gull, Dec., N.J.**

**5 First winter (right) and nonbreeding with Sanderling (center) and first winter Western Sandpiper (left), Nov., N.J.**
• Western is like a miniature Dunlin but is longer legged and whiter chested

**6 Breeding, with Sanderling (center and lower left) and Red Knot, May, N.J.**
• Which is larger, Dunlin or Sanderling? Does the same hold true in other photos?

**7 Breeding and first summer, June, N.J.**
• many one-year-old birds that oversummer on the wintering grounds never attain any breeding plumage and may begin the molt to second-winter plumage as early as late April

**8 First summer, Aug., N.Y.**
• Some one-year-olds attain partial breeding plumage in spring. These birds may molt to nonbreeding on a more adultlike schedule than the bird in photo #7. This bird still retains extensive breeding plumage. Molting adults look similar, but this individual showed excessively worn outer primaries, indicating retained juvenal feathers.

## Dunlin Subspecies Comparision

**9 Breeding *C. a. hudsonia,* June, Manitoba**
• breeds in cen. Canada; winters on Atlantic and Gulf coasts
• large and long billed; more heavily streaked on breast, flanks, and undertail coverts than *pacifica* or *arcticola*

**10 Breeding *hudsonia* with Short-billed Dowitchers, May, N.J.**
• breast-streaking results in indistinct white band between breast and belly at long range
• molts to nonbreeding on breeding grounds

**11 Breeding *pacifica* with Western Sandpiper, June, Alaska**
• breeds in w. Alaska; winters on Pacific Coast
• large and long billed; cleaner white breast-band, flanks, and undertail coverts than *hudsonia*
• molts to nonbreeding on the breeding grounds

**12 Breeding *pacifica* with Western Sandpipers, May, Alaska**
• grayish head and breast with distinct white band between breast and belly at long range
• note that all subspecies are variable, so subspecies identification is not always possible

**13 Breeding *arcticola,* June, Alaska**
• breeds in nw. Canada and n. Alaska; winters in se. Asia
• smaller and shorter billed than *pacifica,* with whiter head and breast
• molts to nonbreeding on breeding grounds; note fresh inner, worn outer primaries

**14 Breeding *arctica* (center) with *schinzii,* Aug., U.K.**
• breeds in Greenland; winters in w. Africa; has been reported on the U.S. Atlantic Coast
• small (slightly larger than Western Sandpiper), short billed and dull with dull silvery-gold scapulars; molts to nonbreeding mostly on wintering grounds (compare with photo #8)

**15 Breeding, May, Fla.**
• Which subspecies is this?
• flight call is a buzzy, drawn-out *jeeezp,* all on one pitch; all subspecies sound similar

**16 Nonbreeding, Feb., N.J.**
• Dunlin regularly roost on coastal rock jetties during cold winter months.

**17 Nonbreeding, with two other species, Oct., N.J.**
• The Dunlins are the long-billed, gray-headed birds. What are the other two species? Consider size, head pattern, wing pattern, and bill shape. See caption #5 for clues.

# Stilt Sandpiper

***Calidris himantopus*** **p. 426**

**Size:** **L 8–$9\frac{1}{4}$ in. (20–23 cm); WS $17\frac{1}{4}$–$18\frac{3}{4}$ in. (43–47 cm); WT 50–70 g**
**about equal to Dunlin; smaller than dowitcher or yellowlegs**

**Structure:** **long legs; long, fine-tipped, slightly drooped bill; oddly puffy rear crown; longer legged, shorter billed than dowitcher; shorter legged, longer billed than Lesser Yellowlegs**

**Behavior:** **walks steadily with neck out, bill down; probes deeply or picks at surface in shallow water, often belly-deep; shallow ponds; breeds on tundra**

**Status:** **fairly common migrant through Great Plains in spring and fall, Atlantic Coast in fall; scarce elsewhere**

**1 Juvenile, Aug., N.J.**
• all plumages with long yellowish legs; long, slightly drooped, dark bill, bold supercilium; holds bill mostly perpendicular to water while feeding
• aged by uniformly fresh plumage with crisply pale-fringed upperparts imparting scaly look
• fresh juveniles show a pinkish buff wash on breast

**3 Juvenile (left), with juvenile Short-billed Dowitchers, Semipalmated Sandpipers, and Lesser Yellowlegs (background), Sept., N.J.**

- longer legged and taller than dowitchers, but smaller head and body with shorter, darker, finer-tipped bill
- many scapulars have been replaced by plain gray nonbreeding feathers

**4 First winter (left), with Lesser Yellowlegs, Nov., Tex.**

- smaller than Lesser Yellowlegs with longer drooped bill, slightly shorter legs, bolder supercilium, and more capped appearance; often feeds with dowitcher-like probing motion, unlike yellowlegs
- most of upperparts are like nonbreeding, but juvenal wing coverts, tertials, and primaries remain

**5 Nonbreeding with another species, Mar., Tex.**

- note unique head shape with sloping forehead and rounded rear crown
- nonbreeding birds show plain gray upperparts
- What is the bird in the background?

**6 Molting adult (left), with breeding Short-billed Dowitcher, Apr., Tex.**
• most easily identified by size, structure, and behavior
• spring migrants often show a mix of breeding and nonbreeding; full plumage acquired by late April or early May

**7 Breeding, July, Alaska**
• breeds in moist tundra near coast

**8 Molting adults (right) with Short-billed Dowitchers, late July, N.Y.**
• due to longer legs and shorter bill, Stilt Sandpipers tilt up higher than dowitchers when feeding
• often associates with dowitchers but much more active feeding style

**9 Molting adult (left) and juvenile, Aug., N.J.**
• basically similar appearance; aged by pattern on wing coverts, tertials, and scapulars

**10 Molting adults, Apr., Tex.**
• narrow wings with white center of underwing; long bill and long foot projection

**11 Juvenile (right) with Lesser Yellowlegs, Sept., N.J.**
• smaller, longer billed, and slightly shorter legged than Lesser Yellowlegs; wings are shorter, narrower, more swept back; wingbeats quicker
• Can you see the different tail pattern and leg color?
• flight call is a low, muffled *pewf*

# Buff-breasted Sandpiper

***Tryngites subruficollis*** **p. 431**

**Size:** L 7¼–8 in. (18–20 cm); WS 17¼–18¾ in. (43–47 cm); WT 43–94 g
same as a small Pectoral Sandpiper

**Structure:** sleek and attenuated with small squarish head, thin neck, short straight bill; upright stance

**Behavior:** walks quickly with peculiar pigeonlike (head-bobbing) gait; picks at surface; short-grass pastures, plowed fields, dry sandflats; nests on tundra

**Status:** uncommon migrant through Great Plains though sometimes in large flocks; small numbers (virtually all juveniles) reach Atlantic and Northwest Coasts in fall

**1 Juvenile, Sept., Conn.**
• all plumages buffy overall with yellowish legs, short, straight, black bill; plain face with prominent dark eye; plain breast with spots at breast-sides
• only juvenile Ruff has similarly plain head and buffy breast: note differences in size and structure (especially bill shape and primary projection); compare with Baird's and Pectoral Sandpipers
• aged by uniformly fresh plumage with relatively small, crisply pale-fringed scapulars, coverts and tertials

**2 Juveniles with Pectoral Sandpiper (right), Sept., N.J.**
• shorter-billed than Pectoral but structure otherwise similar; breast pattern distinctive

**3 Breeding male and (presumably) female during spring migration, May, Neb.**
• differs from juvenile by longer, more loosely arranged scapulars, coverts, and tertials; buff, not white, fringes above
• all plumages show white underwing with prominent dark "comma" at wrist
• male (left) is performing double-wing courtship display at spring staging area; males exhibit wing-flash displays on spring migration at breeding grounds and occasionally in fall (by juveniles)

**4 Breeding male exhibiting single-wing courtship display, June, Alaska**
- uniquely plain, even-toned upperparts
- males use striking underwing pattern to attract females to lek
- males are polygamous, copulating with multiple females at lek; males do not participate in nesting or brood rearing
- females often copulate with more than one male, thus improving gene pool

**5 Molting adult, Sept., N.J.**
- fresh nonbreeding scapulars show broad, smudgy brown fringes; full nonbreeding plumage is rarely if ever seen in North America

**6 Juvenile (right) with Pectoral Sandpipers, Sept., N.J.**
- note differences in underwing and breast patterns
- flight call is a soft, sneezing *gewt,* similar to Pectoral Sandpiper but lower and softer; seldom heard

# Short-billed Dowitcher

***Limnodromus griseus*** p. 434

**Size:** L 9¼–10 in. (23–25 cm); WS 18–20½ in. (45–51 cm); WT 65–154 g
equal to or larger than Lesser Yellowlegs; usually smaller than Long-billed Dowitcher

**Structure:** chunky but attenuated with short legs, long snipelike bill, horizontal stance; usually slimmer and flatter backed than Long-billed in relaxed feeding pose; bill averages shorter and thicker than Long-billed (much overlap in bill shape and length with Long-billed, but extremes diagnostic)

**Behavior:** probes deeply and rapidly ("sewing-machine" style) in shallow water; feeds with flock in one small area; tidal mudflats and nearby pools; nests on taiga/tundra transition

**Status:** common in coastal areas; generally much scarcer inland than Long-billed Dowitcher

**1 Juvenile, Aug., N.Y.**
- in all plumages, bill relatively broad based and blunt-tipped with subtle downward kink near tip; females longer billed than males; bill usually pale-based, often with distinct yellowish tinge
- aged by uniformly fresh plumage with relatively small, crisply pale-fringed scapulars, coverts, and tertials; newly molted gray nonbreeding scapular is larger and rounder-tipped than juvenal scapulars
- more colorful overall than juvenile Long-billed, with complex internal markings on tertials and greater coverts

**3 Juveniles and breeding adult (right) *caurinus* with Surfbirds, Aug., Alaska**
• juvenile *L. g. caurinus* (Pacific Coast) are typically darker overall than other subspecies, with more heavily marked flanks, more limited internal markings on tertials and greater coverts

**4 Juvenile, probably *griseus*, late Aug., N.Y.**
• juvenile *griseus* (primarily Atlantic Coast) average duller than other subspecies, often with just a hint of buff on the breast

**5 Juvenile, probably *hendersoni*, Aug., N.J.**
• juvenile *hendersoni* (Gulf and s. Atlantic coasts and prairies) average brighter than other subspecies, often with a bright orange-buff wash on breast and upperparts and broader fringing above

**6 Nonbreeding with four other species, Oct., N.J.**
• The dowitchers are the long-billed birds with mottled flanks. Can you find a dowitcher with its bill hidden? Can you identify the other species? Focus especially on size, bill shape, and head/breast/flank patterns.

**7 Nonbreeding probably *hendersoni*, Dec., Fla.**
• nonbreeding *hendersoni* average paler than other subspecies, often with thinner barring on flanks and tail, more spotted look to neck and breast, and distinct white fringing to fresh wing coverts (subtle on this bird due to mostly worn, brownish-tinged wing coverts)
• nonbreeding *griseus* is probably intermediate between *caurinus* and *hendersoni*

**8 Nonbreeding *caurinus*, Feb., Calif.**
• nonbreeding *caurinus* average darker than other subspecies, with less white fringing on coverts, more solidly dark breast, broader dark tail barring
• from Long-billed by rounded crown with short "snout," thinner, more chevron-shaped flank barring, dark spots "spilling" into belly (cleaner white belly on Long-billed)

**9 Breeding *hendersoni* (the largest bird) with Dunlin, Stilt Sandpiper (back, center), and Western Sandpiper (the smallest bird), Apr., Tex.**
• spring *hendersoni* shows extensive orange markings on the upperparts and a pale orange wash below, with evenly distributed dark spotting along flanks; orange of underparts brightens by May

**10 Breeding adults and first summers, early June, N.J.**
• breeding *griseus* (right) shows pale orange neck, white belly; breeding *hendersoni* (center) shows orange underparts fading to paler belly, broader orange markings on upperparts than *griseus*
• first-summer birds are mostly gray with a variable scattering of "breeding" scapulars, tertials, and coverts; pale coloration, white-fringed wing coverts, and broad orange markings on "breeding" feathers suggest *hendersoni*

**11 Breeding *caurinus*, Apr., Calif.**
• breeding *caurinus* show evenly orange-washed underparts with extensive spotting and variably barred flanks; relatively dark overall with narrow pale markings on upperparts and usually narrow white tail bars; variable number of "breeding" wing coverts

**12 Breeding *hendersoni*, June, Manitoba**
• breeding *hendersoni* show orange-washed underparts, brightest on neck; flanks and breast-sides with variable but evenly distributed spots and short bars; foreneck and belly unspotted; bright upperparts with broad markings; variable number of "breeding" wing coverts

**13 Breeding *griseus*, May, N.J.**
• breeding *griseus* show variably orange-washed neck, mostly white belly; usually heavy barring on flanks and spotting on breast-sides and neck; dull upperparts with narrow pale markings; relatively few "breeding" wing coverts
• bill relatively heavy and blunt-tipped; relatively long bill suggests female

**14 Breeding Long-billed Dowitcher, early July, Alaska**
• salmon orange underparts, evenly bright past legs; dark markings heaviest at breast-sides, light or absent (when worn) on flanks; upperparts dark with thin markings and numerous "breeding" wing coverts resulting in strong wing-to-belly contrast
• bill subtly thinner based, finer-tipped than Short-billed; short bill indicates male

**15 Breeding *griseus* (left) and *hendersoni* with Lesser Yellowlegs, July, N.J.**
• body size similar to Lesser Yellowlegs
• at long range, all subspecies typically show orange of underparts brightest on neck

**16 Worn breeding *caurinus*, July, Calif.**
• richly colored below with narrow pale markings above like Long-billed, but note spotting along flanks (limited or absent in Long-billed) and slimmer body; though variable in *caurinus,* this individual shows fewer "breeding" wing coverts than Long-billed

**17 Molting adults, July, N.J.**
• To what subspecies do these birds belong?

**18 Molting adult *hendersoni*, early Aug., N.J.**
• on *hendersoni,* fresh nonbreeding feathers are very pale, even-toned gray; new wing coverts are crisply white-fringed; retained breeding tertials and greater coverts have broad pale markings; underparts are washed with pale orange down to belly
• bill relatively broad-based, blunt-tipped, with a subtle kink near the tip, most visible on lower edge

**19 Molting adult *griseus*, early Aug., N.J.**
• fresh nonbreeding scapulars are darker than on *hendersoni* and pale markings on retained breeding scapulars and tertials narrower
• note distinctly yellowish bill-base
• pattern of tail barring highly variable on *griseus* but usually with broader dark bars than on *hendersoni*

**20 Juveniles with Stilt Sandpipers and Laughing Gull, Sept., N.Y.**
• larger than Stilt Sandpiper, with different structure (longer, heavier bill, shorter legs, longer and broader wings); also note bolder face-pattern, grayer underwing, very different rump and tail patterns from above

**21 Breeding *hendersoni* with Least Sandpiper, July, N.J.**
• clear pale-orange underparts with evenly distributed spotting along sides indicate *hendersoni*
• note evenly patterned underwing, typical of all subspecies

**22 Molting adult *hendersoni,* Aug., Fla.**
• in all subspecies, primary molt takes place only in coastal areas, mostly near the wintering grounds

**23 Breeding adult *hendersoni* (the larger birds) with another species, May, Neb.**
• white wedge up back common to all dowitchers
• flight call is a quick, whistled *keu-tu-tu,* lower than all calls of Long-billed
• What is the other species?

# Long-billed Dowitcher

***Limnodromus scolopaceus*** p. 437

**Size:** **L $9^1/_2$–$10^1/_2$ in. (24–26 cm); WS $18^1/_4$–$20^1/_2$ in. (46–52 cm); WT 90–114 g**
**usually slightly larger than Short-billed Dowitcher**

**Structure:** **chunky and rounded with short legs; long snipelike bill; feeding birds often show rounded back and belly as if they had "swallowed a grapefruit"; bill averages longer, thinner based, finer-tipped than Short-billed, but much overlap; many females are distinctively large and long billed**

**Behavior:** **probes deeply and rapidly ("sewing-machine" style) in shallow water; feeds with flock in one small area; freshwater pools, sheltered tidal lagoons; breeds on tundra**

**Status:** **common in the South and West, less common in the Northeast (rare there in spring); much more likely to be seen inland than Short-billed Dowitcher**

**1 Juvenile, Sept., N.Y.**
- in all plumages, bill relatively slim based and fine-tipped, often with very subtle downward arch through outer half; males shorter billed than females and overlap with Short-billed Dowitcher
- aged by uniformly fresh, neatly arranged plumage with relatively small, thinly pale-fringed upperparts
- much less colorful than juvenile Short-billed, with overall appearance much like nonbreeding; coverts and tertials lack internal markings
- fresh individuals show a pinkish buff wash on breast

**2 Juvenile (center), with juvenile Short-billed Dowitchers and Western Willet, Sept., N.J.**
• duller than Short-billed, with no internal markings on tertials or greater coverts
• also note Long-bill's slightly larger size, with longer bill and legs, bolder white crescent under eye; often darker, more greenish tinge to legs and bill-base; often appears to have longer "snout," taller crown, thinner supercilium

**3 Nonbreeding, Feb., Calif.**
• dark overall with smooth gray neck and breast, thick smudgy bars on flanks, subtly dark-centered scapulars, relatively thin supercilium, and greenish-tinged legs and bill-base
• relatively thin, fine-tipped bills and long sloping forehead

**4 Nonbreeding (right) with Short-billed Dowitcher (probably *hendersoni*), Aug., Del.**
• in relaxed feeding pose, Long-billed usually looks quite round-bodied, as if it had "swallowed a grapefruit"; Short-billed is slimmer and flatter-backed; differences obscured when birds are more active or alert
• Short-billed, especiallly *hendersoni*, is often distinctly paler than Long-billed, with white-fringed coverts, thinner, more chevron-shaped flank barring, more speckled neck and breast, and more even-toned scapulars

**5 Nonbreeding with *hendersoni* Short-billed Dowitcher, early Apr., Tex.**
• Which is which?

**6 Fresh breeding (right) with *hendersoni* Short-billed Dowitcher, mid Apr., Tex.**
• Long-bill's upperparts are darker overall with thinner, darker rufous markings and bolder white tips; underparts are darker rufous; a few nonbreeding feathers remain on upperparts
• note Long-bill's longer legs, more upright stance while at rest (due to more front-heavy proportions; a useful distinction with sleeping flocks)

**7 Fresh breeding, late Apr., Neb.**
• darkish upperparts with thin rufous markings and bold white tips; many "breeding" wing coverts; underparts brightest on belly, with heavy barring at breast-sides, finer barring on flanks; white tips on upperparts and underparts wear off by summer
• relatively thin, fine-tipped bill is gently arched through outer half (not kinked in one spot like Short-billed)

**8 Breeding, June, Alaska**
• by June, white tips have worn off; upperparts become very dark; underparts become evenly salmon orange past legs; neck finely speckled; belly unmarked; dark markings along flanks (usually bars, sometimes spots) heaviest at breast-sides, sparser along flanks

**9 Worn breeding, late Aug., N.J.**
• by late summer many dark markings on underparts have worn off, leaving a cluster of spots or bars at breast-sides
• salmon on underparts is paler but solid, often contrasting with white rear flanks
• note strong back-to-belly contrast, rounded body, and long legs

**10 Worn breeding, with *hendersoni* Short-billed Dowitcher, Aug., N.J.**
• the Long-billed is the big fat one
• upperparts on *hendersoni* are brighter and paler with more complex markings; dark markings are evenly distributed along breast-sides and flanks; lacks chevrons or bars on breast

**11 Molting adult, mid Aug., N.J.**
• note structure
• head molt starts earlier than body molt, so many late-summer birds have a distinctive gray-headed, salmon-bellied look
• retained breeding scapulars and tertials are dark with thin, mostly parallel markings; fresh nonbreeding scapulars are slightly darker toward centers

**12 Nonbreeding, early Aug., N.Y.**
• This bird attained full nonbreeding plumage by early August, an unusually early date. There are various reasons why an individual may molt early, including failed breeding resulting in early migration, or poor health preventing migration to the breeding grounds. Unlike Short-billed Dowitchers, virtually all first-summer Long-bills migrate to the Arctic, so all are on the same molt schedule in fall.

**13 Molting adults with American Avocet (left) and Greater Yellowlegs (right), Sept., Del.**
• note upright stance on sleeping birds and strong wing-to-belly contrast on reddish-bellied bird

**14 Breeding, early May, N.J.**
• lesser coverts on underwing often contrastingly whitish

**15 Breeding *hendersoni* Short-billed Dowitcher, early Aug., N.Y.**
• underwing usually more uniform

**16 Molting adults, Oct., Del.**
• note dark gray hoods and whitish underwing patch
• flight call is a sharp, whistled *pseep,* always higher pitched than calls of Short-billed

**17 Molting adults, Sept., Del.**
• white wedge up back as in Short-billed; tail barring variable but typically even, with broad dark bars (matched by many *griseus* and *caurinus* but a useful distinction from *hendersoni*)
• flight feather molt takes place at coastal or inland stop-over sites, a useful distinction from Short-billed in many areas

# Wilson's Snipe

***Gallinago delicata*** p. 441

**Size:** **L 10–11½ in. (25–28 cm); WS 17¼–19½ in. (43–48 cm); WT 79–146 g**
**slightly smaller than dowitchers**

**Structure:** **chunkier than dowitchers, with shorter wings, tail, and legs, larger head; often crouches low to the ground**

**Behavior:** **secretive; probes deeply in shallow water or mud; stays well camouflaged in wet grassy fields, marshes; often not seen until flushed; breeds in fresh marshes and bogs**

**Status:** **common and widespread but harder to see than other shorebirds**

**1 Juvenile, Aug., N.Y.**
• all plumages cryptically patterned with striped crown and back, barred flanks
• aged with caution by thin, even-width fringes to coverts (often divided at tips); these feathers may be replaced quickly so first-year birds often very difficult to age by September

**2 With Long-billed Dowitchers (front left) and Greater Yellowlegs (back right) Jan., Fla.**
• similar shape and feeding style to dowitchers but more compact, with boldly striped upperparts

**3 Nonbreeding, Jan., Fla.**
• differs from juvenile by relatively broad pale tips (often divided) to coverts; adult and first winter not usually separable in the field

**4 Breeding, May, N.Y.**
• in summer, fresh scapulars contrast with worn wing coverts

**5 Jan., Fla.**
• reflections change the color of the water but not that of the birds

**6 May, N.Y.**

**7 May, N.Y.**

• "winnowing" sound in display flight created by wings pushing air past narrow outer rectrices of spread tail

**8 Oct., N.J.**

• shorter wings and legs, darker underwing, more twisting flight than dowitchers

• flight call is a tearing *skaaip,* often given by birds as they flush

# American Woodcock

***Scolopax minor*** **p. 444**

**Size: L 10–12½ in. (25–31 cm); WS 16½–20¼ in. (41–51 cm); WT 135–211 g**
**similar length but nearly twice the weight of dowitchers**

**Structure: fat rounded body with large head, thick neck, long bill, short legs**

**Behavior: largely nocturnal; does not associate with other shorebirds; probes deeply in damp soil with slow rhythmic motion; well camouflaged in thick forest with leaf litter; breeds in damp second-growth forest, bogs, overgrown fields**

**Status: common and widespread but secretive and seldom seen except when flushed or during twilight courtship display**

**1 Dec., N.J.**
- all plumages with rich orange-buff underparts, cryptically patterned upperparts
- age determination not usually possible under normal field conditions; primaries and primary coverts crisply buff-tipped on juvenile, diffusely gray-tipped on adult; wing coverts relatively small with distinct buff fringe in fresh juvenile, but most coverts are replaced by larger, diffusely gray-fringed feathers (like adult) before fall migration

**2 Jan., N.J.**
• when food is scarce during harsh weather, may exhibit aggressive posture to defend feeding territory

**3 Jan., N.J.**
• feeds on earthworms by probing deeply in damp soil; during severe cold spells may seek thawed ground along warm sunny field-edges or roadsides

**4 Feb., N.J.**
• broad rounded wings, fat body, large head, and long bill are distinctive in flight; flight is slow, clumsy, and twisting • flushed birds produce a distinctive wing whistle; low nasal *beent* call is given from ground during display

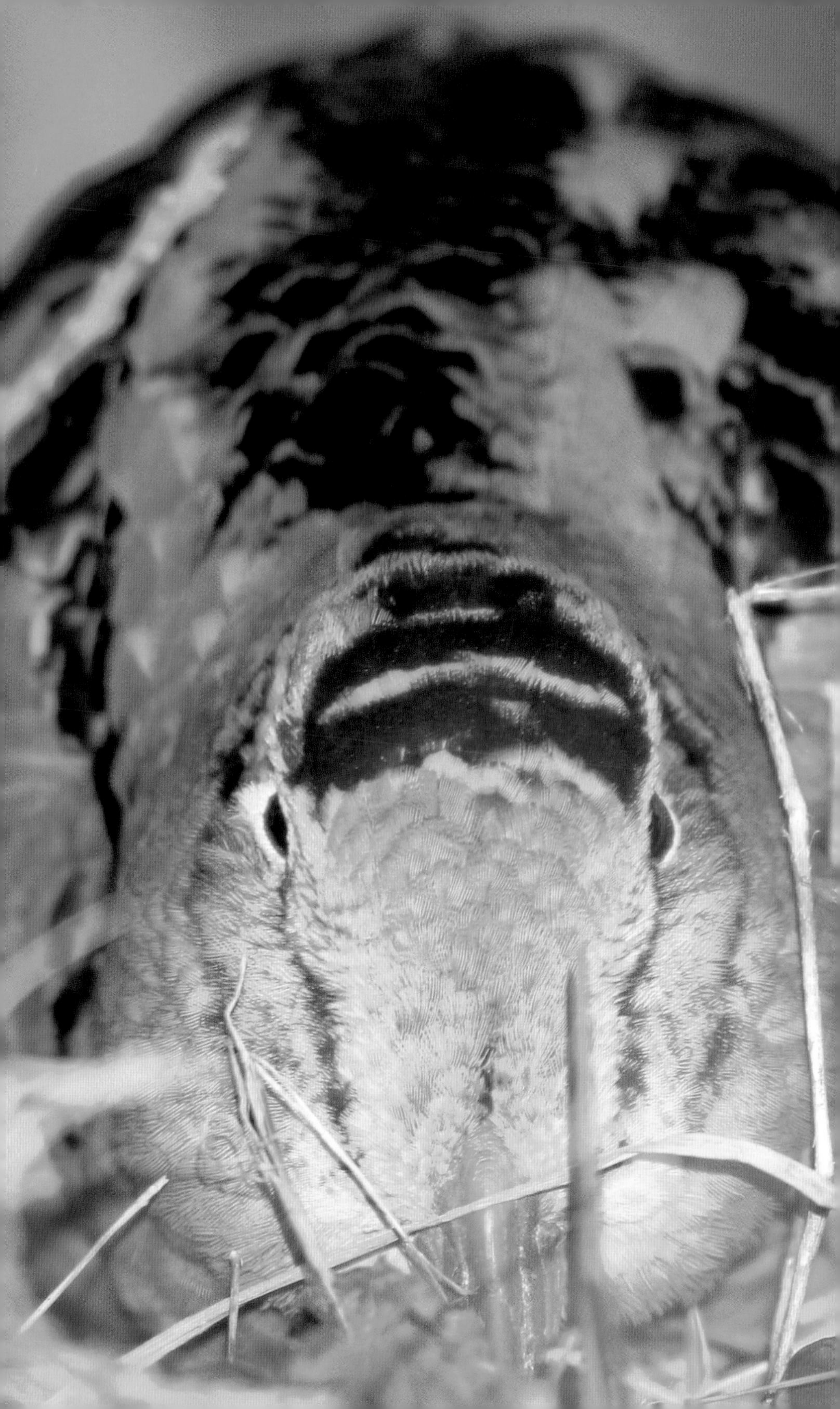

# Wilson's Phalarope

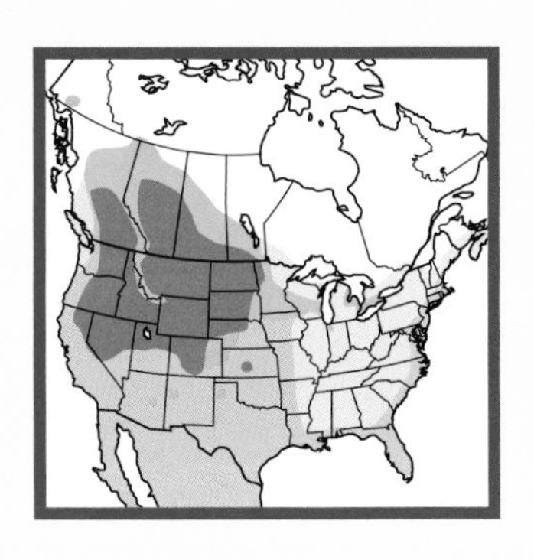

***Phalaropus tricolor*** p. 445

**Size:** **L 8¾–9½ in. (22–24 cm); WS 15½–17¼ in. (39–43 cm); WT 30–85 g**
**close to Stilt Sandpiper; smaller than Lesser Yellowlegs; same or slightly larger than Red Phalarope**

**Structure:** **plump rounded body, small head, long thin neck; medium-length, needle-thin bill; relatively short legs with slightly lobed toes (as in other phalaropes)**

**Behavior:** **active; spins in water, picking at surface, or walks quickly on mud, picking and chasing with head low to the ground; most terrestrial phalarope; shallow grassy ponds, adjacent muddy shorelines; breeds in fresh marshes, potholes**

**Status:** **common, particularly inland; scarce in the East**

**1 Molting juvenile, early Aug., Calif.**
- in all plumages, bright white below with needle-thin, medium-length black bill
- juvenile and nonbreeding show plain whitish face with pale supercilium, prominent dark eye; yellow legs
- aged by dark upperparts with crisp pale fringes; note numerous pale gray nonbreeding scapulars; molts out of juvenal plumage early, so most migrants show variable pale gray upperparts but retain juvenal tertials and wing coverts

**2 Molting juveniles with Western Sandpiper (right), July, Ariz.**
• proportions of head and bill similar to Lesser Yellowlegs, but plump body and short legs distinctive

**3 First winter (right) with juvenile Lesser Yellowlegs, early Aug., N.Y.**
• smaller and whiter than Lesser Yellowlegs
• often feeds by spinning in water, creating vortex that draws insect larvae to surface

**4 First winter, late Sept., Fla.**
• very pale, smooth gray above; mostly white face; dark tertials and coverts are retained from juvenal plumage

**5 Nonbreeding with Red-necked Phalarope (right), early Aug., Calif.**
• Red-necked is slightly smaller and more compact with shorter neck, tail, and bill; bolder ear-patch

**6 Breeding female, May, Nebr.**

**7 Breeding pair, June, Mont.**
• breeding males are drab with pale pinkish wash on neck; breeding females show bold dark stripe through face and neck, bright salmon wash on neck; sexual roles are reversed, as with Red and Red-necked

**8 Breeding males, July, Colo.**
• distinctively plain above and below

**9 Breeding males and females, May, Neb.**
• white stripe up nape, most obvious on females, is unique

**10 Breeding males (2 at center) with White-rumped, Semipalmated, Least, and Stilt Sandpipers, May, Neb.**
• in flight looks plump, thick-chested, thick-necked, with slightly raised head and contrasting white throat; wings are plain, blunt-tipped, pushed forward; wingbeats snappy
• Can you pick out the other species?

**11 Breeding female, May, Neb.**
• white rump
• flight call is a nasal, froglike *werpf*

# Red-necked Phalarope

***Phalaropus lobatus*** p. 447

**Size:** L 7¼–7½ in. (18–19 cm); WS 12¾–16¼ in. (32–41 cm); WT 20–48 g
larger than "peep"; slightly smaller than Red Phalarope

**Structure:** compact with slim neck, small head, needle-thin bill, short legs with lobed toes

**Behavior:** active; spins in water, picking at surface; seaweed lines in pelagic waters; also lakes, rivers, impoundments, nearshore waters in West; nests on tundra

**Status:** common offshore; locally common inland in West; rare inland in East

**1 Juvenile, Apr., England**
• all plumages show small head, slim neck, and relatively short, needle-thin black bill
• juvenile and non-breeding show broad, solid dark ear-patch
• aged by uniformly fresh plumage with blackish, buff-striped upperparts and crisply buff-fringed tertials and coverts; gray-buff wash on neck when fresh
• unlike juvenile Red Phalarope, retains extensive juvenal plumage throughout fall migration

**2 Molting juvenile, Aug., Calif.**
• numerous pale gray, white-edged scapulars and mantle feathers have molted in; buff tones have faded or worn off, resulting in much colder appearance than fresh juvenile

**3 Juveniles, Aug., Calif.**
• starkly black-and-white appearance distinctive

**4 Nonbreeding, Nov., Calif.**
• small head, slim neck, needle-thin bill; slight body rides low in the water, with relatively flat back
• pale gray upperparts with indistinct white lines

**5 Breeding male, June, Alaska**
• male duller overall with pale supercilium; female brighter, darker, more clean-cut
• nests in wet tundra with ponds and lakes

**6 Breeding female, June, Alaska**
• as with other phalaropes, sexual roles are reversed: female does courting, and male does incubation and brood rearing

**7 Breeding female (back center) with Wilson's Phalaropes, Apr., Nev.**
• smaller, more compact, and darker than Wilson's, with more strongly contrasting white throat, dirtier flanks

**8 Molting adult, early Aug., Calif.**
• unlike Red Phalarope, the molt to nonbreeding plumage takes place largely during migration, so fall migrants often show a mix of breeding and nonbreeding plumage

**9 Breeding, May, Calif.**
• narrow, Dunlinlike wing stripe
• breeding birds show dark head with strongly contrasting white throat

**10 Breeding and molting adults, Apr., Hong Kong**
• small and compact with short wings and tail, flat belly, angular chest, small raised head
• underwing with dark markings
• flight call is a hard, squeaky *pwit,* lower and huskier than Red

# Red Phalarope

***Phalaropus fulicarius*** p. 449

**Size:** **L 8–8¾ in. (20–22 cm); WS 16–17½ in. (40–44 cm); WT 36–77 g**
**similar to Sanderling; 40 percent heavier than Red-necked**

**Structure:** **bulkier than Red-necked, with heavier chest, thicker neck, larger head, longer wings and tail, heavier bill; short legs with lobed toes**

**Behavior:** **active; spins in water, picking at surface; pelagic waters, often farther offshore than Red-necked; rarely on inland lakes, rivers, nearshore waters after storms; nests on tundra**

**Status:** **common far offshore; rare nearshore or inland**

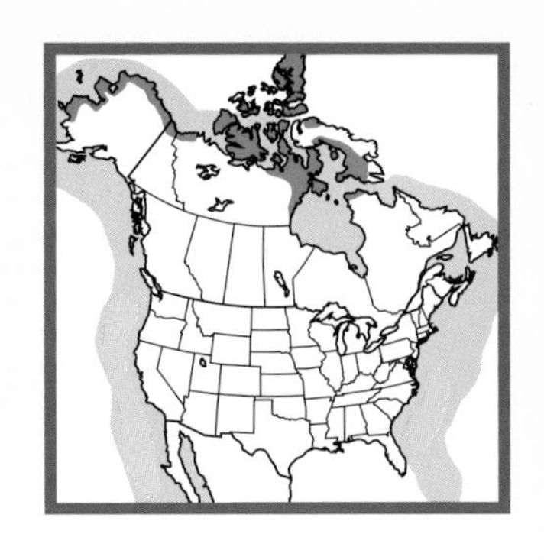

**1 Molting juvenile, Sept., N.J.**

- in all plumages note short, sturdy, often yellow-based bill and short grayish or pinkish legs
- juvenile and nonbreeding show dark ear-patch (sometimes distinctively smudgy)
- aged by blackish tertials and coverts with crisp buff fringes
- molts out of juvenal plumage early, so migrants show variable pale gray upperparts but retain juvenile tertials and coverts; often retains distinctive pinkish wash on throat; full juvenile lacks buffy back stripes

**2 Nonbreeding, Nov., Calif.**
• bulky with uniform pale gray upperparts
• differs from first winter by gray coverts and tertials

**3 Nonbreeding, Jan., England**
• yellow bill-base often prominent

**4 Nonbreeding (right) with juvenile Red-necked Phalarope, Sept., Calif.**
• Red looks bulkier overall with larger head, thicker neck; foreparts often ride higher in the water (not so flat backed) due to bulky chest; wings and tail longer, often more prominently cocked upward

**5 Molting adult, early Apr., N.J.**
• spring migrants may show patchy mixture of plumage; full breeding usually attained by late April or early May

**6 Breeding male with chicks, July, Alaska**
• breeding males duller overall; as with other phalaropes, sexual roles are reversed: female does courting, and male does incubation and brood rearing

**7 Breeding female, June, Alaska**
• spinning in water creates vortex that draws food to surface

**8 Breeding, May, Calif.**
• broad, Sanderling-like wing stripe; clean white underwing
• flight call is a sharp *psip* similar to Long-billed Dowitcher; higher and clearer than Red-necked

**9 Nonbreeding with juvenile Red-necked (top right), Aug., Calif.**
• bulkier than Red-necked, with longer wings and tail, heavier chest, broader wing stripe
• smaller head and bill, less white in primaries than Sanderling

# Rarities and Regional Specialties

# Double-striped Thick-knee

***Burhinus bistriatus*** **p. 319**

**Size:** **L 15–19¼ in. (38–48 cm); WS 34½–40 in. (86–100 cm); WT 780 g slightly larger than American Oystercatcher**

**Structure:** **bulky body; long legs; long neck; short stout bill; large eye**

**Behavior:** **largely nocturnal; ploverlike run-stop-pluck feeding style; dry grasslands, pastures, agricultural fields**

**Status:** **resident Neotropical species; accidental; one record in Texas (Dec. 1961)**

**1 Adult, Mar., Costa Rica**
• size and structure indicate thick-knee family

**2 Adult, Nov., Venezuela**
• differs from other thick-knees by combination of bold white supercilium highlighted above by black, and by strong division between breast and belly
• white patch at base of primaries visible in flight
• juvenile similar but with black stripe on nape below supercilium
• primary call, given most often at dusk or on moonlit nights, is a barking *keh-keh-keh . . .*, often in a prolonged series

# Northern Lapwing

***Vanellus vanellus*** **p. 319**

Size: **L 9¼–12½ in. (23–31 cm); WS 32¾–34¾ in. (82–87 cm); WT 192–254 g close to Black-bellied Plover**

Structure: **Killdeerlike but stockier, shorter tailed; unique paddle-shaped wings and wispy crest**

Behavior: **run-stop-pluck feeding style; pastures, plowed fields, marshes**

Status: **Eurasian species; casual, primarily in late fall and early winter in Northeast**

**1 With European Golden-Plover (center) and Black-headed Gulls, Aug., England**

**2 June, England**

**3 Molting juvenile, Aug., England**
• aged by short crest, gray crown, finely buff-spotted fringes to upperparts; larger, prominently buff-tipped scapulars are fresh nonbreeding feathers

**4 Nonbreeding, Nov., Japan**
• from juvenile by long crest, black crown, prominently buff-tipped scapulars and coverts

**5 Breeding male, June, Scotland**
• female similar but with shorter crest, white mottling on throat

**6 Molting adults, Aug., England**
• unique paddle-shaped wings, broadest in adult male, narrowest in juvenile; pigeonlike flight

**7 Molting adult, Aug., England**
• flight call is a shrill, mewing *pway-vich* (known locally as "Pewit")

# European Golden-Plover

***Pluvialis apricaria*** **p. 322**

**Size:** **L 10½–11½ in. (26–29 cm); WS 26¾–30½ in. (67–76 cm); WT 140–312 g intermediate between Black-bellied Plover and American Golden-Plover**

**Structure:** **rounder body, shorter legs, less attenuated than American; small, fine-tipped bill; usually three or four primaries past tertials**

**Behavior:** **run-stop-pluck feeding style; pastures, plowed fields, tidal flats**

**Status:** **European species; casual spring migrant in e. Canada, occasionally in flocks**

**1 Juvenile, Oct., England**
- compared to American and Pacific, all plumages show: shorter legs; shorter, finer-tipped bill; smaller feathers with more and smaller spots on upperparts; brighter gold upperparts than American
- juvenile and nonbreeding show plain face with prominent dark eye, similar to Pacific; coarser breast markings contrasting with cleaner white belly
- aged by uniformly fresh plumage with neatly arranged upperparts and neatly mottled breast

**2 Nonbreeding, Dec., England**
- differs from juvenile by less neatly arranged upperparts with slightly longer scapulars; blurrier markings on breast and flanks; often very difficult to age
- note stocky body, short legs, short fine-tipped bill

**3 Molting adults with Northern Lapwings and Black-headed Gulls, Aug., England**
- black underparts on breeding birds quite narrow at breast; breast mostly white on some females

**4 Southern breeding female, June, England**
• females show variable black on belly; whitish undertail coverts, flanks, and breast-sides; thin pale supercilium and neck stripe; northern females show more black, approaching male

**5 Breeding male, July, Norway**
• whitish undertail coverts and flanks; broad white at breast-sides; thin white supercilium and neck stripe

**6 Molting adults, Aug., England**
• relatively broad wings with prominent white underwing
• flight call is a clear mournful whistle, *puuii;* lower-pitched than American

# Lesser Sand-Plover

***Charadrius mongolus*** **p. 326**

**Size:** **L 7½–8½ in. (19–21 cm); WS 18–23¼ in. (45–58 cm); WT 39–79 g close to Wilson's Plover**

**Structure:** **slightly longer legged, longer and heavier billed than Semipalmated Plover**

**Behavior:** **run-stop-pluck feeding style; beaches, tidal flats, muddy fields**

**Status:** **Asian species; rare but regular migrant on islands off w. Alaska; casual along Pacific Coast; accidental elsewhere**

**1 Juvenile, Aug., Japan**
• all plumages show gray-brown upperparts, olive gray legs; no white collar
• differs from Greater Sand-Plover by smaller size, smaller head, shorter legs (especially tibia) and smaller bill (no longer than distance from base of bill to back of eye)
• aged by uniformly fresh, crisply pale-fringed upperparts

**2 Molting to nonbreeding, Oct., Australia**
• from juvenile by less neatly arranged upperparts with smudgy pale fringes
• differs from Greater Sand-Plover by size and structure
• differs from Mountain Plover by smaller size, more compact structure, darker legs, more contrasting face-pattern

**3 Breeding, Apr., Hong Kong**
• differs from Greater by size and structure, more extensive rufous breast-band, cold gray-brown upperparts without rufous highlights; hair-thin black line between throat and breast
• white forehead, dusky flank smudging, and rich coloration indicate Siberian subspecies *(mongolus)*, to which all American records refer

**4 Breeding pair, June, Russia**
- breast-band color in males is brick red when fresh, much paler when worn
- brightly marked females may approach males in appearance but usually have paler mask and breast-band

**5 Worn breeding female, Sept., Japan**
- much-faded plumage may recall central Asian populations of Lesser Sand-Plover (*atrifrons* group, unrecorded in North America), but note dark smudging on flanks, whitish forehead, well-defined supercilium
- molt to nonbreeding takes place largely on wintering grounds, so fall migrants are in worn breeding plumage

**6 Sept., Japan**
- whitish patch on inner primaries averages narrower than Greater, but variable
- usually no toe projection past tail
- flight call is a low, hard *kurrip,* reminiscent of male Northern Pintail

# Greater Sand-Plover

***Charadrius leschenaultii*** **p. 328**

**Size: L 7½–10 in. (19–25 cm); WS 21¼–24 in. (53–60 cm); WT 65–103 g**
**averages 20 percent larger than Lesser, but with overlap**

**Structure: slightly larger headed, larger billed, and longer legged than Lesser Sand-Plover**

**Behavior: run-stop-pluck feeding style; beaches, tidal flats, sometimes dry coastal grasslands**

**Status: central Asian species; accidental in California. (Jan.–Apr. 2001)**

**1 Juvenile, Aug., Hong Kong**
- all plumages show gray-brown upperparts, olive legs; no white collar (compare Wilson's Plover)
- all plumages from Lesser Sand-Plover by larger size, larger head, longer legs (especially tibia) and longer, heavier bill; bill tip slightly more pointed; legs average, brighter but variable
- aged by uniformly fresh, crisply pale-fringed upperparts

**2 Breeding, May, Hong Kong**
- differs from Lesser by size and structure, narrower and paler rufous breast-band, cleaner white flanks, paler gray-brown upperparts with faint rufous highlights; no black highlight on breast-band

**3 Nonbreeding (center left) with Terek and Curlew Sandpipers, Apr., Australia**
- differs from juvenile by less neatly arranged upperparts with smudgy pale fringes

**4 Nonbreeding, Apr., Australia**
• long legs (especially tibia) and long heavy bill (greater than distance from base of bill to back of eye)
• upright stance due to front-heavy proportions; sometimes looks long necked
• molt to nonbreeding takes place at or near breeding grounds, so fall migrants are in nonbreeding plumage

**5 Breeding, Mar., Hong Kong**
• toes project past tail tip
• prominent white patch at base of inner primaries averages broader than on Lesser
• flight call is a deep, trilled *trrr-uk-uk;* lower-pitched, more Ruddy Turnstone–like than Lesser

# Collared Plover

***Charadrius collaris*** **p. 329**

**Size: L 5½–6¼ in. (14–16 cm); WS 14½–15½ in. (36–39 cm); WT 28–35 g**
**smallest plover; slightly smaller than Snowy**

**Structure: similar to Snowy Plover but smaller headed, longer legged**

**Behavior: run-stop-pluck feeding style; beaches, river bars, mudflats, shrimp-farm dikes**

**Status: Central and South American species; accidental in Tex. (May 1992)**

**1 Adult (left) with Wilson's Plover, early Oct., Veracruz, Mex.**

- all plumages differ from Snowy Plover by pinkish legs, lack of white collar
- adults show black breast-band and head markings, variable wash of rufous on nape
- juveniles lack black and rufous on head but have variable black breast-band

**2 Adult, early Oct., Veracruz, Mex.**

- some (probably females) show less rufous on nape
- fresh upperparts show rusty tips

**3 Adults, Jan., Jalisco, Mex.**

- faint "sandplover" wing stripe at primary bases; narrow white tail edges
- flight call is a sharp *pweet* resembling Wilson's

# Common Ringed Plover

***Charadrius hiaticula*** p. 333

**Size:** L 7¼–8 in. (18–20 cm); WS 19¼–22¾ in. (48–57 cm); WT 39–84 g
slightly larger than Semipalmated Plover, but with overlap

**Structure:** slightly chunkier body, more thinly attenuated rear than Semipalmated; bill slightly longer, slimmer; toes unwebbed

**Behavior:** like Semipalmated

**Status:** primarily Eurasian species; breeds in remote sections of Arctic Canada, occasionally w. Alaska; winters in Old World; casual migrant on w. Alaska islands; accidental elsewhere

**1 Juvenile Semipalmated Plover, Oct., N.J.**
- pinched lores (pale line between lores and gape); pale orbital-ring; thicker-based bill

**2 Juvenile, Aug., England**
- differs from Semipalmated by broader dark lores extending down to gape; dark mask more even-width through cheeks; little or no white below back of eye; darker orbital-ring; breast-band broader, more often divided (variable); upperparts average colder gray-brown
- differences from Semipalmated subtle; always check toe-webbing, voice, structure
- aged by uniformly fresh, crisply pale-fringed upperparts

**3 Nonbreeding Semipalmated Plover, Sept., Tex.**

**4 Nonbreeding, Feb., Portugal**
- differs from juvenile by uniformly unpatterned upperparts
- differences from Semipalmated as with juvenile but face-pattern somewhat more variable; orbital-ring color more useful

**6 Breeding male Semipalmated Plover, Aug., N.J.**
• pinched lores; yellow orbital-ring; dull supercilium

**7 Breeding male, May, Cyprus**
• from Semipalmated by broader dark lores, blacker cheeks, bold white supercilium; white forehead comes to a point below eye; dark orbital-ring; longer, thinner bill with smaller black tip (usually less than half black); broader breast-band (variable); upperparts average colder gray-brown

**8 Semipalmated Plover, Sept., N.J.**
• diagnostic partial web between middle and outer toes is surprisingly obvious in close view; best seen head-on or from behind

**9 Worn breeding female, Aug., England**
• females show duller black markings, often darker bill than males, but face pattern still distinctive
• note unwebbed toes

**10 Worn breeding female, Aug., England**

**11 Worn breeding male (front) and females, Aug., England**

**12 Breeding, with Dunlin, Aug., England**

- typically bolder wing stripe than Semipalmated; blacker cheek (on males) and whiter supercilium results in more contrasting mask visible even in flight
- flight call is a distinctive, mellow *poo-eep;* lower, less rising, and with longer first syllable than Semipalmated
- Dunlin are of the race *C. a. schinzii;* smaller than American Dunlin

# Little Ringed Plover

***Charadrius dubius*** **p. 337**

**Size:** **L 5½–6¾ in. (14–17 cm); WS 16¾–19¼ in. (42–48 cm); WT 26–53 g slightly smaller than Semipalmated Plover (overlap)**

**Structure:** **more delicate than Semipalmated with smaller head, slimmer bill, longer legs and tail; no primary projection past long tertials; toes unwebbed**

**Behavior:** **run-stop-pluck feeding style; muddy shorelines along coast or inland; often in water; seldom with flocks of other birds**

**Status:** **Eurasian species; casual spring vagrant to w. Aleutian Is., Alaska**

**1 Juvenile, Aug., Japan**
• differs from Semipalmated by structure, indistinct supercilium, often brighter orbital-ring
• aged by uniformly fresh, pale-fringed upperparts; juvenile with unique pattern of thin dark subterminal line within broad buff fringe

**2 Nonbreeding, Jan., Sri Lanka**
• differentiated from Semipalmated by structure, indistinct supercilium, striking yellow orbital-ring
• from juvenile by less neatly arranged upperparts with larger scapulars, faint pale fringes

**3 Breeding male, June, Japan**
• differs from Semipalmated by structure, bold yellow orbital-ring; white highlight above mask; mostly dark bill; usually pinkish legs
• female similar but mask and breast-band suffused with brown

**4 Breeding male, Apr., Japan**

**5 Juvenile, Aug., Japan**
• very faint wing stripe
• flight call is a clear, descending whistle, *peeu,* similar to Lapland Longspur's whistle call

# Eurasian Dotterel

***Charadrius morinellus*** p. 341

**Size:** **L 8–9½ in. (20–24 cm); WS 22¾–25½ in. (57–64 cm); WT 86–200 g smaller than golden-plovers**

**Structure:** **golden-plover-like but more compact, with heavier chest, smaller bill; upright stance**

**Behavior:** **run-stop-pluck feeding style; arid uplands with short or sparse vegetation, gravel; agricultural fields; remarkably tame and approachable**

**Status:** **Eurasian species; rare visitor and sporadic breeder in w. Alaska; casual along Pacific Coast, primarily in September**

**1 Juvenile, Sept., Scotland**
- all plumages show bold supercilia joining at nape; pale breast-band, yellowish legs
- aged by uniformly fresh plumage with bold pale fringing on upperparts, coarse streaking on breast

**2 Nonbreeding, May, England**
- differentiated from juvenile by less neatly arranged upperparts with thinner pale fringes; smoother, grayer breast

**3 Breeding female, Scotland**

**4 Breeding male, July, Sweden**
• males paler and less richly colored than females; less solidly gray foreneck

**5 Juvenile, Sept., Scotland**
• unpatterned wings and tail
• flight call is a low, rolling *pjjrt;* generally quiet

# Eurasian Oystercatcher

***Haematopus ostralegus*** p. 342

**Size:** L 16–18½ in. (40–46 cm); WS 32–34½ in. (80–86 cm); WT 300–745 g same as American Oystercatcher

**Structure:** somewhat slimmer, smaller billed, shorter legged, longer winged than American

**Behavior:** similar to American but also occurs in coastal pastures and meadows, on river bars

**Status:** Eurasian species; accidental in Newfoundland (Apr., May)

**1 Juvenile, Aug., England**
• differentiate all plumages from American by darker upperparts, white rump
• aged by dark eye and bill tip; uniformly fresh plumage with subtle pale fringing on upperparts

**2 Breeding, June, Scotland**
• breeding adults show red eye and orbital-ring; solid orange bill; darker upperparts than juvenile

**3 Breeding and first summer with Bar-tailed Godwit, Aug., England**
• first summer shows white throat-band like nonbreeding; duller bill and eye than adults; dark bill tip like juvenile; upperparts less neatly arranged than juvenile

**4 Breeding, summer, England**
• differs from American by blacker upperparts, bright red eye and orbital ring

**5 Adults, Aug., England**
• white rump, broader wing stripe, longer wings and tail than American

# Black-winged Stilt

***Himantopus himantopus*** p. 346

**Size:** **L 14–16 in. (35–40 cm); WS 26¾–33¼ in. (67–83 cm); WT 137–289 g same as Black-necked Stilt**

**Structure:** **like Black-necked Stilt**

**Behavior:** **like Black-necked Stilt**

**Status:** **Eurasian species; accidental in w. Alaska**

**1 Juvenile, Sept., Hong Kong**
- all plumages differ from Black-necked Stilt by paler, less contrasting head-pattern
- aged by uniformly fresh plumage with thin pale fringing on upperparts, white trailing edge to wing (dark in adult); legs duller than adult
- voice like Black-necked Stilt

**2 Adults, Sept., Hong Kong**
- variable head-pattern; always less black than Black-necked Stilt

# Northern Jaçana

***Jaçana spinosa*** **p. 349**

**Size:** L 6¾–9½ in. (17–24 cm); WS 17–22 in. (43–56 cm); WT 82–161 g slightly larger than Lesser Yellowlegs

**Structure:** chunky body and small head; long legs with extremely long toes

**Behavior:** walks on floating vegetation and picks at surface of water or plants; may use long toes to flip over lily pads; freshwater marshes with abundant floating vegetation; flooded grassy fields, roadside ditches

**Status:** Mexican and Central American species; rare, irregular visitor to s. Texas, mostly Oct. to May; has bred; casual in s. Arizona

**1 Juvenile, Nov., Costa Rica**
- all plumages identified by unique structure and behavior, bold yellow wing-pattern

**2 Adult, Aug., Ariz.**
- bold yellow flight feathers; often holds wings up momentarily after landing
- yellow "spur" at bend of wing

**3 Adult, Aug., Cozumel, Mex.**
- primary call, usually given in flight, is a screeching, rail-like chatter, *swee swee swee swee swee . . .* ; interspersed with clucks

# Common Greenshank

***Tringa nebularia*** **p. 350**

**Size:** **L 12–13½ in. (30–34 cm); WS 27¼–28 in. (68–70 cm); WT 128–270 g same or slightly larger than Greater Yellowlegs**

**Structure:** **similar to Greater Yellowlegs; slightly longer bill**

**Behavior:** **like Greater Yellowlegs**

**Status:** **Eurasian species; rare migrant on w. Alaska islands; casual in California and e. Canada**

**1 Juvenile with Northern Lapwing, Aug., England**

- all plumages from Greater Yellowlegs by greenish legs, white wedge up back
- aged by uniformly fresh, brownish-tinged upperparts with distinctly pale-fringed, moderately pointed scapulars and coverts

**2 Nonbreeding, Sept., Hong Kong**

- differs from juvenile by paler, grayer, more rounded scapulars with crisp pale fringes

**3 Breeding, May, Russia**
• breeding birds show a variable scattering of black feathers above; lack spotted upperparts and flank barring of Greater Yellowlegs

**4 Nonbreeding, Oct., Hong Kong**
• mostly white tail and white wedge up back in all plumages; some nonbreeding birds look very white-faced
• similar structure to Greater Yellowlegs with bulky, angular chest and sturdy, slightly upturned bill
• flight call is a loud, ringing *kEeu-teu-teu,* very similar to Greater Yellowlegs but more liquid, less strident
• the very similar Nordman's Greenshank (*T. guttifer*), unrecorded in North America, has clean white underwing, shorter foot projection past tail, and more distinctly two-toned bill

# Marsh Sandpiper

*Tringa stagnatilis* p. 355

**Size:** L 8¾–10 in. (22–25 cm); WS 22–23½ in. (55–59 cm); WT 43–120 g
equal or slightly smaller than Lesser Yellowlegs

**Structure:** similar to Lesser Yellowlegs but smaller head, longer, thinner bill, longer legs

**Behavior:** walks steadily, picking at surface; often in belly-deep water; marshes, flooded fields, ponds; seldom in saltwater habitats

**Status:** Eurasian species; accidental in w. Alaska (Aug., Sept.)

**1 First winter, early Sept., Japan**
- all plumages from Lesser Yellowlegs by structure, greenish legs, white wedge up back, whiter face
- aged by fresh, crisply white-fringed non-breeding scapulars contrasting with more worn, browner coverts and tertials with buff fringes and internal markings

**2 Nonbreeding, Dec., Gambia**
- upperparts pale gray with white fringes
- very thin bill and white face may recall Wilson's Phalarope

**3 Breeding, Apr., Lesbos, Greece**
• unique black scrawls on upperparts in breeding plumage

**4 Breeding, Apr., Hong Kong**
• prominent white wedge up back
• flight call is a soft, whistled *teu,* similar to Lesser Yellowlegs

**5 Nonbreeding with Common Greenshank (right), Oct., Hong Kong**
• similar tail and rump pattern to Common Greenshank but half the bulk with smaller head, slimmer bill, slimmer chest, longer legs

# Common Redshank

***Tringa totanus*** **p. 356**

**Size:** **L 10¾–11½ in. (27–29 cm); WS 23½–26½ in. (59–66 cm); WT 105–184 g between Greater and Lesser Yellowlegs**

**Structure:** **more compact than yellowlegs with larger head, shorter legs; longish straight bill**

**Behavior:** **walks actively, picking and shallowly probing; marshes, tidal flats, sheltered lagoons, flooded fields**

**Status:** **Eurasian species; casual in Newfoundland**

**1 Juveniles (3 at left) with molting adults, Aug., England**

- all plumages darkish with prominent eye-ring, indistinct supraloral stripe, striking orange-red legs
- juveniles aged by uniformly fresh, brownish plumage with crisply spotted fringes to upperparts, coarsely streaked breast, duller legs, and dull bill-base
- molting adults show brighter bill-base, less neatly arranged upperparts; bird in foreground has acquired extensive plain gray upperparts and white underparts of nonbreeding plumage

**2 Molting adults, Aug., England**

- all plumages show bold white trailing edge to wing, white wedge up back
- flight call is a rich, whistled *teu-dr-dr,* intermediate in character between Lesser Yellowlegs and Willet

**3 Breeding, June, Scotland**

- smaller and more compact than Spotted Redshank with shorter, straighter bill, different plumage pattern

# Spotted Redshank

***Tringa erythropus*** **p. 357**

**Size:** **L $11\frac{1}{2}$–$12\frac{1}{2}$ in. (29–31 cm); WS $24\frac{1}{2}$–$26\frac{3}{4}$ in. (61–67 cm); WT 121–205 g same or smaller than Greater Yellowlegs**

**Structure:** **similar to Greater Yellowlegs but with smoother, less-sculpted contour; shorter legs; longer, straighter bill with subtle droop at tip**

**Behavior:** **active; walks steadily, picking at surface; often dashes after fish; marshes, sheltered lagoons, flooded fields; often in deep water; regularly swims**

**Status:** **Eurasian species; very rare migrant on w. Alaska islands; casual on both coasts; accidental inland**

**1 Juvenile, Oct., Japan**
• all plumages show unique bill shape with orange-red base to lower mandible
• juvenile and non-breeding show bright orange-to-red legs, bold dark loral stripe
• aged by uniformly fresh plumage with heavily barred under-parts; dark brownish upperparts with crisp pale notching along fringes

**2 First winter with Greater Yellowlegs, Dec., N.Y.**
• differentiated from Greater Yellowlegs by structure, leg and bill color, darker loral stripe
• aged by contrast between fresh, pale gray nonbreeding upper scapulars and more worn, brownish juvenal lower scapulars, tertials, and coverts; tertials show distinct pale notches
• size difference obvious here, but some Spotted Redshanks are as large as Greater Yellowlegs

**3 First winter, Mar., Japan**
• more advanced molt than the bird in photo #2; aged by contrast between fresh nonbreeding scapulars and worn (brownish) juvenal coverts; juvenal wing feathers will be retained until the annual complete molt in late summer
• nonbreeding adult similar, but tertials and coverts same color and pattern as scapulars

**4 Nonbreeding (right) with Marsh Sandpiper, Oct., Hong Kong**
• all plumages show barred rump and tail, white wedge up back, white underwing
• flight call is an emphatic, whistled *chu-IT;* similar to Semipalmated Plover but lower

**5 Molting adult, Apr., Japan**
• molting birds may show a mottled appearance below

**6 Breeding male, Apr., Japan**
• legs become dark in breeding season
• female similar but with white feather tips

# Wood Sandpiper

*Tringa glareola* p. 359

**Size:** L 7½–8½ in. (19–21 cm); WS 22½–22¾ in. (56–57 cm); WT 41–98 g slightly larger than Solitary Sandpiper

**Structure:** similar to Solitary Sandpiper but less attenuated, shorter billed, longer legged

**Behavior:** active; walks steadily, picking and probing at surface; often chases insects; marshes, muddy or grassy shorelines, stagnant pools; avoids tidal areas

**Status:** Eurasian species; uncommon to fairly common spring migrant, scarce fall migrant and rare breeder on w. Alaska islands; accidental outside Alaska

**1 Juvenile, Sept., England (inset, juvenile Lesser Yellowlegs, Aug., N.J.)**
- all plumages from Solitary Sandpiper by structure, larger spots above, bolder supercilium, white rump, pale underwing
- all plumages from Lesser Yellowlegs by smaller size, less attenuated shape with shorter primary projection, shorter, duller legs, bolder supercilium
- aged by uniformly fresh, dark brownish upperparts with large buff notches and relatively small scapulars

**2 Juvenile, Aug., England**
- paler, more diffuse breast pattern than Solitary Sandpiper

**3 Juvenile with juvenile Ruff, Aug., England**
- capped appearance and eye-ring may recall Sharp-tailed Sandpiper, but note spotted upperparts
- the Ruff is probably a female, being scarcely twice the bulk of Wood Sandpiper

**4 Nonbreeding, Oct., Japan**
• differentiated from juvenile by smoother gray breast and grayer, less neatly arranged upperparts; scapulars are larger, with white fringes broken by dark bars

**5 Breeding, May, Israel**
• white rump and strongly barred tail in all plumages; paler underwing than Solitary Sandpiper
• flight call is a high, whistled *pseu-hu-hu;* similar to Greater Yellowlegs but weaker and with the shrillness of Solitary Sandpiper

**6 Breeding, May, Greece**
• more compact than Lesser Yellowlegs, with shorter, stouter bill, shorter legs, shorter primary projection

# Green Sandpiper

*Tringa ochropus* p. 360

**Size:** L 8½–9½ in. (21–24 cm); WS 22¾–24½ in. (57–61 cm); WT 53–112 g slightly larger than Solitary Sandpiper

**Structure:** very similar to Solitary but with shorter wing point, slightly shorter primary projection; fuller, more rounded body; slightly straighter bill

**Behavior:** like Solitary Sandpiper

**Status:** Eurasian species; casual in spring to w. Alaska islands

**1 Juvenile, Aug., Cyprus (inset, juvenile Solitary Sandpiper, Aug., N.J.)**
- all plumages from Solitary by structure, darker overall plumage, slightly weaker eye-ring, white rump
- aged by uniformly fresh, neatly arranged upperparts with smallish scapulars
- juvenile with more coarsely streaked breast than juvenile Solitary

**2 Nonbreeding, Nov., Portugal**
- differs from juvenile by grayer, less neatly arranged upperparts with larger scapulars; white dots bleed slightly into feather centers

**3 With Common Sandpiper, Aug., England**
- round-bodied and very dark, with clean-cut division to white belly

**4 Nonbreeding, Jan., Japan**
- often looks very dark with faint spotting

**5 Molting adult, early Sept., Japan**
- note short primaries and very straight bill
- breeding birds have coarsely streaked breast

**6 Breeding, Mar., Israel**
- white rump and bold tail barring; wings are darker above and below and slightly broader than Solitary's
- flight call is a clear, ringing *tuEEt-wit-wit;* lower, stronger, less shrill than Solitary Sandpiper

# Gray-tailed Tattler

***Tringa brevipes*** **p. 368**

**Size:** **L 10–11½ in. (25–28 cm); WS 18–21½ in. (45–54 cm); WT 98–140 g same as Wandering Tattler**

**Structure:** **very much like Wandering but slightly longer neck, shorter nasal groove; tarsus scaling scutellated (overlapping like tiles); Wandering's are reticulate (non-overlapping)**

**Behavior:** **foraging behavior and motions similar to Wandering; rocky coasts and islands, mudflats, beaches; not so tied to rocky habitats as Wandering**

**Status:** **Asian species; uncommon to rare migrant on w. Alaska islands; casual on Alaska mainland; accidental along Pacific Coast**

**1 Juvenile (front) with juvenile Wandering, Sept., Alaska**
- all plumages paler overall than Wandering with bolder supercilium, more contrasting dark lores; tail often distinctly paler than back
- aged by uniformly fresh, neatly arranged upperparts; smallish scapulars and coverts have pale-dotted fringes
- differs from juvenile Wandering by cleaner white flanks, pale-dotted fringes to upperparts

**2 Juvenile, Sept., Japan**
- paler overall, so lores contrast more; broader supraloral; more prominent rear supercilium
- nasal groove extends just over halfway to bill tip; bill-base averages paler

**3 Juvenile Wandering Tattler, Nov., Calif.**
- darker overall, so lores contrast less; thinner supraloral; less prominent rear supercilium
- nasal groove well over halfway to bill tip

**4 Nonbreeding with Red-necked Stint, Nov., Australia**
• differs from Wandering by pale overall coloration, mostly whitish flanks, contrastingly dark lores, and pale tail
• differentiated from juvenile by unpatterned, less neatly arranged upperparts with larger scapulars and coverts

**5 Breeding and first summer, Apr., Australia**
• differs from breeding Wandering by paler overall coloration, finer barring below, with extensive white belly, finely barred uppertail coverts; face-pattern differences as with juvenile and nonbreeding; generally more gregarious than Wandering
• first-summer bird (back left) holds nonbreeding plumage through the summer

**6 Juvenile, Sept., Hong Kong**
• pale with contrastingly dark lores and underwing
• flight call is a whistled *tuu-eet* or *tuwi-di-di,* much like Common Ringed-Plover; very different from Wandering

# Common Sandpiper

***Actitis hypoleucos*** **p. 368**

**Size:** **L 7½–8½ in. (19–21 cm); WS 15¼–16½ in. (38–41 cm); WT 33–70 g same as Spotted Sandpiper**

**Structure:** **like Spotted, but has longer tail and subtly longer, straighter bill**

**Behavior:** **like Spotted Sandpiper**

**Status:** **Eurasian species; rare but regular migrant on w. Alaska islands; casual on Alaska mainland**

**1 Juvenile, Aug., Cyprus (inset, juvenile Spotted Sandpiper, Aug., N.J.)**
• aged by uniformly fresh, neatly arranged, pale-fringed upperparts
• differs from juvenile Spotted by longer tail, more grayish green legs, less prominently barred look to coverts, pale-spotted fringes to tertials and greater coverts; slightly straighter, grayer-based bill; darker breast

**2 Nonbreeding, Jan., Sri Lanka**
• differs from juvenile by lack of pale fringing above
• from nonbreeding Spotted by structure, leg and bill color, dark spotted edges on tertials and greater coverts, darker breast

**3 Breeding, May, Japan**
• streaked breast; dark markings on upperparts

**4 Aug., England**
• longer, broader wing stripe connects with white flanks; tail slightly longer with more prominent white fringe
• both species fly with distinctive shallow, vibrant wingbeats mixed with stiff-winged glides
• flight call is a quick, shrill *swEE-dee-dee-dee;* quality similar to Solitary Sandpiper but faster

**5 Spotted Sandpiper, July, N.J.**
• shorter, narrower wing stripe does not connect with white flanks; shorter tail with less white

# Terek Sandpiper

***Xenus cinereus*** **p. 371**

**Size: L 8¾–9½ in. (22–24 cm); WS 22¾–23½ in. (57–59 cm); WT 58–108 g slightly larger than Dunlin**

**Structure: similar to Spotted Sandpiper but stockier, with longer neck, longer wings, shorter tail, and unique, long upturned bill**

**Behavior: active; walks steadily, picking and probing, often with head held low; bobs tail like Spotted Sandpiper; mudflats, beaches, tide pools, reefs**

**Status: Eurasian species; rare migrant on w. Alaska islands; casual on Pacific Coast**

**1 Juvenile, Aug., Japan**
• all plumages mostly gray above with yellow legs and unique long upturned bill
• aged with close view by uniformly fresh, neatly arranged upperparts with relatively small, thinly buff-fringed scapulars and coverts; shorter black markings on upper scapulars; blurry streaks on neck and breast-sides

**2 Nonbreeding, Dec., Japan**
• differentiated from juvenile by plain gray, less neatly arranged upperparts with relatively large scapulars and coverts; smudgy breast pattern; indistinct stripe on scapulars

**3 Breeding and first summer, Apr., Australia**
• breeding birds show finely streaked head, neck, and breast, bold scapular stripe, darker bill
• first-summer birds (back right) hold nonbreeding plumage through the summer and undergo their annual complete molt (indicated here by fresh, crisply pale-fringed primaries) earlier than adults

**4 Breeding, June, Finland**

**5 With Curlew Sandpipers, Apr., Hong Kong**
• distinct white trailing edge to the wing
• flight call is a quick series of plaintive whistles, *pri-pi-pi*

# Little Curlew

***Numenius minutus*** **p. 374**

**Size:** **L 11¾–12 in. (28–30 cm); WS 27¼–28½ in. (68–71 cm); WT 119–210 g same as Upland Sandpiper**

**Structure:** **similar to Whimbrel but more delicate, with slimmer neck, much shorter, slimmer bill, longer legs; primaries usually fall even with tail tip**

**Behavior:** **walks steadily, picking and probing; dry grasslands, pastures, beaches**

**Status:** **Asian species; casual on Pacific Coast**

**1 Adult with Whimbrel (left), Apr., Japan**
• size and structure distinctive (but see Eskimo Curlew, p. 268); all plumages similar
• much smaller and buffier than Whimbrel, with pale lores, often less distinct crown stripes; much shorter, slimmer bill; dull pinkish legs

**2 Fresh adult, Apr., Japan**
• fresh birds are paler and buffier overall with fainter crown stripes; upperparts have broader buff fringes and more-distinct internal markings

**3 Worn adult, Sept., Calif.**
• worn birds are darker overall with bolder crown stripes; upperparts have narrower, paler fringes, more solidly dark centers

**4 Adult, Mar., Australia**
• very slim bill
• prominent eye on plain face recalls Upland Sandpiper

**5 Mar., Australia**
• though this is an uncommon species, wintering birds may gather in large flocks

**6 Worn adult, Sept., Calif.**
• brown underwing like Whimbrel

**7 Worn adult, Sept., Calif.**
• flight call is a rapid, whistled *twi-ti-ti-ti;* higher and weaker than Whimbrel

# Eskimo Curlew

***Numenius borealis*** **p. 375**

**Size:** **L 12¾–14¾ in. (32–37 cm); WS 28¾–33¼ in. (72–83 cm); WT 270–454 g slightly to distinctly larger than Little Curlew**

**Structure:** **like Little Curlew but longer wing point, longer bill, shorter legs**

**Behavior:** **foraging habits probably like Little Curlew; pastures, plowed fields in spring; heath barrens, salt marshes in fall**

**Status:** **Presumed extinct; formerly bred in nw. Canada, n. Alaska; passed through Great Plains in spring, Northeast in fall; no documented sightings since early 1960s**

**1 Adult Eskimo Curlew, Mar. 1962, Tex.**
• primaries project well past tail tip; longer bill, shorter legs, more heavily marked underparts than Little Curlew
• in flight, shows cinnamon underwing, plain gray flight feathers

**2 Adult Little Curlew, Oct., Japan**
• slightly smaller than Eskimo Curlew; primaries usually equal with tail tip (shorter here due to molt); shorter bill, longer legs, more sparsely marked underparts than Eskimo Curlew
• in flight, shows brown underwing

**3 Adult Upland Sandpiper, Sept., Del.**
• smaller than Eskimo Curlew; shorter, straighter bill, tail projects well past primaries; weak head-pattern

**4 Juvenile Whimbrel, Aug., Conn.**
• usually distinctly larger and longer billed than Eskimo Curlew, but runts overlap in both characters; note bulkier body, bolder head-pattern, heavier bill
• in flight, shows brown underwing, checkered flight feathers

# Bristle-thighed Curlew

***Numenius tahitiensis*** **p. 378**

**Size: L 16–17½ in. (40–44 cm); WS 30½–35½ in. (76–89 cm); WT 310–800 g same as Whimbrel**

**Structure: same as Whimbrel**

**Behavior: walks steadily, picking and probing; beaches, reefs, grassy areas in winter; breeds on open, hilly tundra**

**Status: rare and local endemic breeder in w. Alaska; casual along Pacific Coast**

**1 Nonbreeding, Nov., Midway Is., Hawaii**
• all plumages similar to Whimbrel but buffier, with larger spots on upperparts, buffy tail, bold buffy rump-patch, cleaner underparts
• fresh fall adults are bright buff with lightly streaked breast; bill often extensively pink
• primaries are molted rapidly after the wintering grounds are reached, often rendering birds flightless for two weeks; this individual shows no primaries past tertials indicating that molt is still underway

**2 Breeding, Mar., Midway Is., Hawaii**
• spring birds are darker with more heavily streaked breast, darker bill

**3 Breeding, June, Alaska**
• differentiated from Whimbrel by streaked breast contrasting with buff belly, larger spots on upperparts, orange-buff tail

**4 Breeding, June, Alaska**
• rich orange-buff rump and tail diagnostic
• flight call is a clear, easily imitated whistle, *ee-o-eet;* similar to Black-bellied Plover but faster, higher, and repeated more frequently

# Far Eastern Curlew

***Numenius madagascariensis*** **p. 380**

**Size:** **L 21 1/4–25 1/2 in. (53–64 cm); WS 35 1/4–41 1/2 in. (88–104 cm); WT 600–800 g slightly larger than Long-billed Curlew**

**Structure:** **similar to Long-billed Curlew**

**Behavior:** **walks steadily, picking and probing; beaches, tidal flats, marshes**

**Status:** **Asian species; casual on w. Alaska islands; accidental in British Columbia**

**1 Juvenile with Black-tailed Gull, Sept., Japan**
• all plumages similar; best identified by size, structure, and plain brown overall color without white rump; less buffy than Long-billed Curlew without rich buffy wing pattern • aged by uniformly fresh, relatively small scapulars, tertials, and coverts; coverts often contrastingly paler than scapulars; bill often distinctly shorter than adult's; underparts more sparsely marked

**2 Nonbreeding, Nov., Australia**
• differs from juvenile by less neatly arranged upperparts with larger scapulars, tertials, and coverts; underparts more heavily marked; very long bill may be extensively pink in nonbreeding season

**3 With Eurasian Curlews (right and foreground), Aug., Japan**
• plain brown underwing distinctive among large curlews
• Eurasian Curlew looks colder brown with contrastingly whitish belly and has bolder face-pattern with more pronounced supercilium

**4 Breeding, Apr., Japan**
• breeding birds have rufous highlights in upperparts

**5 With Eurasian Curlews, June, Hong Kong**
• plain brown plumage distinct from Eurasian Curlew's bold white rump and underwing
• flight call is a clear, mournful whistle, *cuu-ree;* flatter than Long-billed; often repeated

# Slender-billed Curlew

***Numenius tenuirostris*** **p. 380**

Size: **L 14½–16½ in. (36–41 cm); WS 32–36¾ in. (80–92 cm); WT 255–360 g close to Whimbrel**

Structure: **similar to Whimbrel but with slightly longer legs and shorter, slimmer bill**

Behavior: **walks steadily, picking and probing; shallow inland marshes, tidal flats, dry grasslands**

Status: **Eurasian species; nearly extinct; accidental in Ontario**

**1 Nonbreeding, Jan. 1988, Morocco**

- all plumages similar; note uniquely slim bill, bright white ground color to neck, breast, and belly, weak crown stripes
- nonbreeding birds show large, teardrop-shaped spots on flanks; juveniles have streaked underparts without spots; breeding birds show streaked underparts with a few flank spots

**2 Nonbreeding, Jan. 1988, Morocco**

- bright white underwing, rump, and tail
- flight call is a shrill *kew-EE;* like Eurasian Curlew but higher and faster

# Eurasian Curlew

***Numenius arquata*** **p. 382**

**Size:** **L 20–24 in. (50–60 cm); WS 36–40 in. (80–100 cm); WT 500–1360 g slightly larger than Long-billed Curlew**

**Structure:** **similar to Long-billed Curlew**

**Behavior:** **walks steadily, picking and probing; rocky or sandy beaches, tidal flats, marshes**

**Status:** **Eurasian species; casual on Atlantic Coast in fall and winter**

**1 Juvenile, Sept., England**
- all plumages similar; best identified by size, structure, and white rump
- aged by uniformly fresh, relatively small scapulars, tertials, and coverts; bill often distinctly shorter than adult's

**2 Adult *N. a. arquata,* Aug., England**
- differentiated from juvenile by less neatly arranged upperparts with larger scapulars, tertials, and coverts
- western *arquata,* which accounts for North American records, is more heavily streaked below (variable)

**3 Adult *N. a. orientalis* with Dunlin, Jan., Japan**
- eastern *orientalis* is more sparsely streaked below but shows much variation; some individuals are intermediate

**4 Molting adult *arquata*, Aug., England**
• white rump and lightly marked white underwing
• these birds are mid-way through primary molt, so worn outer primaries contrast with fresh and still-growing inners
• flight call is a rich, whistled *courrr-E,* slightly lower and flatter than Long-billed; also a melodic *coy-coy-coy* of a similar cadence to Spotted Sandpiper but much lower and richer

**5 Juvenile *orientalis,* Oct., Hong Kong**
• usually cleaner white underwing and more sparsely marked belly than *arquata*

# Black-tailed Godwit

***Limosa limosa*** p. 384

**Size:** **L 16–17½ in. (40–44 cm); WS 30½–35½ in. (70–82 cm); WT 160–500 g slightly larger than Hudsonian Godwit**

**Structure:** **somewhat stockier than Hudsonian, with longer legs and longer, heavier, straighter bill**

**Behavior:** **walks steadily, probing deeply in mud; tips up more than other godwits while feeding; marshes, tidal flats, flooded fields, impoundments**

**Status:** **Eurasian species; rare spring migrant on w. Alaska islands; casual on Atlantic Coast**

**1 Juvenile *L. l. melanuroides,* Sept., Japan**
• all plumages best identified by size, structure, and wing and tail patterns
• aged by uniformly fresh, neatly arranged plumage with prominent buff fringes to upperparts; buff-washed neck distinctive
• eastern *melanuroides,* which occurs in Alaska, is duller with finer-tipped, often subtly drooped bill

**2 Juvenile, presumed *L. l. islandica,* Aug., England**
• *islandica*, which breeds in Iceland, Norway, and Scotland, and is casual in North America, is the brightest subspecies overall with most heavily barred greater coverts and tertials
• nominate *limosa* of Europe to c. Asia is intermediate in plumage

**3 Nonbreeding *melanuroides,* Oct., Hong Kong**
• nonbreeding plumage is plain gray

**4 Nonbreeding *islandica* or *limosa,* Aug., England**

**5 Breeding male *melanuroides,* July, Russia**
• *melanuroides* shows the dullest rufous underparts, usually confined to neck
• note slightly drooped, fine-tipped bill

**6 Breeding male, presumed *islandica,* with Common Redshanks, Apr., England**
• *islandica* shows the brightest, most extensive rufous underparts, but some overlap with *limosa;* females of all races duller, longer billed than males

**7 Juvenile, presumed *islandica* with Little Egret, Aug., England**
• all plumages show bold wing and tail patterns; wing stripe averages broader in *limosa* and *islandica*

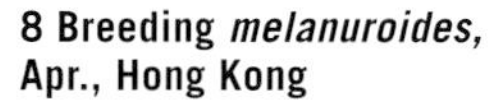

**8 Breeding *melanuroides,* Apr., Hong Kong**
• white underwing
• flight call is a nasal, creaking *ke-weeku,* often repeated

# Bar-tailed Godwit

***Limosa lapponica*** **p. 388**

**Size:** **L 14¾–15½ in. (37–39 cm); WS 28–32 in. (70–80 cm); WT 233–455 g between Marbled and Hudsonian Godwits**

**Structure:** **somewhat stockier than Hudsonian with shorter neck, shorter legs, longer bill**

**Behavior:** **walks steadily, probing deeply in mud; marshes, tidal flats, sheltered beaches; nests on open rolling tundra**

**Status:** **primarily Eurasian species; uncommon breeder in w. Alaska from the Y-K Delta to Barrow; casual along both coasts**

**1 Juvenile, Sept., Japan**
- aged by uniformly fresh, neatly arranged plumage with prominent buff checkering on upperparts
- short legs and checkered upperparts unique among godwits

**2 Juvenile with Marbled Godwits, Sept., Wash.**
- shorter legged and less buffy than Marbled, with bolder supercilium, different pattern above
- averages smaller than Marbled but with overlap

**3 Nonbreeding, Nov., Gambia**
- nonbreeding is plain pale gray-brown with dark feather centers, giving a unique streaked appearance above

**4 Breeding male, June, Alaska**
• rich chestnut on head and underparts; unlike other godwits, bill becomes mostly dark in the breeding season

**5 Breeding female, June, Alaska**
• larger and longer billed than males, with more variable plumage; most show some pale chestnut wash or barring on underparts, but some are as plain as non-breeding

**6 Breeding males and females, Apr., Australia**

**7 Breeding *L. l. baueri* with Black-tailed Godwits and Red Knot, Apr., Australia**

- eastern *baueri,* a breeder in Alaska, shows plain brown rump and underwing in contrast to Black-tailed's vivid wing and tail patterns

**8 Breeding female *baueri,* Oct., Hong Kong**

- *baueri* shows heavily marked underwing in all plumages

**9 Breeding *L. l. lapponica,* Aug., England**

- western *lapponica,* casual on the Atlantic Coast, shows a bold white rump and white underwing in flight
- flight call is an emphatic, slightly nasal *kIEY-did* or *kIEY-dey-did;* deeper than Hudsonian

# Great Knot

***Calidris tenuirostris*** p. 396

**Size:** **L 10½–11¼ in. (26–28 cm); WS 24¾–26½ in. (62–66 cm); WT 130–230 g slightly larger than Red Knot**

**Structure:** **similar to Red Knot but longer billed, longer legged, longer winged, and smaller headed**

**Behavior:** **like Red Knot**

**Status:** **Asian species; casual migrant (mostly spring) on w. Alaska islands; accidental in Oregon**

**1 Juvenile, July, Japan**
• all plumages most easily identified by size and structure
• aged by uniformly fresh, neatly arranged upperparts with thin pale fringing; note mostly dark mantle and scapulars and dark subterminal "anchors" on coverts and tertials

**2 Nonbreeding (front) with Red Knots and Curlew Sandpipers, Nov., Australia**
• longer bill and longer legs than Red Knot are quite striking in direct comparison; upperparts have subtly darker centers and paler fringes, so has more streaked, less flat gray look than Red Knot; face-pattern often less contrasting; underparts often more coarsely marked

**3 Breeding, with Whimbrel, Bar-tailed Godwit, Gray-tailed Tattler, and Ruddy Turnstone, Apr., Australia**
• size and plumage pattern similar to Surfbird, but structure very different

**4 Breeding, Apr., Australia**
• some individuals with much less rufous on scapulars

**5 Breeding, with Ruddy Turnstone, Apr., Hong Kong**
• longer, narrower wings than Red Knot; fainter wing stripe; darker primary coverts; paler rump
• flight call is a muffled *nyut-nut,* similar to Red Knot; seldom heard

# Red-necked Stint

***Calidris ruficollis*** **p. 406**

**Size: L 5¼–6½ in. (13–16 cm); WS 14½–15¼ in. (35–38 cm); WT 16–51 g about the same as Semipalmated Sandpiper**

**Structure: much like Semipalmated but more attenuated; shorter legs; shorter, slimmer, finer-tipped bill; unwebbed toes; usually horizontal stance**

**Behavior: like Semipalmated and often seen with it**

**Status: primarily Asian species; uncommon migrant and rare breeder in w. Alaska; rare but annual migrant elsewhere in North America, particularly along Atlantic and Pacific Coasts**

**1 Juvenile, Sept., Japan**
• aged by uniformly fresh, crisply pale-fringed upperparts imparting scaly look
• note long primary projection and tail; rufous-edged scapulars contrasting with relatively plain gray coverts and tertials; relatively pale head with contrasting dark lores and often a subtle "split supercilium"; smudgy breast-band with faint blurry streaks

**2 Juvenile, Sept., Japan**
• smudgy breast-band with faint blurry streaks; often clean-cut white throat; often looks quite pale-headed with contrasting dark lores
• flight call is a slightly rough *kiirp,* similar to Semipalmated Sandpiper but usually higher

**3 Nonbreeding, Nov., Japan**
• from Semipalmated and Western by unwebbed toes, finer bill, shorter legs, more attenuated shape; may show prominent shaft streaks on rear scapulars and coverts
• cleaner, grayer upperparts, whiter breast-sides, shorter legs, longer tail than Little Stint

**4 Breeding, Apr., Hong Kong**
- note horizontal stance, accentuated by short legs and attenuated shape
- fresh spring birds may be frosted with white above

**5 Breeding, May, Japan**
- clean rufous neck contrasts with spotted breast-sides; extensive rufous on mantle and scapulars but not on tertials or coverts

**6 Molting adult, Aug., Japan**
- molting birds often hold a telltale wash of clean rufous on neck; note shorter legs than Little; contrasting dark lores

**7 Molting adult (center) with Semipalmated and Least Sandpipers, late July, N.J.**
- most easily picked out from Semipalmateds by rufous-washed neck, though often looks paler-headed, shorter legged, and more attenuated

# Little Stint

***Calidris minuta*** **p. 407**

**Size:** **L 5½–6¼ in. (14–17 cm); WS 13½–14¾ in. (34–37 cm); WT 17–44 g same or just smaller than Semipalmated Sandpiper**

**Structure:** **more potbellied than Semipalmated, with smaller head, slimmer, finer-tipped, often drooped bill; unwebbed toes; legs sometimes look particularly slim and long**

**Behavior:** **like Semipalmated and often seen with it**

**Status:** **Eurasian species; casual in w. Alaska and along both coasts**

**1 Juvenile, Sept., Ireland**

- aged by uniformly fresh, crisply pale-fringed upperparts, imparting scaly look
- note long primary projection; blackish-centered coverts and tertials; prominent mantle lines; buffy breast-side patches with distinct dark streaks; contrasting gray nape; "split supercilium"

**2 Juvenile, Sept., Japan**

- "split supercilium" is highlighted by dark central crown ridge
- edges on upperparts vary from bright rusty to duller buff, but most birds are distinctively contrasty
- flight call is a very short, squeaky *chit* or *tik*, much shorter and harder than Semipalmated

**3 First winter, Nov., Cyprus**

- aged by contrast between fresh scapulars and worn coverts and tertials
- note unwebbed toes, long legs, fine bill, dirty (brownish gray) upperparts with extensive dark feather centers, extensively dirty breast-sides

**4 Nonbreeding, Jan., Sri Lanka**

- breast-sides look dirty without a clean intrusion of white; often a complete breast-band
- primaries often project well past tail tip

**5 Fresh breeding, May, Uzbekistan**
• rufous-edged upperparts including tertials and coverts; breast-sides with fine spots over rufous wash; fresh birds are frosted with white and usually brightest rufous on cheeks

**6 Fresh breeding, May, Uzbekistan**
• most breeding birds show prominent mantle lines

**7 Breeding, July, Norway**
• more-worn birds attain brighter rufous face and breast-sides, but throat still white

**8 Worn breeding, with Semipalmated Sandpiper, early July, N.J.**
• worn birds have pale heads and look brightest on the nape; this is a particularly large, heavy-billed individual

# Temminck's Stint

***Calidris temminckii*** p. 408

**Size:** **L 5¼–6 in. (13–15 cm); WS 13½–14¾ in. (34–37 cm); WT 15–36 g slightly smaller than Semipalmated Sandpiper**

**Structure:** **chunky body with long wings and tail, short legs; slim, fine-tipped, slightly drooped bill**

**Behavior:** **similar to Least Sandpiper; creeps slowly on bent legs; avoids open mudflats**

**Status:** **Eurasian species; rare migrant on w. Alaska islands; accidental on Pacific Coast**

**1 Juvenile, Aug., Cyprus**
• all plumages relatively plain and uniform with solid wash across breast; contrasting eye-ring; long wings and tail; yellow-olive legs
• aged by uniformly fresh upperparts with subtle, crisp buff fringes and dark subterminal "highlights" (unique among "peep")

**2 Juvenile, Sept., Japan**
• all plumages have long tail with bold white outer feathers

**3 Breeding, June, Finland**
• flight call is a crisp, ringing trill, *tidididup;* recalls Smith's Longspur rattle but higher and more rapid; often repeated

**4 First winter, Nov., Cyprus**
• aged by contrast between fresh, plain scapulars and worn coverts with thin pale fringes and dark subterminal "highlights"
• very plain gray-brown with smooth hood; white eye-ring stands out on plain face

**5 Breeding, May, Greece (inset, molting adult Least Sandpiper, Aug., N.J.)**
• breeding birds show a variable scattering of scapulars and coverts with blackish centers and dull rufous edges
• some Least Sandpipers may approximate this plumage (inset), but note longer bill, less attenuated shape, bolder supercilium and cheek spot, weaker eye-ring, less solid breast-band

# Long-toed Stint

***Calidris subminuta*** p. 409

**Size:** L $5^{1}/_{4}$–$5^{1}/_{2}$ in. (13–14 cm); WS $13^{1}/_{4}$–14 in. (33–35 cm); WT 23–37 g same as Least Sandpiper

**Structure:** similar to Least Sandpiper but longer necked and more potbellied; longer legs and toes; shorter, finer-tipped bill; more tilted over when feeding; more upright when alert

**Behavior:** similar to Least Sandpiper; awkward gallinule-like gait due to long toes

**Status:** Asian species; fairly common spring migrant on outer Aleutian Is.; casual elsewhere in w. Alaska; accidental in Oregon and California

**1 Juvenile, Sept., Japan**
• aged by uniformly fresh, crisply pale-fringed upperparts
• all plumages from Least Sandpiper by structure; dark crown continues to bill and lores; loral stripe fades toward eye; often faint dark cheek-patch; often pale-based lower mandible
• on juveniles, also note wing coverts contrastingly duller than scapulars and with pale fringes broken at tips; streaking extending down flanks

**2 Nonbreeding, Nov., Hong Kong**
• differentiated from Least Sandpiper by structure, face-pattern (though more subtle than on juvenile), broad dark centers on scapulars; often blander face-pattern than Least, with obscure cheek-patch

**3 Breeding, May, Russia**
• face-pattern and structure as juvenile
• often much brighter than Least, with broader orange fringes above; scapulars sometimes noticeably larger; breast more sharply streaked, often with bright orange wash
• due to long legs, must tilt down while feeding

**4 Breeding, Apr., Hong Kong**
• some individuals are strikingly bright orange throughout head, breast, and upperparts, recalling Little Stint; note structure, face-pattern, and yellow legs

**5 Worn breeding, Sept., Japan**
• note structure, face-pattern, and pale base to lower mandible
• though worn and dull, several scapulars show broad orange fringes; single fresh nonbreeding scapular shows broad dark center

**6 Least Sandpiper, Dec., Calif.**
• toes usually equal with tail tip

**7 June, Russia**
• toes project beyond tail tip
• flight call is a rich, trilled *churrd,* similar to Pectoral Sandpiper; much lower than Least

# Juvenile Stint Comparison

**1 Juvenile Red-necked Stint, Sept., Japan**
• slim and attenuated; short legs; short, fine-tipped bill; long primary projection and tail
• reddish-edged upper scapulars contrast with paler, grayer lower scapulars and coverts; smudgy breast-band

**2 Juvenile Little Stint, Aug., Cyprus**
• potbellied; short, fine-tipped bill; long primary projection
• contrasty upperparts with prominent mantle lines; prominent buffy breast-side patch with distinct dark streaks

**3 Juvenile Semipalmated Sandpiper, Aug., N.J.**
• straight, blunt-tipped bill; short primary projection
• same color buff tinge on crown, cheeks, breast-sides, upperparts; diffuse streaks at breast-sides

**4 Juvenile Western Sandpiper, Sept., N.J.**
• large head and thick neck (front-heavy), long legs and bill (often drooped tip); short primary projection
• reddish upper scapulars contrast with duller lower scapulars, coverts, and crown; very pale face; small cluster of streaks at breast-sides

**5 Juvenile Long-toed Stint, Aug., Cyprus**
• longer yellowish legs, toes; longer neck
• dark crown continues to bill and lores; loral stripe fades toward eye; faint dark cheek-patch; wing coverts duller than scapulars; pale-based lower mandible

**6 Juvenile Least Sandpiper, Aug., N.J.**
• shorter yellowish legs, toes; shorter neck
• pale forehead; strong dark loral stripe and cheek-patch; wing coverts as bright as scapulars; all-dark bill

# Worn Breeding Red-necked and Little Stint and ID Pitfalls

**1 Breeding Red-necked Stint, Aug., Japan**
• slim and attenuated; short, fine-tipped bill; long primary projection and tail
• clean rufous-washed face and neck; spotted breast-sides; rufous-based scapulars contrast with plain coverts and tertials

**2 Breeding Little Stint, July, Japan**
• potbellied; short, fine-tipped bill; long primary projection
• rufous wash throughout head and upperparts, including coverts, tertials, and breast-sides

**3 Breeding Sanderling, July, N.J.**
• molting birds often show rufous-washed neck like Red-necked Stint, but are much larger with heavier body, larger head, longer, heavier bill, no hind toe; usually plainer face-pattern

**4 Breeding Western Sandpiper, July, N.J.**
• large head and thick neck; long legs and bill; short primary projection
• rufous-based upper scapulars contrast with plain wings; no rufous on neck; spotted underparts

**5 Breeding Sanderlings and Semipalmated Plovers with Semipalmated Sandpipers (smallest birds), late July, N.J.**
• Sanderlings are much larger than "peeps"

**6 Juvenile Least Sandpiper (right) with Semipalmated Sandpipers, early Aug., N.J.**
• juvenile Leasts often look strikingly bright next to Semipalmated and have been mistaken for adult Little Stints; note stubby rear end, short, yellowish legs, dark cheek-patch; it is critical to correctly age any potential vagrant peep

# Sharp-tailed Sandpiper

***Calidris acuminata*** p. 417

**Size:** L 6¾–8 in. (17–20 cm); WS 16¾–19¼ in. (42–48 cm); WT 45–114 g slightly smaller than Pectoral Sandpiper; males 25–30 percent larger than females

**Structure:** similar to Pectoral but slightly rounder body, longer legs, shorter, slimmer bill

**Behavior:** like Pectoral; most sightings are with that species

**Status:** Asian species; casual in spring, fairly common in fall in w. Alaska; accidental in spring, rare in fall along Pacific Coast; casual in fall elsewhere

**1 Juvenile (left) with juvenile Pectoral Sandpiper, Oct., Japan**
• aged by uniformly fresh, relatively small scapulars and coverts with crisp rufous fringes and white outer tips
• differs from Pectoral Sandpiper by structure, bright buff breast with limited streaks, brighter rufous cap, bolder whitish supercilium, usually bolder eye-ring
• flight call is a mellow, whistled *treeip,* often doubled; recalls Barn Swallow; very different from Pectoral Sandpiper

**2 Juvenile, Oct., Japan**
• bill often darker than Pectoral's
• supercilium broadens behind eye, so cap looks "tilted forward"

**3 Nonbreeding, Nov., Australia**
• note rounder body and longer legs than Pectoral
• upperparts plain; head and breast pattern much like juvenile but colors more subdued; always lacks heavy breast-streaking of Pectoral

**4 Breeding, Apr., Japan**
• rounded body and small bill
• heavy breast-markings extend in chevrons down flanks, but do not form clean-cut division like Pectoral; rufous cap; bold eye-ring

**5 Molting adult, Aug., Del.**
• molting birds often retain a few telltale chevrons on flanks
• note structure, bold eye-ring, prominent cap

# Curlew Sandpiper

***Calidris ferruginea*** **p. 425**

**Size:** **L 7¼–7½ in. (18–19 cm); WS 16¾–18½ in. (42–46 cm); WT 35–103 g same as Dunlin or Stilt Sandpiper**

**Structure:** **similar to Dunlin but slimmer and more attenuated; longer wings, neck, and legs; finer-tipped, more evenly decurved bill; more upright stance while resting**

**Behavior:** **walks actively, rapidly picking and probing; impoundments, tidal creeks and flats, quiet beaches**

**Status:** **Eurasian species; rare but regular migrant on Atlantic Coast; casual elsewhere**

**1 Juvenile, Aug., Sweden**
• juvenile and nonbreeding differ from Dunlin by structure, paler breast, bolder supercilium; differ from Stilt Sandpiper by structure, black legs
• aged by uniformly fresh plumage with crisp pale fringing on upperparts imparting scaly look; buff wash on breast when fresh

**2 Nonbreeding with Red-necked Stint and Greater Sand-Plover, Nov., Australia**
• differs from juvenile by plain gray upperparts

**3 Breeding male, May, Europe**
• breeding birds show rich brick-red head and underparts, scalloped with white when fresh; females average paler with more dark barring below and have longer, heavier bills than males

**4 Molting female with Semipalmated Plover, July, N.J.**
• when worn, underparts become a paler orange color; molting birds retain a few telltale rufous feathers on underparts late into the season

**5 Molting to nonbreeding with Sanderling and Western Sandpipers, Aug., N.J.**
• upright stance while resting

**6 Molting to nonbreeding, July, N.J.**
• all plumages show white rump and wing stripe
• flight call is a low, trilled *chrreep,* with a quality like Pectoral but a pattern recalling Least

# Spoon-billed Sandpiper

***Eurynorhynchus pygmeus*** p. 428

**Size:** **L 5½–6½ in. (14–16 cm); WS 14–15¼ in. (35–38 cm); WT 26–39 g same as Semipalmated Sandpiper or Red-necked Stint**

**Structure:** **similar to Red-necked Stint but larger head, longer wings, and longer legs; unique bill is heavy and long with broadly spatulate tip; roosting birds stand more upright**

**Behavior:** **walks steadily, swishing bill from side to side in shallow water or wet mud ("vacuum-cleaner" style); marshes, tidal flats; often with Red-necked Stint**

**Status:** **Asian species; casual migrant in w. and n. Alaska; accidental in British Columbia**

**1 Juvenile, Sept., Japan**
• structure and behavior distinctive in all plumages
• aged by uniformly fresh, crisply pale-fringed upperparts imparting scaly look
• juveniles show bright, contrasty upperparts and breast-side patches similar to juvenile Little Stint; white mantle lines; very white face and variably prominent "split" supercilium
• flight call is a quiet, rolled *preep* or a shrill *wheet*

**2 Juvenile, Sept., Japan**

**3 Nonbreeding with Red-necked Stints, Broad-billed Sandpipers, and Lesser Sand-Plovers, Apr., Hong Kong**
• Can you find the Spoon-billed Sandpiper?
• nonbreeding birds are plain gray above with extensive white forehead and distinct "split" supercilium
• breeding plumage usually acquired relatively late, so spring migrants may look gray while Red-necked Stints are beginning to show color

**4 Breeding, with Red-necked Stint (back), June, Alaska**
• similar plumage pattern to Red-necked Stint, but note larger head, longer legs, heavier bill (spoon-shaped tip may be difficult to see from side view); streaking often extends well down flanks; beware of "peeps" with mud on bill!

# Broad-billed Sandpiper

***Limicola falcinellus*** **p. 429**

**Size:** **L $6^{1}/_{2}$–$6^{3}/_{4}$ in. (16–17 cm); WS $14^{3}/_{4}$–$15^{1}/_{2}$ in. (37–39 cm); WT 24–49 g between "peep" and Dunlin**

**Structure:** **similar to Dunlin but slimmer body, longer wings, shorter legs; heavier bill, kinked at tip**

**Behavior:** **walks slowly and probes dowitcher-like in soft mud; sheltered mudflats, tidal creeks, wet meadows**

**Status:** **Eurasian species; casual fall migrant on Aleutian Is., Alaska; accidental in New York**

**1 Juvenile, Sept., Japan**
- all plumages with uniquely thick bill, kinked down at tip; dull olive legs; bold head pattern with distinct "split" supercilium
- aged by uniformly fresh plumage with relatively small, crisply rust- and white-fringed scapulars, coverts, and tertials; crisply streaked breast

**2 First winter, Sept., Japan**
- aged by contrast between plain gray scapulars and darker, distinctly pale-fringed coverts and tertials retained from juvenal plumage
- nonbreeding adults show plain gray upperparts, paler crown though still with distinct "split" supercilium

**3 Fresh breeding, Apr., Australia**
- frosted look with pale feather fringes in early spring
- flight call is a drawn-out, buzzy, Dunlin-like *jrrreeeit,* with a distinct rising inflection

**4 Breeding, May, Greece**
- much darker by late spring
- from juvenile by more heavily marked breast, less neatly arranged upperparts

# Ruff

***Philomachus pugnax*** **p. 432**

**Size:** **L 8–12 in. (20–30 cm); WS 19¼–23¼ in. (48–58 cm); WT 70–268 g males average 20 percent larger than females; male close to Greater Yellowlegs; female slightly larger than Lesser Yellowlegs**

**Structure:** **chunky, rounded body; longish legs and neck; small head with short, slightly drooped bill; hunches over while feeding with "loose" mantle feathers sticking up; upright when alert**

**Behavior:** **active; wanders continuously, picking and shallowly probing; impoundments, marshes, flooded fields; often with yellowlegs**

**Status:** **Eurasian species; rare but regular migrant, mostly along Atlantic and Pacific coasts, Upper Midwest, and w. Alaska; rare in winter in California; casual elsewhere in winter**

**1 Juvenile, Sept., Japan**
• variable; in all plumages, leg color varies from greenish to yellowish to orange to pinkish; bill-base either the same color (especially in males) or dark
• juvenile and nonbreeding distinctively plain buff to grayish on head and breast; from Buff-breasted Sandpiper by size and structure (note particularly long tertials and short or lacking primary projection)
• aged by uniformly fresh plumage with crisp pale fringing on upperparts imparting scaly look

**2 Nonbreeding female, Mar., Japan**
• differs from juvenile by less neatly arranged upperparts with softer pale fringes
• females (sometimes called "Reeve") are smaller, usually dark-billed

**3 Nonbreeding male, Feb., England**
• males average 20 percent larger, often with orange- or pink-based bill; sometimes white on face and neck

**4 Juvenile (second from left) with Short-billed Dowitcher and Lesser Yellowlegs, Sept., N.Y.**
- size similarity to dowitcher and Lesser Yellowlegs indicates a female
- combination of size, longish legs, short bill, and plain buffy face and breast distinctive

**5 Juvenile and nonbreeding males and females with European Golden-Plover (left), Aug., England**
- size dimorphism striking

**6 Molting adult male, May, Uzbekistan**
- breeding males highly variable; fancy head tufts and neck ruffs held in spring and early summer; colors may vary from white to black to chestnut with many combinations; most have black-splotched bellies; this individual has partially developed white neck ruffs

**7 Breeding, May, Finland**
• in full breeding plumage, males have elaborate neck ruffs and head tufts, bare facial skin, usually all pinkish or orange bill
• Ruffs are unique among shorebirds for their striking sexual dimorphism and male polymorphism

**8 Breeding females, May, Greece**
• breeding females have upperparts similar to males and variable dark markings on breast; bills are mostly dark
• males and females typically segregated by sex in wintering areas and migration

**9 Breeding female, May, Finland**

**10 Breeding males at lek, June, Russia**
- grouselike lek display performed by groups of males at stop-over sites and on the breeding grounds; leks occasionally seen in North America
- note white underwing

**11 Molting adult, July, England**
- molting birds often retain patches of black on belly

**12 Juvenile with Dunlin, Aug., England**
- white V-shaped rump-patch unique
- mostly silent

# Jack Snipe

***Lymnocryptes minimus*** **p. 439**

**Size:** **L 6¾–7½ in. (17–19 cm); WS 15¼–16¾ in. (38–42 cm); WT 33–106 g two-thirds the bulk of Wilson's Snipe**

**Structure:** **slimmer and stubbier than Wilson's, with larger head, shorter neck; shorter, more pointed tail; shorter legs; shorter, more rounded wings; much shorter bill**

**Behavior:** **highly secretive; flushes only on very close approach, then flies weakly before dropping a short distance away; body rocks rhythmically as it probes in soft mud; usually among dense cover in fresh marshes, wet meadows**

**Status:** **Eurasian species; accidental in Alaska, California, Washington, and Labrador**

**1 Oct., England**

- best identified by size, structure, and behavior
- all plumages similar; back is green-glossed with thick buffy stripes; dark central crown with bold buffy lateral crown-stripes; flanks streaked, not barred
- mostly silent away from breeding grounds

**2 Oct., England**

# Common Snipe

***Gallinago gallinago*** p. 440

**Size:** **L 10–10¾ in. (25–27 cm); WS 17½–18¾ in. (44–47 cm); WT 72–163 g averages slightly larger than Wilson's Snipe**

**Structure:** **like Wilson's Snipe but 14, not 16, rectrices; single pair of broad outer rectrices separated from rest of tail during display (on Wilson's 2 pairs of more slender outer rectrices)**

**Behavior:** **like Wilson's Snipe**

**Status:** **Eurasian species; uncommon migrant and rare breeder in w. Aleutian Is., Alaska; rare migrant elsewhere on w. Alaska islands**

**1 Juvenile, Sept., Japan**
- averages larger, paler, buffier than Wilson's, often with weaker flank bars, but much overlap; tail and wing patterns diagnostic
- aged with caution by thin, even-width fringes to coverts (often divided at tips); very difficult to age after early fall
- voice very similar to Wilson's

**2 Adult, May, Japan**
- differs from juvenile by relatively broad pale tips (often divided) to coverts; first winter similar

**3 Spread tail, June, Russia**
- usually 14 tail feathers compared to 16 on Wilson's Snipe; outer pair (r7) broad; due to broad outer tail feathers, "winnowing" sound is hollower, more humming, and much lower than that of Wilson's

**4 Jan., Japan (top inset, Common Snipe underwing, June, Russia; bottom inset, Wilson's Snipe, Dec., Fla.)**
• axillaries and underwing coverts usually with extensive white unlike Wilson's, which is more evenly checkered; secondaries usually broadly tipped white; some show a pattern closer to Wilson's

**5 Dec., Oman (inset, Wilson's Snipe, Dec., Fla.)**
• secondaries usually with broad white trailing edge unlike Wilson's; some show narrower white edge closer to Wilson's

# Pin-tailed Snipe

***Gallinago stenura*** **p. 442**

Size: **L 10–10¾ in. (25–27 cm); WS 17½–18¾ in. (44–47 cm); WT 84–155 g averages slightly larger than Wilson's Snipe**

Structure: **similar to Wilson's or Common but slightly shorter bill, steeper forehead, larger eyes; shorter, blunter wings and shorter tail, so less attenuated; longer foot projection past tail in flight; tail with 24–28 feathers, the outer 7–9 pair extremely narrow and pinlike**

Behavior: **similar to Wilson's Snipe but also tolerates drier habitats; heavier, more direct flight**

Status: **Asian species; accidental on w. Aleutian Is., Alaska**

**1 Sept., Japan (inset, spread tail, Sept., Japan)**
• 7–9 pairs of pinlike outer rectrices diagnostic; other differences from Wilson's more subtle; all plumages similar; very difficult to age
• averages paler, buffier than Wilson's or Common, usually with more contrasting pale internal markings; outer edge to scapulars averages thinner, inner edge more distinct (but variable); averages broader supraloral, thinner loral stripe
• mostly silent away from breeding grounds

**2 With Common Snipe (left), Sept., Japan**
• paler, buffier with less pronounced back stripes than Common; shorter-billed and less attenuated; often shorter tail projection past wings
• in flight, wings dark like Wilson's but often with contrasting pale panel on upper secondary coverts; toes project farther past tail

# Eurasian Woodcock

***Scolopax rusticola*** p. 443

**Size:** **L 13¼–14 in. (33–35 cm); WS 20–24 in. (50–60 cm); WT 144–382 g larger than American Woodcock**

**Structure:** **similar to American Woodcock but smaller head and more pointed wings**

**Behavior:** **like American Woodcock**

**Status:** **Eurasian species; formerly casual visitor to Northeast; no recent records**

**1 Oct., England**

- all plumages similar; age determination not usually possible in the field
- similar to American but much duller and more uniform overall; lacks gray mantle stripes and rich orange-buff color below; underparts heavily barred
- in flight, wings much more pointed than American; occasionally gives snipelike call on takeoff

**2 Feb., Netherlands**

# Oriental Pratincole

***Glareola maldivarum*** **p. 451**

Size: **L 9¼–9½ in. (23–24 cm); WS 24–26¼ in. (60–65 cm); WT 75–136 g close to Black Tern**

Structure: **very long ternlike wings; short, stout, slightly decurved bill; shallowly forked tail; wings project well past tail tip**

Behavior: **walks actively in ploverlike fashion; forages mostly on flying insects with ternlike flight; meadows, shallow pools, cultivated fields; upright stance when alert**

Status: **Asian species; accidental in w. Alaska**

**1 Juvenile, July, Japan**
• size, structure, behavior unique to pratincoles
• all plumages from other pratincoles by a combination of characters: primaries project well past tail tip; lower breast moderately buff-washed; chestnut wing-linings and all-dark secondaries visible in flight
• aged by crisply pale-fringed upperparts, including primaries

**2 Nonbreeding, Sept., Japan**
• differentiate from juvenile by plain upperparts

**3 Breeding, May, Japan**
• differentiate from nonbreeding by black lores and throat-band

**4 Breeding, July, England**
• ternlike flight on long pointed wings
• bold white rump, all-dark secondaries, shallowly forked tail; wing-linings chestnut
• flight call is a sharp, ternlike *kewp* or *kuw-ik-ik*

# Hybrid Shorebirds

Hybrid shorebirds are extremely rare, much rarer than hybrid gulls or waterfowl. However, they have been found with increasing frequency in recent years, perhaps the result of an increasing number of birders who look carefully at birds and are armed with better identification information. Although it is often not possible to identify hybrids with 100 percent certainty, a little detective work and a detailed knowledge of variation in each species often reveals a very likely hybrid combination. Usually a hybrid will look superficially like one species but with a number of characters that are outside that species' range of variation. The key is to figure out what other species could contribute those "outside" characters without otherwise changing the appearance of the bird. As with any identification process, identifying hybrids must begin with a careful assessment of size and structure.

**1–3 Juvenile, presumed American x Pacific Golden-Plover (with Black-bellied Plover, left, fig. 1), mid-Oct., N.J.** This bird was well studied by the authors over a period of several days. During its entire stay, the bird remained on a beach, an expected behavior for Pacific but very unusual for American. Structurally, it has the "Black-bellied Plover look" typical of Pacific: relatively large squarish head, chunky rounded body, and relatively short rear end. However, it possesses the characteristic long primary projection of American with four primaries past the tertials, plus it has relatively short legs like American. In color, it is brightest (yellow-washed) on the fore-face, lores, crown, and nape, duller on the back and breast. This color distribution mostly fits Pacific, though the latter is usually brighter overall, with yellow extending into the breast. The overall plumage pattern is better for Pacific, with more contrastingly streaked face, neck, and breast, more contrastingly spotted nape, and more sparsely marked flanks and belly than on American. Perhaps the most compelling support of the hybrid theory is voice: the bird gave typical calls of both species.

**4 Adult Black-necked Stilt x American Avocet, Mar., Calif.**
Some hybrids are relatively straightforward to identify. This relatively large, long-legged shorebird shows the rusty-buff head and neck, white scapulars, and grayish legs of American Avocet. However, it is more extensively black above (including on the crown, nape, and cheeks), its bill is too straight for an avocet's, and its legs are too long and slightly tinged pink.

**5 Adult American x Black Oystercatcher (center) with Black Oystercatchers, Feb., Calif.**
Birds like this, showing characters intermediate between American and Black, are occasionally seen on the s. Pacific Coast. Interestingly, Pacific Coast American Oystercatchers *(H. p. frazari)* often show extra flecks of black below the hood, perhaps a result of past hybridization with Black Oystercatcher.

**6–7 Juvenile presumed Little x Temminck's Stint, Sept., Holland.**
At first glance, this small *Calidris* looks crouched, short-legged, and rather dark, recalling Least Sandpiper. However, many characters are wrong for Least, most notably the long primary projection. Upon close scrutiny there are several characters that indicate Temminck's Stint as one parent, including short legs, complete smooth brownish breast-band, and, most indicative of all, dark subterminal lines on lower scapulars. For the other parent, there are many characters in favor of Little Stint. These include long primary projection, pale mantle lines, extensively blackish centers to upper scapulars and tertials, whitish outer tips to lower scapulars and greater coverts, contrast between gray nape and more colorful breast-sides, and generally pale face with extensive white throat and faint upper (split) supercilium. The legs are a drab olive color, supportive of a crossing between a yellow-legged and dark-legged species.

**8 Juvenile, presumed White-rumped x Buff-breasted Sandpiper, early Nov., Newfoundland.**
This bird was seen with a flock of White-rumped Sandpipers and appeared slightly larger. General impression is similar to juvenile Buff-breasted, with plain buffy face fading to paler belly, yellowish legs, and scaly upperparts with feather centers darker subterminally. However, the structure is a bit off, with slightly shorter neck, shorter legs, longer bill, and longer wings. All of these characters point to White-rumped or Baird's Sandpiper as the other parent. Two features which support White-rumped over Baird's are the breast streaks extending down the flanks (lacking in both Baird's and Buff-breasted) and a subtle hint of brownish at the base of the lower mandible, a character unique to White-rumped. Circumstantial support of White-rumped is provided by the fact that the bird was seen with a flock of juvenile White-rumps during their peak migration season. Both Baird's and Buff-breasted pass through Newfoundland much earlier.

**9–10 Juvenile, possible Baird's x Buff-breasted Sandpiper, Sept., Mass.**
Like the previous hybrid, this bird has the general distinctive impression of a juvenile Buff-breasted, with plain buffy face fading to paler belly and scaly upperparts with a strong contrast between mantle and head. However, the bill is too long, the legs too short and dark (though not quite black), and the body is a bit too slight compared to the head for Buff-breasted. What is the other parent? Two characters which point to Baird's are the very straight, slim bill and a ghost of the Baird's loral pattern. Baird's fits otherwise, as well, though the bird is perplexingly pale on the head and breast, almost suggesting Sanderling as the other parent, though nothing else supports that supposition.

**11 Juvenile, presumed Baird's Sandpiper x Dunlin, Oct., Me.**
Seen here with a juvenile Semipalmated Sandpiper, this bird looks considerably larger, approaching Sanderling or Dunlin in size. Structurally the bird is clearly Dunlin-like, with a chunky, thick-chested, thick-necked frame and a relatively long, slightly drooped bill. However, it looks slightly short-legged, short-billed, and long-winged for Dunlin (all appropriate for Baird's). In terms of plumage, the bird looks mostly wrong for Dunlin but right for Baird's. The overall buffy appearance, brightest on the head, along with a very scaly look to the upperparts, is perfect for juvenile Baird's. The head-pattern fits Baird's, being rather plain with prominent dark lores and pale supraloral. The breast-streaking is moderately heavy, appropriately intermediate between Baird's and Dunlin. The fact that the bird is holding full juvenal plumage in October is a point in favor of Baird's. Most Dunlin have molted many head and body feathers by October.

**12–13 First winter, presumed Purple Sandpiper x Dunlin, Nov., England.**
This bird looks very much intermediate between Dunlin and Purple Sandpiper. Its very dark gray plumage and boldly white-fringed coverts are matched only by first-winter Purple (and Rock) Sandpiper. Dunlin's parentage is seen in the longer, drooped, all-dark bill, slightly longer dark legs, mostly whitish flanks contrasting with dark hood, bolder supercilium, and whiter underwing. Aside from range (which is a major consideration!), the only character pointing away from Rock Sandpiper instead of Purple is the relatively narrow wing stripe.

**14 Juvenile, presumed Pectoral x Curlew Sandpiper ("Cox's Sandpiper"), Sept., Mass.**
This much-celebrated hybrid combination was once thought to be a separate species, wintering in very small numbers in Australia and presumed to breed in Siberia. Seen here with a juvenile Semipalmated Sandpiper, the bird most closely resembles Pectoral Sandpiper in size, structure, and plumage pattern, but has longer, darker legs, a longer, darker, finer-tipped bill, longer primary projection, and a slightly plumper, more rounded body. All these characters are appropriate for Curlew Sandpiper. The breast-streaking is too sparse for Pectoral but too heavy for Curlew or Sharp-tailed Sandpipers. The head pattern is appropriate for Curlew or Pectoral Sandpipers and lacks the rich chestnut cap and very bold supercilium that would be expected from a Sharp-tailed Sandpiper parent.

# Aberrant Shorebirds

**1 Leucistic juvenile Least Sandpiper, Aug., Ohio.** Rarely, shorebirds may show diluted pigment and appear mostly white. This juvenile Least is identifiable by structure and aged by uniformly fresh plumage with relatively small, neatly arranged, crisply buff-fringed coverts and scapulars.

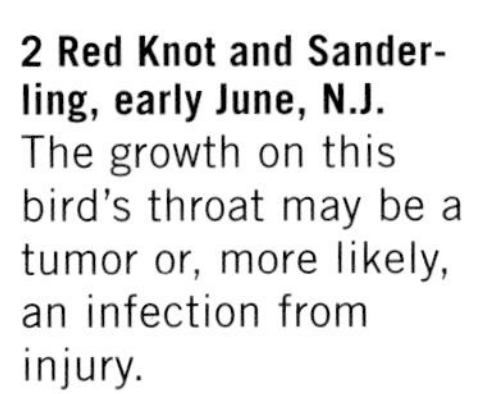

**2 Red Knot and Sanderling, early June, N.J.** The growth on this bird's throat may be a tumor or, more likely, an infection from injury.

**3–4 Ruddy Turnstones, Apr., N.J.**
Occasionally a bird's bill may grow abnormally. Observers should be prepared for odd variations and never rely on a single field character to make an identification.

# Species Accounts

## THICK-KNEES (FAMILY BURHINIDAE)

### Double-striped Thick-knee, p. 226

***Burhinus bistriatus***

**STATUS**

**Neotropical species.** Accidental in Texas (Dec. 1961). Locally common to uncommon within range. Resident in lowland areas from s. Veracruz, s. Oaxaca, and Hispaniola (and recently in the Bahamas) south to Colombia and n. Brazil. Slowly expanding its range because of the conversion of forest to pasture.

**TAXONOMY**

Four subspecies are recognized, but differences are slight. Nominate *bistriatus* occurs from s. Mexico to Costa Rica, *pediacus* in n. Colombia, *vocifer* in Venezuela and n. Brazil, and *dominicensis* in Hispaniola. Nominate *bistriatus* is the largest (especially wing and tarsus) and is relatively dark with a grayish breast; *pediacus* is the palest and has a cinnamon-buff tone to upperparts; *vocifer* is relatively dark and has a brownish breast; *dominicensis* is the smallest. The Texas record pertains to the nominate subspecies.

**BEHAVIOR**

Largely nocturnal. Occurs in arid to semiarid open country such as grasslands, pastures, and agricultural fields. Forages in ploverlike fashion, running short distances then plucking prey from the ground. Eats primarily invertebrates, particularly insects, worms, and mollusks, but will also take small lizards and mammals. Generally found in pairs or family groups. Spends much of the day roosting on the ground, where cryptic patterning makes it difficult to see. Seen most often on dirt roads at night when its large eyes strongly reflect headlights. Usually runs rather than flies to flee danger. Sometimes semi-domesticated for its watchdog qualities and to control household vermin.

Breeds primarily in the dry season, January to April in Costa Rica, April and May in Hispaniola, March to May in Colombia.

**MIGRATION**

Essentially sedentary.

**MOLT**

**First life year:** The molt out of juvenal plumage is partial, including most head and body plumage. This molt may be protracted for up to a year.

**Adult:** Molt patterns of adults are poorly known but appear to be gradual.

**VOCALIZATIONS**

Primary call is a far-carrying, barking *keh-keh-keh* . . . , often in a prolonged series or extended into a sputtering trill; sounds froglike at a distance. Most vocal at dusk or on moonlit nights.

## LAPWINGS AND PLOVERS (FAMILY CHARADRIIDAE)

### Northern Lapwing, p. 227

***Vanellus vanellus***

**STATUS**

Eurasian species. Casual, primarily in late fall and early winter, in the northeastern states and provinces. Most records are of single birds, though major flights occurred in December 1927 and January 1966, during which hundreds of birds appeared, apparently swept across the n. Atlantic by strong storms. Common and widespread in the Palearctic. Breeds from Scandinavia and Spain east to se. Russia and China. Winters from the Faroes and n. Africa east to Japan and Taiwan.

**TAXONOMY**

No subspecies recognized.

**BEHAVIOR**

Breeds in a variety of open habitats, particularly pastures and agricultural fields. During migration and winter, uses similar habitats, including pastures, plowed fields, damp meadows, and marshes.

Forages visually, employing run-stop-scan foraging technique. When found, prey is quickly plucked from the surface. Eats a variety of invertebrates, particularly earthworms and insects. Feeds by day or night, the latter apparently preferred under full moon. In the Palearctic, usually seen in large flocks, frequently numbering in the hundreds or thousands. Often associates with European Golden-Plover. Birds in North America often associate with Killdeer.

Nests early March to early July, mostly April/May.

#### MIGRATION

Partial to intermediate-distance migrant. Migrates primarily by day.

**Spring** migration somewhat variable, depending on weather. Departure from southern wintering areas begins as early as January and continues through April, with peak movement in March. Arrival on breeding areas in central Europe as early as January but generally in early March. Arrival in Scandinavia occurs from mid-February to early April. **Adults** arrive before **one-year-olds.**

**Fall** migration takes place between late May and December, mostly in a southwesterly direction. Freezing weather late in the fall and into the winter may provoke large movements. Early-season movements leisurely by comparison. **Adults** depart the breeding grounds between late May and September (failed breeders first) and pass through central Europe from June, with peak passage in September/October. Arrival at southern wintering areas generally in late October to early November. Southbound **juveniles** depart northern breeding grounds from July to October and pass through central Europe primarily in October/November. Juveniles generally winter farther south than adults.

#### MOLT

**First life year:** Molt out of juvenal plumage takes place from July to December. This molt includes head and body, inner secondaries, some wing coverts, and a variable number of tail feathers. From late January to early April there is a limited molt to breeding plumage involving head, neck, breast, and some wing coverts and inner secondaries. This plumage resembles breeding adult but is distinguished by retained juvenal primaries and outer secondaries.

**Second life year:** The complete molt to nonbreeding plumage takes place from about May (of first life year) to August/September, after which second-year birds are indistinguishable from adults.

**Adult:** From late January to early April there is a limited molt to breeding plumage involving head, neck, breast, and some wing coverts and inner secondaries. The complete molt to nonbreeding plumage takes place from May/June to August/September, occasionally October. flight feather and body molt occur simultaneously during migration.

#### VOCALIZATIONS

Highly vocal. Flight call is a shrill, mewing *pay-ee* or a more vibrant, cracking *pway-vich.*

## Black-bellied Plover, p. 29

***Pluvialis squatarola***

#### STATUS

Common and widespread. Occurs nearly around the globe, breeding in the high Arctic from w. Russia east to Baffin I. and wintering in coastal areas from s. Canada and the United States to s. South America, w. Europe, Africa, se. Asia, and Australia. Generally scarce inland except at selected sites during migration. The world population is estimated at 498,000, with 200,000 in North America.

#### TAXONOMY

No subspecies recognized.

**BEHAVIOR**

Breeds in moist to dry grassy and lowland heath tundra north of the treeline, often near the coast. Nest site is typically on dry, gravelly, or lichen-covered ridges or mounds among wetter or more vegetated tundra. In migration and winter primarily coastal, preferring tidal flats and beaches as well as plowed fields and pastures within a few miles of the coast. Migrants occasionally seen on interior lake shores and agricultural fields, particularly in the n. Great Plains. Like other plovers, forages visually in characteristic run-stop-pluck style with very upright posture. Feeds during both day and night. On the breeding grounds eats primarily insects and other invertebrates but also occasionally seeds and berries. During nonbreeding season feeds primarily on worms, bivalves, crustaceans, and, in some areas, insects. Typically seen in loose flocks, but often seen in large communal roosts numbering in the hundreds.

Nest initiation primarily late May in much of Alaska and mid-June in Canada and Alaska's North Slope. Egg dates mid-May to mid-July. Incubation 27 days. Both adults share parental duties, but female deserts family after about 12 days. Probably single-brooded. Peak hatching late June to mid-July. Young can fly after about 35–43 days. The male's display flight is high and zigzagging with slow, butterfly-like wingbeats.

**MIGRATION**

Intermediate to long-distance migrant. Migrates during both day and night.

**Spring** migrants begin to depart wintering areas in early April. Migrants are most common in coastal areas and rarer inland. Peak numbers pass through the Gulf and Pacific coasts in late April and the s. Alaskan and mid-Atlantic coasts in May. Most birds arrive on the breeding grounds between late May and mid-June and commence nesting within about one week. Some **one-year-old** birds remain on the wintering grounds through their first summer, while others migrate partway or all the way north to the breeding grounds. Hence small numbers of Black-bellied Plovers are regularly seen through much of the species' range in midsummer. Spring migration of first-summer birds averages later than that of adults.

In **fall**, southbound migration of **adults** begins by mid- to late July and continues into December. Peak numbers of adults pass through the mid-Atlantic and central Pacific coasts in late August and early September and through the Gulf Coast in mid-September. **One-year-old** birds typically move south well ahead of adults. Southbound **juveniles** depart the breeding grounds in mid- to late August, and head primarily toward coastal areas. The first juveniles arrive along the mid-Atlantic and mid-Pacific coasts in late August with peak passage there in October. Arrival on the wintering grounds takes place between October and December.

**MOLT**

**First life year:** The partial molt out of juvenal plumage is highly variable in extent and timing and involves some or all head and body feathers. Some birds begin to molt by late September, while many retain full juvenal plumage until December or January and a few until spring. Between about April and June there is a variable head and body molt. The resulting plumage may resemble anything from adult nonbreeding to adult breeding. Birds of this age are distinguished from adults by retained worn juvenal primaries and wing coverts.

**Second life year:** The complete molt to nonbreeding plumage takes place between about June and October, after which second-year birds are usually indistinguishable from adults. The spring molt to breeding plumage is as adult's.

**Adult:** Between about March and May there is a partial molt into breeding plumage involving most head and body feathers. This molt may commence either before or after birds leave the wintering areas. The complete molt to nonbreeding plumage takes place primarily between July and November. Many migrants seen between mid-August and mid-September are in heavy flight feather and body molt. This molt is generally completed upon arrival on the wintering grounds. However, molt timing is variable, depending on wintering areas. Birds that make the long overwater flight to South America do most of their molting on the wintering grounds, saving energy for migration. Other birds, presumably shorter-distance migrants, molt at high latitudes and do most of their southward migration in nonbreeding plumage.

#### VOCALIZATIONS

Flight call is a distinctive, far-carrying whistle, typically a slurred, three-syllable *plEE-uu-ee* with the middle note lower; sometimes shortened to one or two syllables. Distinctly lower pitched than calls of golden-plovers. Flight call is given frequently by standing and flying birds and is given in nocturnal migration. Flight song, given by male on the breeding grounds, is a similar three-syllable whistle but is softer and with the middle note highest: *wu-DEE-uu.* This is followed by a melodious trilled song as the bird descends.

## European Golden-Plover, p. 230
### *Pluvialis apricaria*

#### STATUS

Breeds from e. Greenland and Iceland to w. Siberia. Winters from w. Europe to n. Africa and the Middle East. Casual spring migrant in Newfoundland and Labrador with occasional large incursions in April and May. Accidental in Alaska.

#### TAXONOMY

No subspecies recognized. Northern and southern birds were once treated as separate subspecies based on wide variation in the amount of black on the face, neck, and breast in breeding plumage. Southern birds, on average, show less extensive black.

#### BEHAVIOR

Breeds primarily in dry heath tundra with scattered hummocks. During migration and winter, uses a variety of habitats including pastures, plowed fields, and tidal flats. Forages visually, employing run-stop-scan foraging technique, often focusing on one small area. When prey is found, the bird will quickly lean over and pluck it from the surface, or shallowly probe to extract it. In winter eats mostly earthworms and beetles. On the breeding grounds eats a variety of invertebrates including adult and larval insects, spiders, small mollusks, and crustaceans. Also eats berries, seeds, and leaves. Feeds mostly during the day. In Europe usually seen in flocks, often with Northern Lapwings.

Nests mid-April to early July.

#### MIGRATION

Partial to intermediate-distance migrant. Some birds in the British Is. are resident. Migrates by day or night.

**Spring** migration takes place between February and early June. Passage takes place through the Mediterranean region in February and early March, and through the British Is. between March and May, though some arrive as early as mid-February. Arrival on the breeding grounds takes place between mid-April (Iceland) and mid-June (Scandinavia).

**Fall** migration takes place between July and November. **Adults** depart the breeding grounds from early July in South to September in North, with peak passage through the British Is. mostly in late September and October. Southbound **juveniles** depart the breeding grounds by October/November and pass through the British Is. mostly in November.

#### MOLT

**First life year:** The partial molt out of juvenal plumage takes place from August/October to October/December. The molt includes most head and body feathers, some tertials, coverts, and tail feathers. The partial molt to breeding plumage takes place mostly during spring migration, between March and May. It includes a variable number of head and body feathers, wing coverts, and tail feathers and is distinguished from adult breeding by retained juvenal primaries, secondaries, and tail feathers.

**Second life year:** Molt patterns as in adult, but spring molt tends to take place slightly later and fall molt slightly earlier (beginning mid-May to mid-June).

**Adult:** The molt to nonbreeding plumage begins mid-June to mid-July and is completed mid-September to early November. Secondaries and some coverts are usually molted only every two years, but the molt is otherwise complete. In continental birds this molt begins with a limited head, body, and inner primary molt on the breeding grounds, is suspended during migration, and is completed on the wintering grounds. In Icelandic birds the molt is completed near the wintering grounds before migration. The partial molt to breeding plumage takes place mostly during spring migration, between late February and early May. It includes a variable number of head and body feathers, wing coverts, and tail feathers. This molt tends to be less complete in southern populations, which breed earlier and therefore stop molt earlier.

#### VOCALIZATIONS

Flight call is a clear, mournful whistle, *puuii;* much lower and simpler than American's. Song is a whistled *prr-pEE-u,* repeated monotonously as the bird performs its aerial display. Other more complex whistles are sometimes given during display.

## American Golden-Plover, p. 34

***Pluvialis dominica***

#### STATUS

Locally common migrant in spring in the Great Plains. Less common in fall in the Northeast. Generally scarce along the s. Atlantic and Pacific coasts. Rare elsewhere. Breeds in subarctic and Arctic regions of Alaska east to Baffin I. and n. Manitoba. Winters mostly from s. Peru and central Brazil to c. Argentina, rarely north to the Gulf and s. Atlantic coasts. Usually more common inland than Black-bellied Plover. The world and North American population is estimated at 150,000.

#### TAXONOMY

No subspecies recognized. Until 1993 this species and the Pacific Golden-Plover were considered conspecific.

#### BEHAVIOR

Breeds primarily in dry rocky tundra with sparse or low vegetation. Where sympatric with Pacific in Alaska, tends to prefer sparse rocky slopes. During migration and winter prefers upland areas with sparse or low vegetation such as native prairie, pastures, sod farms, and plowed fields. To a lesser degree also uses tidal flats, beaches, and salt-marsh pools. Forages visually, employing run-stop-scan foraging technique characteristic of plovers. When prey is found, the bird will quickly lean over and pluck it from the surface. Eats a variety of invertebrates including adult and larval insects, spiders, small mollusks, and crustaceans. Also eats berries, seeds, and leaves. Feeds mostly during the day. In prime habitat, migrants often seen in moderate flocks of about 30–50. Otherwise seen singly or in small flocks. Often associates with Killdeer or Black-bellied Plover.

Nest initiation primarily late May and early June. Egg dates late May to mid-July. Incubation 26–27 days. Both adults share parental duties, but male generally tends young longer than female. One brood per season but will renest after failure. Peak hatching late June to mid-July. Young can fly after 21–22 days. Display flight, performed by male, involves a silent ascent to a height of 30–330 ft. (9–100 m), followed by the display song given with slow, deep, butterfly-like wingstrokes.

**MIGRATION**

Long-distance migrant. Most follow an elliptical route, passing through the Great Plains in spring and over the Atlantic Ocean in fall. Smaller numbers regularly use other routes. Migrates by day or night.

**Spring** migration takes place between February and May. Arrivals along the Gulf Coast begin in late February, with peak passage from mid-March to late April. Peaks in the n. Great Plains in early to mid-May. Small spring passage takes place through New England and the mid-Atlantic Coast between mid-March and mid-May and through the central Pacific Coast in April and May. Arrival on the breeding grounds takes place from early May at southernmost sites to mid-June at high latitudes. A very few **one-year-old** birds remain at wintering sites through the summer, but most migrate north to breed.

**Fall** migration takes place between late June and late November. **Adults** depart the breeding grounds from late June (failed breeders) to early August. Peak numbers of adults pass through e. Canada in August and early September. Most depart e. Canada and head south on a direct flight over the Atlantic to wintering grounds in South America. Smaller numbers pass through New England and the mid-Atlantic, peaking from late August to mid-September, before heading offshore. Small numbers move south down the Atlantic, Central, and Pacific flyways, peaking between late August and mid-September. Southbound **juveniles** depart the breeding grounds by mid-August, with stragglers remaining into September. Peak numbers pass through e. Canada in September. Smaller numbers pass through the Pacific Northwest between late August and mid-October and through much of the rest of the continent between late September and late October. Stragglers often linger in coastal areas to late November, occasionally December.

**MOLT**

**First life year:** Molt out of juvenal plumage takes place between about October/December and December/March, mostly on the wintering grounds, and includes most body feathers and primaries. Most wing coverts and some tertials and rectrices are retained. The molt to breeding plumage takes place during spring migration between February and early May, though most birds are in largely nonbreeding plumage until mid- to late April. This molt involves head and body feathers, tertials, and most wing coverts and rectrices. This molt produces a very adult-like plumage in most birds, though birds remaining on the wintering grounds apparently molt in fewer breeding plumage feathers.

**Adult:** The complete molt to nonbreeding plumage begins during incubation in June and is completed by January. The head-and-body molt is completed by October. The primary molt takes place on the wintering grounds between about September and January. As with first-year birds, the molt to breeding plumage takes place during spring migration between February and early May, though most birds are in largely nonbreeding plumage until mid- to late April.

#### VOCALIZATIONS

Flight call is an urgent, whistled *pleE-dle* or *que-eed,* higher and more slurred than Pacific's with less change in pitch. Song is a shrill, whistled *klee-EEp,* repeated monotonously as the bird performs its aerial display. Other more complex whistles are sometimes given during display.

### Pacific Golden-Plover, p. 40

***Pluvialis fulva***

#### STATUS

Fairly common within limited U.S. breeding range. Uncommon along West Coast and central California in migration and winter. Breeds in Arctic and subarctic regions from the Yamal Pen. in w. Siberia east to w. Alaska. Winters primarily in coastal areas from Somalia east to s. China, New Zealand, and Oceania. A few winter along the U.S. West Coast, primarily in s. California. The world population is estimated at 125,000, with 16,000 in Alaska.

#### TAXONOMY

No subspecies recognized. Until 1993 this species and the American Golden-Plover were considered conspecific.

#### BEHAVIOR

Breeds primarily in tundra with dense vegetation and few rocks. Where sympatric with American in Alaska, tends to prefer lower, lusher areas near the coast or river valleys. During migration and winter uses a wide variety of habitats including pastures, lawns, plowed fields, salt marshes, tidal flats, beaches, and mangroves. Forages visually, employing run-stop-scan foraging technique characteristic of plovers. During "stop" will often lift up one foot, a behavior not typical of American. When prey is found, the bird will quickly lean over and pluck it from the surface. Will occasionally dig to pursue prey. Eats a variety of invertebrates including adult and larval insects, spiders, small mollusks, and crustaceans. Also eats small vertebrates, berries, seeds, and leaves. Feeds mostly during the day. On the West Coast usually seen singly or in small flocks. Premigratory roosts in Hawaii may number up to several hundred.

Nest initiation primarily late May and early June. Egg dates late May to mid-July. Incubation 25 days. Both adults share parental duties, but male generally tends young longer than female. One brood per season but will renest after failure. Peak hatching late June to mid-July. Young can fly after 26–28 days. Display flight, performed by male, involves a fluttering ascent to 30–330 ft. (9–100 m) followed by deep, butterfly-like wingstrokes with display song given throughout.

#### MIGRATION

Long-distance migrant. Many birds in eastern part of range make long overwater flight to w. Pacific islands. These birds average longer winged than western birds that follow transcontinental migration route. Migrates by day or night.

**Spring** departure from wintering areas begins as early as February in southern parts of the range and in early to mid-April in California. Peak numbers pass through Japan between late April and mid-May, n. Oceania in late April and early May, and the U.S. Pacific Coast in April and early May. Arrival on the breeding grounds takes place from late

April at southernmost sites to mid-June at high latitudes. Many **one-year-olds** and some **two-year-olds** remain at wintering sites through the summer, while others migrate north to breed.

**Fall** migration takes place between late June and late November. **Adults** depart the breeding grounds from late June (failed breeders) to mid-July, sometimes as late as early August. Adults pass through the West Coast between early July and mid-September, peaking in August, somewhat earlier than American Golden-Plover. Arrives in Hawaii in August. Southbound **juveniles** depart the breeding grounds by mid- to late August, with stragglers remaining until early October. Juveniles pass through the West Coast mostly between mid-August and mid-November, peaking in September and October. Juveniles arrive in Hawaii in October.

#### MOLT

**First life year:** Molt out of juvenal plumage begins in October, after birds have reached n. Oceania, and is completed on the wintering grounds by early December. The molt includes most body feathers and some wing coverts, but juvenal primaries are retained, unlike American Golden-Plover. Most wing coverts, some tertials, and rectrices also retained. The molt to breeding plumage begins on the wintering grounds in late January or February and is completed during migration by early May. The extent of this molt is highly variable in first-year birds. In some birds this molt involves head and body feathers, tertials, and most wing coverts and rectrices, producing an adultlike plumage. Other birds molt into a partial breeding plumage or remain in full nonbreeding. Those birds with the least advanced plumage are most likely to remain on the wintering grounds. One-year-old birds differ from one-year-old American Golden-Plover in retained juvenal primaries.

**Second life year:** The complete molt to nonbreeding plumage begins during incubation in June and is completed by January. About half of birds are in nearly full nonbreeding body plumage before departing the breeding grounds. The primary molt begins on the breeding grounds in July, is suspended during migration, and is completed on the wintering grounds by January. Fall migrants differ from American Golden-Plover in their fresh inner primaries. This is particularly visible in second-year birds with contrastingly very worn juvenal outer primaries. As with first-year birds, the molt to breeding plumage takes place mostly during spring migration, between late January and early May. Most second-year birds attain full breeding plumage, though some do not attain full plumage until their **third life year.**

**Adult:** Molt patterns as in second life year.

#### VOCALIZATIONS

Flight call is a clear, whistled *ku-EEid,* lower pitched and more clearly disyllabic than American's with emphasis on second syllable. Often given by overhead migrants. Song is a clearly enunciated low whistle, *pEE-prr-EE,* middle note lower. This phrase is repeated monotonously as the bird performs its aerial display. Other more complex whistles are sometimes given during display.

## Lesser Sand-Plover, p. 232

***Charadrius mongolus***

#### STATUS

Asian species. Rare but regular migrant on islands off w. Alaska. Casual in mainland Alaska (where it has bred) and along Pacific Coast. Accidental elsewhere in North America. Alaska records primarily May to September; Lower 48 records primarily July to October. Breeds in several disjunct populations from e. Kazakhstan, n. India, and w. China to the Chukotski and Kamchatka Pens. in e. Russia. Winters in coastal regions from South Africa and Saudi Arabia east to Taiwan and Australia.

**TAXONOMY**

Five subspecies recognized, falling into two distinct groups, which some authorities consider separate species. The mongolus (northern) group ("Mongolian Plover") is comprised of *mongolus,* which breeds in interior e. Russia, and *stegmanni,* which breeds in Kamchatka, Commander Is., and Chukotski Pen. Both races winter in Taiwan and Australia. All North American records refer to the latter group. The *altifrons* (southern) group ("Lesser Sand-Plover") is comprised of three subspecies: *pamirensis* breeds in s. central Russia and winters from Africa to w. India; *atrifrons* breeds in Himalayas and s. Tibet and winters from India to Sumatra; and *schaeferi* breeds in e. Tibet and winters from Thailand to Greater Sundas. Northern birds differ from southern in larger size, more bulbous-tipped bill, darker and colder upperparts, usually mottled (compared to clean white) rear flanks, narrower white rump-sides, and contrastingly dark tail. Additionally, breeding male northern birds have extensively white foreheads and more brick-colored (compared to pale orange) breast-bands, usually with dark upper borders.

**BEHAVIOR**

Breeds on barren Arctic tundra and mountain steppe and basins up to 18,000 ft. (5,486 m). Nest is usually on bare or sparsely vegetated sand or gravel near water. In migration and winter prefers sand or gravel beaches, mudflats, and muddy fields. Forages visually in typical plover fashion, employing a run-stop-pluck technique. Also employs "sewing-machine" style probing more than other plovers. Often wades in shallow water. Eats a variety of invertebrates including insects, crustaceans, mollusks, and marine worms.

Breeds primarily from May to July.

**MIGRATION**

Intermediate- to long-distance migrant. E. Siberian birds migrate coastally and overland. Most central Asian birds make nonstop flight over mountains to coastal wintering areas. Many nonbreeders summer in wintering areas, while others return north.

In **spring,** northbound **adults** depart wintering areas primarily in early April. Southern breeding grounds are reached by mid-April to early May and Arctic breeding grounds by late May or early June. A few birds, presumably **one-year-olds,** occasionally oversummer on the wintering grounds.

In **fall, adults** begin to flock near the breeding grounds by early July and depart through early to mid-August. Arrival at wintering areas is primarily in late August and September. **Juveniles** depart the breeding grounds by early September and arrive at the wintering grounds by mid-September to November.

**MOLT**

Unlike Greater Sand-Plover, the complete molt to nonbreeding plumage begins on the wintering grounds.

**First life year:** The molt out of juvenal plumage may be partial or complete. It takes place from late September to March and usually includes all but flight feathers, greater coverts, and outer tail feathers. Between late December and April, older nonbreeding feathers begin to be replaced by new nonbreeding feathers. In April/May, some breeding plumage appears on head and body. Flight feather molt is highly variable. Some birds molt no flight feathers, while others begin flight feather molt between December and May and complete it from May onward (see second life year). First-year birds in spring are similar to breeding adults but are less fully developed and are usually further distinguished by some or all retained worn juvenal primaries and outer tail feathers.

**Second life year:** The complete molt to nonbreeding plumage takes place from July/August to September/October. For birds that completed flight feather molt earlier, flight feathers are molted for a second time, often before first cycle is completed. Birds that commenced flight feather molt in late spring (of first life year) may not molt them again in summer. Between early April and early May there is a partial molt to breeding plumage as in adult.

**Adult:** Between early April and early May adults undergo a partial molt to breeding plumage, including most head and body feathers. The complete molt to nonbreeding plumage begins in mid- to late August after the wintering grounds are reached, though in some birds the forehead may begin to change during migration. Body molt is completed by October, and flight feather molt is completed by December to February.

#### VOCALIZATIONS

Flight call is a low, hard *kurrip,* reminiscent of a male Northern Pintail's call.

## Greater Sand-Plover, p. 234

### *Charadrius leschenaultii*

#### STATUS

Asian species. Accidental in California (Jan. to Apr., 2001). Breeds in desert and semidesert lowlands and plateaus from Turkey and Jordan east to s.-central Russia (L. Baikal), Mongolia, and central China. Winters in coastal regions from South Africa and Egypt east to Taiwan and Australia.

#### TAXONOMY

Three subspecies are recognized. *C. l. columbinus* breeds from Turkey and Jordan east to Caspian Sea; *crassirostris* breeds from Caspian Sea to e. Kazakhstan; nominate *leschenaultii* breeds in w. China, Mongolia, and s.-central Russia. Of the three, *columbinus* is the smallest with the shortest bill (similar in proportion to Lesser Sand-Plover) and, in breeding plumage, has the most extensive rufous in the upperparts and along the foreflanks; *crassirostris* has the longest bill and wing; and *leschenaultii* has intermediate-length bill and wing but the deepest bill measurements. The California record pertains to the nominate race.

#### BEHAVIOR

Breeds in arid lowlands and plateaus up to 9,800 ft. (3,000 m). Nest is usually on hard, flat, often salt-encrusted ground or saltpan with sparse shrubby vegetation, usually near water. In migration and winter prefers beaches, mudflats, and sometimes dry coastal grasslands. Forages visually in typical plover fashion, employing a run-stop-pluck technique. Often wades in shallow water. Eats a variety of invertebrates, particularly insects, but also crustaceans and marine worms. Forages singly or with other shorebirds.

Nests late March to June.

#### MIGRATION

Partial to long-distance migrant. Most birds make nonstop flight to coastal areas. Small numbers of nonbreeders summer in wintering areas, while others return north.

In **spring,** northbound **adults** depart wintering areas from late February to mid-April, with arrival in western breeding areas in March and eastern breeding areas in May.

In **fall, adults** begin to flock near the breeding grounds by mid-July and begin to head south by early August, with peak movement in mid- to late August. Arrival at wintering areas is primarily in August and September. **Juveniles** depart the breeding grounds by mid-August and arrive at the wintering grounds beginning in late August, with peak arrival in late September and October.

#### MOLT

Unlike Lesser Sand-Plover, the complete molt to nonbreeding plumage takes place largely at or near the breeding grounds.

**First life year:** The partial molt out of juvenal plumage begins in September and is completed from November to March. The molt is variable and may include all but flight feathers and some coverts and tail feathers, or it may be very limited, with most juvenal plumage retained. In April/May a variable head-and-body molt takes place. On birds returning to the breeding grounds, a partial breeding plumage is attained, and the outer primaries are also replaced. These birds may be distinguished from adults by duller overall plumage and the contrast between fresh outer and worn inner primaries. On birds remaining on the wintering grounds, new nonbreeding feathers appear, producing a plumage much like adult nonbreeding. On these birds the complete molt to nonbreeding plumage may begin as early as June (or even April in *crassirostris*), starting with inner primaries.

**Second life year:** On birds summering on the wintering grounds, the complete molt to nonbreeding plumage begins in April to June (of first life year) and is completed by July to September. Some individuals may molt primaries for a second time starting in July/August, sometimes before the first primary molt has been completed. Birds summering on the breeding grounds are on a schedule like adults, with the complete molt starting in July and completed by November to January. From February to March/April there is a partial molt to breeding plumage as in adult.

**Adult:** Between February and March (rarely April) there is a partial molt to breeding plumage including head, underparts, upper mantle, and sometimes central tail feathers. In some individuals the molt is more extensive, including most of upperparts and some coverts. The molt is typically more extensive in males. The complete molt to nonbreeding plumage begins by about mid-July at or near the breeding grounds, where much of the body plumage as well as inner 0–4 primaries are replaced. Arrives at wintering ground in nearly complete nonbreeding from about mid-August onward. Primary molt takes place mostly at the wintering grounds from August to November/December.

#### VOCALIZATIONS

Flight call is a deep, trilled, Ruddy Turnstone–like *trrr-uk-uk* or a shorter *krreeu.* Lower pitched and more trilled than calls of Lesser Sand-Plover.

## Collared Plover, p. 236

***Charadrius collaris***

#### STATUS

Central and South American species. Accidental in Texas (May 1992). Primarily resident from s. Sinaloa and s. Tamaulipas south to central Chile and central Argentina.

#### TAXONOMY

Two subspecies are sometimes recognized. Northern *gracilis* occurs from Mexico to Brazil and is slightly shorter winged than the nominate.

#### BEHAVIOR

Prefers sand or gravel beaches, river bars, sometimes lake shores and shrimp farms. Forages visually in typical plover fashion, employing a run-stop-pluck technique. Eats a variety of invertebrates. Forages singly or in small flocks with other plovers.

Nesting season primarily April to December in Mexico, March to June in Central America, May to September in South America.

**MIGRATION**

Largely resident, but some populations at least partially migratory. Commoner from April to September in central Chile. Some movement evident February–April and October–November in Central America.

**MOLT**

Molt timing poorly known but appears to vary depending on time of breeding.

**First life year:** The molt out of juvenal plumage appears to begin shortly after fledging (mostly late fall in northern populations) and involves head and body feathers and some wing coverts.

**Adult:** Molt timing and extent is poorly known, but, as in most other shorebirds, there appears to be a complete molt after breeding (mostly winter in northern populations) and a partial molt before breeding (mostly summer in northern populations).

**VOCALIZATIONS**

Flight call is a sharp *pweet* (resembling Wilson's Plover's flight call) and a more rolling *kerrp*.

## Snowy Plover, p. 44

### *Charadrius alexandrinus*

**STATUS**

Widespread throughout temperate regions of North and South America, Europe, Africa, and Asia. Generally uncommon in North America, where it is found at scattered sites throughout the arid western interior as well as along the Pacific Coast north to s. Washington and the Gulf Coast. Some populations are sedentary but most interior birds migrate to the Pacific or Gulf coasts in winter. The world population is estimated at 586,000 with 16,000–21,000 in North America.

**TAXONOMY**

Six subspecies recognized worldwide, with two in North America. *C. a. tenuirostris* occurs in the Caribbean and along the Gulf Coast east of Louisiana and *C. a. nivosus* in w. North America east to Louisiana. *C. a. tenuirostris* is often distinguished by its paler upperparts coloration, but the distinction is sometimes not apparent in the field, and some authorities lump the two subspecies.

**BEHAVIOR**

Breeds primarily in sparsely vegetated sandy beaches, lagoons, salt flats, or river bars. Nests on the open ground with at least some water nearby. During migration and winter, prefers coastal beaches, tidal flats, salt ponds, and wastewater ponds. Forages visually in characteristic run-stop-pluck feeding style, typically running faster and farther than other plovers. Feeds on a variety of terrestrial and aquatic invertebrates. Generally gregarious in winter, when it may form loose flocks of a few dozen up to 300. In breeding season typically seen in pairs or small groups.

Nesting begins as early as early March in California, late March in Florida, and late April to early June in the Great Basin. Last clutch usually laid by mid-July in most of range but as late as early September in Florida. Egg dates range from early March to early September. Incubation 24 days. Both adults share parental duties, though female often deserts after chicks are a few days old. Commonly rears two to three broods and may attempt to renest up to six times after failures. Peak hatching of first clutch ranges from mid-April to late June, depending on region. Young can fly after about 27–31 days.

#### MIGRATION

Short-distance and partial migrant. Migration timing varies regionally. Birds nesting in the Great Plains head primarily to the Gulf Coast, while birds nesting in the Great Basin head primarily toward the Pacific Coast and the Gulf of California. Birds nesting on the California coast may disperse either north or south for the winter. Florida birds are partially migratory, with some local dispersal and some birds departing the state.

**Spring** migration takes place between late February and early June. **Adults** typically arrive at Pacific Coast breeding sites between early March and late April. Some birds breed in more than one location per season in coastal California. These birds may depart the first breeding site as early as late April and arrive at the final site as late as June. Most birds in the interior West arrive between late March and early May.

**Fall** migration of **adults** takes place mostly between late June and November. Birds in California may depart their first breeding site as early as late April but then head to other breeding sites. Departure to wintering areas generally begins from late June to early August. Main departure from interior breeding sites takes place from July to October, with stragglers lingering into November. Migratory **juveniles** may depart the breeding areas any time from early June to mid-October, depending on nest timing.

#### MOLT

**First life year:** The molt out of juvenal plumage begins in late June for early hatchlings and mid-September for latest hatchlings, and is completed by about October/December. This molt involves head and body feathers and usually most wing coverts, tertials, and tail feathers. The partial molt to breeding plumage begins as early as mid-October on coastal birds or as late as March on interior birds and is completed by February/April. This molt includes head and body feathers as well as some wing coverts, tertials, and tail feathers, but is often less complete than in adult. This plumage resembles breeding adult but is distinguished by retained juvenal flight feathers.

**Adult:** The partial molt to breeding plumage begins as early as mid-October on coastal birds or as late as March on interior birds, and is completed by January/March. This molt includes most head and body feathers and some wing coverts, tertials, and tail feathers. The complete molt to nonbreeding plumage is poorly described but takes place mostly between June/July and early October/November. It begins on the breeding grounds but takes place extensively at stop-over sites.

#### VOCALIZATIONS

Flight call is a low, burry *drrrp* and variations, often repeated several times. A drawn-out, rising *toorrEET* is given by both sexes in alarm, aggressive encounters, and courtship. Ground-display song is a whistled *turEEu.* No flight song, unlike Old World Kentish Plover *(C. a. alexandrinus),* which some authorities consider a separate species.

## Wilson's Plover, p. 47

***Charadrius wilsonia***

#### STATUS

Locally common throughout its range in temperate regions of the Americas. Strictly coastal, occurring from Virginia and the Gulf of California south through the Caribbean, Gulf of Mexico, Brazil, and Peru. Resident in many areas, but most birds along the Atlantic and Gulf coasts move south in winter. Rare stray north of breeding range along Atlantic and Pacific coasts. Very rare stray inland. The North American population is estimated at 6,000.

#### TAXONOMY

Three subspecies recognized worldwide, with two occurring in North America. Nominate *C. w. wilsoni* breeds along the Atlantic and Gulf coasts south to Belize and the Caribbean. Strays have occurred along the coast as far north as Nova Scotia and inland north to Minnesota. *C. w. beldingi* breeds in the Gulf of California and along the Pacific Coast from central Baja California south to Peru. Strays have occurred along the coast as far north as Oregon and inland at the Salton Sea (one breeding record). *C. w. beldingi* is very similar to *wilsoni* but is smaller and slightly darker overall, with more extensive dark on lores and forehead and shorter, narrower supercilium. Breeding males have a more rufous-tinged nape.

#### BEHAVIOR

Strictly coastal. Breeds in areas of high salinity and scattered vegetation including salt flats, dunes, and overwashed beaches. Nests on bare ground, often near vegetation clump, drift, or other windbreak. Forages on relatively dry mud or sand flats and wide flat areas of beach with intertidal pools. Like other plovers forages visually in characteristic run-stop-pluck style. When alert, often assumes characteristic upright posture with back at a 70° angle. Also frequently assumes a crouched posture with back parallel to ground. Feeds primarily on crustaceans, especially Fiddler Crabs. Typically seen in pairs but often forms flocks of up to 30 or more in nonbreeding season.

Nest initiation primarily mid-April in south, late May in north. Egg dates early April to mid-July in the south, early May to late June in the north. Incubation 24–25 days. Both adults share parental duties. Typically single-brooded but may have second brood if first is lost. Peak hatching late May to mid-June. Young fly at 21 days.

#### MIGRATION

In **spring,** northbound **adults** begin to depart wintering areas in Middle America and the Caribbean in late February and March. Spring arrivals take place between late February/early March along the Gulf Coast and mid- to late April on the mid-Atlantic Coast.

In **fall,** most **adults** depart the breeding grounds in July, though some gather at nearby beaches through August. Most fall migration of adults takes place in August and September. Some northern birds winter in s. Florida or Texas, while others continue to points south in Middle America and the Caribbean. Southbound **juveniles** on the Atlantic Coast typically depart the breeding areas in August and September, with the peak movement through Florida and the Gulf Coast in October.

#### MOLT

**First life year:** The molt out of juvenal plumage takes place primarily in August to November and involves body feathers and some wing coverts. In February and March there is a partial molt involving body feathers, some wing coverts, and some flight feathers. This plumage resembles adult breeding plumage but can be distinguished by retained juvenal flight feathers. Also, the male's black markings are usually less extensive.

**Adult:** Between January and March there is a head-and-body molt to breeding plumage. Adults undergo a complete molt between July and October.

#### VOCALIZATIONS

Flight call is a short *pip* or *pi-dit,* given by standing and flying birds. Alarm note is a sharp, rising *pweet,* also given by standing and flying birds. On the breeding grounds a short, gurgling rattle, *jrrrrrrid,* is given by both sexes during ground and air chases.

## Common Ringed Plover, p. 237

***Charadrius hiaticula***

**STATUS**

Primarily Eurasian species. Fairly common in its restricted North American range. Breeds on the ne. coast of Baffin I., e. Ellesmere I., and Greenland. Widespread in the Old World, breeding in Arctic and subarctic regions from Europe to Siberia. Regular in spring on St. Lawrence I., Alaska, where it has bred. Winters primarily in w. Europe, Africa, and the Middle East. Virtually unrecorded in North America outside breeding range. The world population is estimated at 442,500, with fewer than 10,000 in North America.

**TAXONOMY**

Two subspecies recognized, but differences subtle and clinal. Nominate *C. h. hiaticula* breeds from ne. Canada to w. Europe. *C. h. tundrae* breeds from n. Scandinavia to Siberia and sometimes Alaska. *C. h. hiaticula* is generally larger and slightly paler (but the latter perhaps only because they have more-worn upperparts in spring/summer), but there is a cline, with southern birds being the largest and palest. Northern *hiaticula* and *tundrae* much more similar. Molt patterns typically differ between the two subspecies.

**BEHAVIOR**

In North America, breeds primarily in Arctic and subarctic areas above treeline. Nest site is typically on bare or sparsely vegetated sand or gravel near shorelines. In migration and winter prefers sand or gravel beaches, mudflats, marshes, muddy fields, and pastures. Forages visually in typical plover fashion, employing a run-stop-scan technique. Captures prey by leaning forward and picking at surface. Also employs a "foot-trembling" feeding method in muddy situations, causing prey to move and become more conspicuous. Eats a variety of invertebrates including insects, crustaceans, and mollusks as well as vegetable matter. Typically forages in well-scattered flocks. In North America seen singly or in small groups.

Nest initiation primarily mid-June. Egg dates range from mid-June to late July in Greenland, probably earlier in Alaska. Incubation 23–26 days. Both adults share parental duties. One or possibly two broods. Will sometimes renest after failure. Peak hatching mid-July in Greenland. Young can fly 25 days after hatching. The male's flight display is performed over its territory on slow, deep, butterfly-like wingstrokes with body alternately tilting from side to side.

**MIGRATION**

Partial to long-distance migrant. Only the Arctic populations are considered here.

In **spring,** northbound **adults** depart wintering areas in w. Africa in early to mid-April. Migrants pass through w. Europe in mid- to late May and arrive in Canada and Greenland by late May or early June. Arrival in w. Alaska takes place at about the same time. Other populations are on different schedules, some returning to breeding areas as early as February/March.

In **fall,** most **adults** depart breeding areas by early August, passing through w. Europe between mid-August and mid-September. Failed breeders may begin to head south earlier. **Juveniles** can fly by early to mid-August, after which they gather at local estuaries until their departure for Europe and Africa in early to mid-September. Peak passage through w. Europe is from mid-September to October.

**MOLT**

**First life year:** In *hiaticula* the molt out of juvenal plumage is partial, including body (but not all scapulars), tail, tertials, and some coverts. It takes place from August to January. As in adults, there is little or no spring molt in *hiaticula,* so one-year-old birds are readily distinguished from adults by retained worn juvenal flight feathers as well as some coverts and body feathers. In *tundrae* the molt out of juvenal plumage is complete, taking place primarily on the wintering grounds from about November/January to March/May. One-year-old *tundrae* are not readily distinguishable from adults.

**Adult:** In February and March *tundrae* undergo a partial molt to breeding plumage, including head, underparts, and some tail feathers, tertials, and coverts. This molt is minimal or lacking in *hiaticula.* The complete molt to nonbreeding plumage is variable, depending on latitude. Southern birds (including most *hiaticula*) molt on or near the breeding grounds, and primaries are molted without suspension from July to October. Most northern birds (including *tundrae* and some *hiaticula*) begin primary molt on the breeding grounds but suspend and complete the molt on or near the wintering grounds, from November to January/April. Siberian birds, those with the longest migration, may molt entirely on the wintering grounds.

#### VOCALIZATIONS

Flight call is a low, mellow *poo-eep;* lower and less rising than Semipalmated Plover, with a longer first syllable. Display song is a repeated, more emphatic version of the flight call, *tu-eea tu-eea tu-eea tu-eea . . .*

### Semipalmated Plover, p. 50
***Charadrius semipalmatus***

#### STATUS

Common throughout most of North America, especially in coastal areas. Breeds primarily in Arctic and subarctic regions from w. Alaska to Newfoundland. Winters in coastal areas from Washington and Virginia south to Chile and Argentina. Regular migrant inland. The world and North American population is estimated at 150,000.

#### TAXONOMY

No subspecies recognized but generally decreases in size from east to west.

#### BEHAVIOR

Breeds primarily in subarctic and low Arctic tundra near or above treeline. Nest site is typically on bare or sparsely vegetated ground in well-drained sand, gravel, or shale areas or dry tundra. Migrants commonest in coastal areas but numerous at favorable interior sites through much of North America. In migration and winter prefers sand or gravel beaches, mudflats, marshes, and fields with mud or short vegetation. Forages visually in typical plover fashion, employing a run-stop-scan technique. Captures prey by leaning forward and picking at surface. Also employs a "foot-trembling" feeding method in muddy situations, causing prey to move and become more conspicuous. Feeds by day and night. Eats a wide variety of aquatic invertebrates as well as flies, beetles, and spiders. Typically forages in well-scattered flocks of a few to several hundred.

Nest initiation primarily mid-May in milder areas to mid-June in the Arctic. Egg dates range from early May to mid-July. Incubation 23 days. Both adults share parental duties. Female abandons young and male after about 15 days. One brood but may renest after failure. Peak hatching late June in south, mid-July in north. Young can fly 22–31 days after hatching. The male's flight display is performed over its territory on slow, deep, butterfly-like wingstrokes with body alternately tilting from side to side.

#### MIGRATION

Migrates during both day and night.

In **spring,** northbound **adults** depart southern wintering areas mostly in March and April. Peak numbers pass through the Gulf Coast in late April, the Pacific Coast and interior West in late April/early May, and the mid-Atlantic Coast and upper Midwest in mid- to late May. Arrival on the breeding grounds takes place in early May in Alaska and British Columbia but mostly late May and early June in Manitoba (males before females). Egg laying begins shortly after arrival. Many **one-year-old** birds summer on the wintering grounds, while some migrate partway or all the way north to the breeding grounds.

In **fall,** most **adults** depart breeding areas by late July and early August (females before males). Failed breeders may begin to head south by mid-June. The first fall migrants appear in the Pacific Northwest in late June and in New England in early July. Peak numbers of adults pass through the Pacific Northwest in late July/early August, through much of the remainder of the continent in early to mid-August, and through the Gulf Coast in late August/early September. More common in New England and the Maritime Provinces in fall than spring. **Juveniles** can first fly by late July/early August and begin their southbound migration within a week or two. Most birds head to the coast, but interior sites are also used. The first juveniles arrive across the northern tier of states in late July, with peak passage there in early September and along the Gulf Coast in late September/early October.

**MOLT**

**First life year:** The molt out of juvenal plumage takes place during late fall, primarily on the wintering grounds, and includes head and body feathers and presumably some tail feathers and wing coverts. Between about April and June, first-year birds undergo a limited head and body molt. The acquired plumage is similar to breeding adult but is less fully developed and is distinguished by retained worn juvenal primaries and wing coverts.

**Second life year:** Molt patterns as adult except the complete molt to nonbreeding plumage begins as early as June.

**Adult:** Between January/February and March/April, **adults** undergo a partial molt to breeding plumage including head and body feathers and presumably some wing coverts and tail feathers. This molt takes place mostly on the wintering grounds. The complete molt to nonbreeding plumage usually begins in mid-July. The body molt takes place during migration and is usually completed by October (though some birds begin body molt as late as October). The primary molt usually begins between July and October (probably earlier with failed breeders) and is completed by December or January.

**VOCALIZATIONS**

Flight call is a strong, whistled *chu-EEp* with a distinct rising inflection on second syllable. Flight call is given frequently by standing and flying birds. Threat call is an accelerating, descending *yrp, yrp, yrp-yrp-yrp-yr-r-r-r-r-r.* Display song is a repeated, more emphatic version of the flight call, *ku-eep ku-eep, ku-eep . . . ,* often followed by a trilled *yr-r-r-r-r-r-yrp.*

## Piping Plover, p. 54

***Charadrius melodus***

**STATUS**

Widespread but uncommon and local in e. and central North America. Globally threatened and endangered. Breeds along the Atlantic Coast from Newfoundland to North Carolina and in the prairies from Alberta and Wisconsin south to the panhandle of Texas. Winters on the Atlantic Coast from North Carolina to Florida and along the Gulf Coast to Texas and the Yucatán. A few winter in the Bahamas and West Indies. Rare vagrant in the West. The world and North American population is estimated at 5,945 as of 2005.

### TAXONOMY

Two subspecies are recognized, including nominate *C. m. melodus* of the Atlantic Coast and *C. m. circumcinctus* of the prairies. On average, *circumcinctus* is darker overall with more contrastingly dark cheeks and lores. Breeding male *circumcinctus* shows more extensive black on forehead and bill-base and more often shows complete breast-bands. Some overlap exists.

### BEHAVIOR

An inconspicuous bird of dry sandy beaches. Breeds in open sand, gravel, or shell-strewn beaches and alkali flats. Nest site is typically near small clumps of grass, drift, or other windbreak. In winter prefers sand beaches and mudflats. Migrants seldom seen inland but occasionally show up at lake shores, river bars, or alkali flats. Forages visually in typical plover fashion, employing a run-stop-scan technique. Captures prey by leaning forward and picking at surface. Also employs a "foot-trembling" feeding method, causing prey to move and become more conspicuous. Feeds by day and night. Eats a wide variety of aquatic and terrestrial invertebrates, including marine worms, insects, mollusks, and crustaceans. Seldom found in large numbers except at a few favored wintering or staging sites, where numbers sometimes reach 100 or more. More typically seen in pairs or in groups of 3 or 4. When approached, more often runs than flies.

Nest initiation primarily late April on the Atlantic Coast, mid-May inland. Egg dates late April to mid-August inland, mid-April to mid-August on the Atlantic Coast. Incubation 26–28 days. Both adults share parental duties. Female sometimes abandons young and male after about 10 days. Generally one brood but will renest several times if earlier attempts fail. Peak hatching mid-June inland, late May to early June on the Atlantic Coast. Young can fly after 21–35 days. The male's flight display is performed over its territory on slow, deep, butterfly-like wingstrokes with body alternately tilting from side to side.

### MIGRATION

Short- to intermediate-distance migrant. Interior birds head primarily to the Gulf Coast and, to a lesser degree, the s. Atlantic Coast, with few birds stopping inland. Atlantic Coast breeders head south and winter primarily along the s. Atlantic Coast, the Bahamas, and West Indies, though some birds cross over to the Gulf Coast. Migrates during both day and night.

**Spring** migration through the Gulf Coast takes place between late February and mid- to late April, with peak numbers in March. Arrival on interior breeding grounds as well as in New England and the Maritime Provinces takes place between mid-April and mid-May. Arrival on mid-Atlantic Coast beaches takes place between early March and mid-April. Males often arrive before females. Some **one-year-old** birds remain on the wintering grounds through the summer, though up to half may return north to breed.

In **fall,** most **adults** depart the breeding grounds by mid-July (females before males), though some birds may leave as early as late June (mid-June for some failed breeders), while others with late nests may remain into September. Peak numbers of adults pass through the mid-Atlantic and Gulf coasts from mid-July to mid-August. Southbound migration of **juveniles** begins between late July and early September. Peak numbers of juveniles pass through New Jersey in August, through Virginia in September/October, and through Texas in October and early November. A few stragglers may remain along the mid-Atlantic Coast into November or December.

**MOLT**

**First life year:** The molt out of juvenal plumage takes place during early fall, primarily on or near the wintering grounds, and includes at least head and body feathers.

**Second life year:** In early spring there is a head-and-body molt to breeding plumage. This plumage is similar to breeding adult but is distinguishable by retained worn juvenal primaries and wing coverts. For birds that did not breed, the complete molt to nonbreeding plumage presumably takes place earlier than in adults.

**Adult:** In February and March there is a partial molt to breeding plumage including at least head and body feathers. The complete molt to nonbreeding plumage usually begins in July and is completed by October. Body molt takes place during migration, while primary molt usually begins after the wintering grounds are reached.

**VOCALIZATIONS**

Flight call is a soft, whistled *peep* given by standing and flying birds. Frequently heard alarm call is a soft *pee-werp,* with the second syllable lower pitched. Male's display song is a repeated, high-pitched *pirp, pirp, pirp, pirp, pirp . . .* or more drawn-out *pooeep, pooeep, pooeep, pooeep, pooeep . . . ,* often repeated 40 or more times per flight.

## Little Ringed Plover, p. 241

***Charadrius dubius***

**STATUS**

Palearctic species. Casual spring vagrant to w. Aleutians. Breeds from Norway and Morocco east to se. Russia and New Guinea. Winters from the n. tropics of Africa east to s. China and New Guinea.

**TAXONOMY**

Three subspecies recognized. Nominate *dubius* breeds from the Philippines to New Guinea; *jerdoni* breeds in India and se. Asia; *curonicus* breeds through remainder of range from Europe and n. Africa to Japan. Differences are very slight and not usually recognizable in the field. Nominate *dubius* has the longest bill, *jerdoni* the shortest bill and wing, and *curonicus* the darkest bill. Alaskan records presumably pertain to *curonicus.*

**BEHAVIOR**

More of an inland species than Ringed or Semipalmated Plovers. Breeds primarily in temperate interior lowlands (up to 6,600 ft. or 2,000 m), usually near fresh water. Nest site is typically in bare or sparsely vegetated areas of gravel or sand, particularly along river or lake shores. Also uses man-made sites such as gravel pits, industrial sites, or gravel rooftops. In migration and winter, uses muddy shores inland and along coast, often wading in shallow water. Forages visually in typical plover fashion, employing a run-stop-pluck technique, though more active than Ringed or Semipalmated Plovers. Also employs a "foot-trembling" feeding method. Eats primarily insects and other small invertebrates. Less gregarious than Ringed or Semipalmated Plovers, seldom joining large flocks of other birds.

Breeding season April to September in Europe, December to June in India.

**MIGRATION**

Partial to intermediate-distance migrant.

In **spring,** northbound **adults** depart wintering areas from late February to early April, with peak movement through w. Europe in April and early May. Northernmost breeding areas are reached from mid-May to early June.

In **fall, adults** begin to disperse from breeding areas in late June and July, with peak passage through Europe in August and through n. Africa in September. Arrives at northern wintering areas by late August and southern wintering areas by early October. **Juveniles** begin to depart the breeding grounds by July, with peak movement through Europe in late August and early September and through n. Africa in late September and October.

#### MOLT

**First life year:** The partial molt out of juvenal plumage takes place from August to December and typically includes most of upperparts, innermost secondaries, and some coverts and tail feathers. Occasionally all plumage is replaced except outer secondaries and primaries. Rarely is the molt complete. Between January and April/May there is a partial molt to breeding plumage, including most of head, body, tertials, and wing coverts. One-year-old birds may be distinguished from adults by typically retained juvenal flight feathers and coverts and some brown nonbreeding feathers in the breast-band.

**Adult:** Between January and April/May, there is a partial molt to breeding plumage, including most of head, body, tertials, and wing coverts. The complete molt to nonbreeding plumage begins in June on the breeding grounds, where some body plumage as well as inner 2 or 3 primaries are replaced before molt is suspended. The molt continues at staging areas in July and August, then is completed on the wintering grounds by November or December.

#### VOCALIZATIONS

Flight call is a clear, far-carrying *peeu,* descending in pitch; reminiscent of Lapland Longspur's whistled flight call.

## Killdeer, p. 58

### *Charadrius vociferus*

#### STATUS

Common, widespread, and familiar throughout most of North America. The most commonly seen shorebird inland. Breeds from se. Alaska to Newfoundland, south through the West Indies and n. Mexico. Winters from coastal British Columbia, s. Nebraska, and e. Massachusetts south to w. South America from n. Chile to Venezuela. Some southern populations are resident, and an isolated population is resident from Peru to n. Chile. The world population is estimated at more than 1 million.

#### TAXONOMY

Three subspecies recognized worldwide, with one occurring north of Mexico. Nominate *C. v. vociferus* breeds throughout North American range and winters south to the Caribbean and n. South America. *C. v. ternominatus* is resident in the West Indies, and *C. v. peruvianus* is resident in Ecuador, Peru, and n. Chile. The latter races are both smaller than *vociferus* with, on average, more rufous in the wing coverts. Note, however, that much variation exists in this character, with southern populations of *vociferus* showing more rufous than northern populations. On average, *ternominatus* has the grayest scapulars of the three and *peruvianus* the most rufous-edged scapulars.

#### BEHAVIOR

Breeds in a wide variety of open situations. Nest site is typically on bare dirt or gravel with little or no vegetation. Often uses graveled rooftops or gravel parking lots and frequently nests close to human activity. Migrants equally common inland as along the coast. In migration and winter prefers short-grass pastures (including turf farms, ball fields, and so on), plowed fields, and mudflats, usually near water.

Forages visually in typical plover fashion, employing a run-stop-scan technique. Captures prey by leaning forward and picking at surface or probing shallowly into mud. Also employs a "foot-trembling" feeding method, causing prey to move and become more conspicuous. Feeds by day and night. Eats a variety of invertebrates, including earthworms, snails, grasshoppers, and beetles. Less frequently eats small vertebrates and seeds. Typically forages in well-scattered flocks of a few to several hundred.

Breeding season protracted in the south, more restricted in the north. Nest initiation primarily early April in south, mid-May in north. Egg dates mid-February to late October in the south, late April to early July in the north. Incubation 24–26 days. Both adults share parental duties. Usually one brood in north, two or three (rarely six) in south. May renest after failure. Peak hatching early May in south, mid-June in north. Young fly after about 40 days. The male's flight display is performed over its territory on slow, deep, butterfly-like wingstrokes. Display flight may last for up to an hour. Injury-feigning display, commonly observed in this species due to its propensity to nest near human habitation, is used as a distraction to lure predators away from eggs or young. This display involves frantic calling while dragging spread wing and tail as if injured.

#### MIGRATION

Migrates during both day and night. Severe cold or snowfall may induce movements in late fall or early winter by birds attempting to winter at more northerly latitudes. Occasionally swept north in late fall by coastal storms.

In **spring,** northbound **adults** depart southern wintering grounds primarily between mid-February and March. Peak numbers pass through much of the continent in March, with arrival at most northerly breeding sites between mid-April and mid-May. Most **one-year-old** birds apparently migrate north to breed.

In **fall, adults,** at least in northern portions of range, withdraw in winter, some moving as far south as the West Indies and n. South America. Postbreeding adults tend to gather in flocks between late June and August, and southbound migration may begin in that time period. Peak migration of adults typically occurs from late August in the north to early October farther south. A hard freeze or heavy snow later in the season may cause the most northerly birds to move farther south or to coastal areas. Southern **juveniles** can fly as early as mid-April, while northern juveniles first fly by early July. Young from later nests in the south may fledge as late as January. Birds at least in northern portions of range withdraw in winter, some moving as far south as the West Indies and n. South America. Peak migration of juveniles typically occurs from late September/early October in the north to November farther south.

#### MOLT

Molt timing variable, depending on region.

**First life year:** In northern birds, the molt out of juvenal plumage takes place between July and November (some completed by mid-September) and includes head and body feathers, some or all rectrices, some wing coverts, and some inner secondaries. Late-hatching southern birds may molt much later.

**Adult:** Between about February and May there is a variable partial molt to breeding plumage. This molt often includes some head and body feathers and some wing coverts, though at times the molt may be skipped. The complete molt to nonbreeding plumage usually takes place between mid-June and early November (primarily August and September) and begins on the breeding grounds.

**VOCALIZATIONS**

Flight call is a strident, drawn-out *deeee* with a rising inflection. Variations of this call, including a variable *dee-dit-dit* or longer trill, may be given in mild alarm. Primary alarm call is the familiar, strident *kill-deea,* often repeated incessantly and typically given by running or flying birds. A variation of this call is given monotonously by male in courtship display flight.

## Mountain Plover, p. 62

### *Charadrius montanus*

**STATUS**

Uncommon and declining. Breeds in the w. Great Plains and Colorado Plateau. Winters primarily in central California, with progressively fewer wintering south and east to s. Arizona, n. Mexico, and s. Texas. Migrants occur throughout the interior West. Seldom found near water. The world and North American population is estimated at 9,000.

**TAXONOMY**

No subspecies recognized.

**BEHAVIOR**

A bird of arid plains, not mountains. Seldom seen near water. Breeds in dry flat short-grass prairie and semidesert areas with short sparse vegetation. Frequently found in the vicinity of prairie dog towns, cattle, or other areas with disturbed soil. Nest site typically in area of reduced vegetation with some bare ground, often near an object such as a cow manure pile. In migration and winter prefers dry plowed fields, heavily grazed grasslands, sod farms, and alkaline flats. Forages visually in typical plover fashion, employing a run-stop-scan technique. Captures prey by leaning forward and picking at surface or probing into cracked soil. In winter also employs a "foot-trembling" feeding method, causing prey to move and become more conspicuous. Eats a wide variety of terrestrial invertebrates, including grasshoppers, beetles, crickets, and ants. In winter, typically forages in scattered flocks of 30 to several hundred. Rarely occurs singly. When approached, will often crouch and "hide" rather than fly.

Nest initiation primarily mid-May. Egg dates mid-April to mid-July. Incubation 28–31 days. One brood per adult per season, but male may raise a first brood while female raises a second. Some pairs raise only one brood. May renest after failure. Peak hatching mid-June. Young can fly after 33–34 days. The male's flight display, performed over its territory, involves a steep rise followed by a "falling leaf" descent with wings held in a deep "V" as the bird rocks back and forth.

**MIGRATION**

Migrates primarily at night.

In **spring, adults** depart the wintering areas primarily between mid-February and early March, and many apparently fly nonstop to the breeding grounds. Arrival on the breeding grounds takes place between early March and mid-April. Egg laying does not usually begin until several weeks after arrival. Virtually all **one-year-old** birds apparently return to the breeding grounds with adults.

In **fall, adults** depart the breeding grounds with juveniles by early August. Migrants occur throughout the interior West in August and September and arrive on the wintering grounds between mid-September and early November. **Juveniles** depart the breeding grounds with adults by late July or early August. Migrants occur throughout the interior West in August and September and arrive on the wintering grounds between mid-September and mid-October.

**MOLT**

**First life year:** The molt out of juvenal plumage is variable. In March and April juveniles undergo a head-and-body molt to breeding plumage. This plumage resembles adult breeding except for some retained juvenal scapulars, wing coverts, and primaries.

**Second life year:** The complete molt to nonbreeding plumage takes place between mid-June and mid-August, primarily on or near the breeding grounds.

**Adult:** In March and April there is a partial molt to breeding plumage, including head and some body feathers and presumably some wing coverts. The complete molt to nonbreeding plumage takes place between mid-June and mid-August, primarily on or near the breeding grounds.

**VOCALIZATIONS**

Generally quiet. Flight call is a rough, grating *kirrp.* Also gives a soft, whistled *hoit.* Flight song is a buzzy, whistled *ji-ji-ji-ji-ji-ji-li-li-li-li-li-li* . . . ; also given during aggressive encounters. In breeding season male gives a low, cow-like *mooo.*

## Eurasian Dotterel, p. 243

### *Charadrius morinellus*

**STATUS**

Palearctic species. Rare visitor and sporadic breeder in w. Alaska. Casual on Aleutians in August and along West Coast, mainly September. Widespread in the Palearctic but generally uncommon in summer, with patchy distribution. Breeds from Scotland and Scandinavia to e. Siberia and at widely scattered high-mountain sites from Italy to Mongolia. Wintering birds concentrated in small area in Spain, n. Africa, and the Middle East.

**TAXONOMY**

No subspecies recognized.

**BEHAVIOR**

Remarkably tame, approachable at all seasons. Breeds on barren tundra and dry, sparsely vegetated plateaus and slopes up to 9,000 ft. (2,743 m) where the landscape is dominated by moss, lichen, and scattered rocks. During winter prefers semidesert regions with sand or gravel and high barren plateaus. Avoids rich, cultivated lands. Like other plovers forages visually, employing run-stop-scan-pluck technique. Eats primarily insects but also takes worms, snails, leaves, berries, and seeds. Usually seen singly or in small flocks in summer, larger flocks in winter.

Nest initiation primarily early to mid-June. Egg dates late May to mid-July. Incubation 23–29 days. Female polyandrous, laying two or more clutches to be hatched and raised by different males. Female may assist in incubation of last clutch but does little brood rearing. Each pair single-brooded but will renest after failure. Peak hatching late June to mid-July. Young can fly after 19–23 days. Display flight, performed by female over territory, involves shallow, shivering wingstrokes at a height of 100–330 ft. (30–100 m) on an often undulating flight path. Calls are given throughout.

**MIGRATION**

Medium to long-distance migrant. Some individuals move at least 6,200 mi. (10,000 km) between breeding grounds in e. Siberia to wintering areas in n. Africa. Remarkably consistent in its time and location of return to spring stopover and breeding areas. Many fall migrants may reach wintering grounds in one flight.

**Spring** departure from wintering areas takes place in March and April. Migrants return to Ireland and Scotland by late April and early May and to the Siberian tundra by early to mid-June.

**Fall** migration takes place between July and December. **Adult** females begin departing the breeding grounds from mid-July (Siberia) to mid-August (n. Europe), with peak numbers passing through n. Europe in late August and early September and reaching n. Africa by September and October. Southbound adult males and **juveniles** follow about 2 weeks after females, departing Siberia by early August and n. Europe by late August. Arrival on the wintering grounds takes place from October to November/December.

**MOLT**

**First life year:** The partial molt out of juvenal plumage begins primarily in late September (some start August to early November) and may be completed as early as mid-October or as late as April/May. The molt includes most body feathers and a few wing coverts and tail feathers. The limited partial molt to breeding plumage takes place between early March and mid-May and includes a variable number of feathers on the underparts. Timing of primary molt not well known, but some apparently molt outer 2 primaries in late winter or spring. Others may molt inner primaries in late spring to begin the complete molt to nonbreeding plumage.

**Second life year:** Molt patterns presumably as in adult, though the complete molt to nonbreeding apparently may begin as early as late spring (of first life year).

**Adult:** The partial molt to breeding plumage takes place between early March and mid-May. This molt involves at least most of the underparts and often extensive head, upperparts, wing coverts, tertials, and tail feathers. The complete molt to nonbreeding plumage begins in early July/early August on the breeding grounds with the replacement of the inner primaries, scattered body feathers, and some coverts, tertials, and tail feathers. The molt is suspended during migration, then completed at or near the wintering grounds. The body molt is completed by late September/late October and the primary molt by late October/December.

**VOCALIZATIONS**

Generally rather quiet. Flight call is low, rolling *pjjrt*. In display flight female gives a soft, ringing *peet-peet-peet* . . . , rapidly repeated. Also trilled variations of this call from agitated birds.

## OYSTERCATCHERS (FAMILY HAEMATOPODIDAE)

### Eurasian Oystercatcher, p. 245

***Haematopus ostralegus***

**STATUS**

Palearctic species. Accidental in Newfoundland (May 1994) and casual spring vagrant to Greenland. Breeds from Norway, Iceland, and France east to w. Russia and nw. China, then disjunctly from Kamchatka to e. China. Winters from Iceland and n. tropical Africa east to se. China.

**TAXONOMY**

Three subspecies are recognized. Nominate *ostralegus* breeds in Europe and coastal w. Russia; *longipes* breeds in central Asia; and *osculans* breeds from Kamchatka to e. China. Bill length increases from west to east. *H. o. longipes* is paler and browner above and has a long nasal groove extending more than halfway down the bill (less than half in *ostralegus* and *osculans*). Nominate has less white in primary webs and shafts (lacks white in outer primaries) than other races. The Newfoundland record is presumed to represent the nominate subspecies.

**BEHAVIOR**

Primarily coastal though also occurs locally far inland along rivers and lakes during breeding season. Breeds primarily on sand, shell, or gravel beaches, dunes, and grassy bluffs along the coast, grassy meadows and gravel river bars inland. Forages in a wide variety of primarily coastal habitats including salt marshes, shellfish beds, tidal flats, rocky shorelines, and coastal pastures and meadows. Along coast eats primarily shellfish but also marine worms and crustaceans. Inland eats primarily earthworms, sometimes insects. Walks steadily while visually searching for prey. Feeds by probing in the ground or stabbing and hammering shellfish. Feeds by day and night. Gregarious outside breeding season, often forming flocks of up to several hundred.

Nests late March to late July. Young may be dependent on parents for up to 26 weeks. Age of first breeding normally 4 years.

#### MIGRATION

Partial to intermediate-distance migrant.

**Spring** migration of **adults** takes place between late January and early May. Most return to breeding sites in March and April, though northernmost breeders arrive on territory as late as early May. Immatures (up to 3 years old) typically remain on the wintering grounds, though some may return to breeding area without nesting.

In **fall,** southbound movement of **adults** and **juveniles** begins in July and continues at least through early November. Staging areas are occupied primarily in August and September. Family groups are usually maintained throughout the migration period and into early winter.

#### MOLT

**First life year:** The partial molt out of juvenal plumage takes place from August/September to December. This molt includes most body feathers and some tertials, coverts, and tail feathers. The partial molt to breeding plumage takes place between about January and May and includes some body feathers, tertials, wing coverts, and tail feathers. The resulting plumage resembles breeding adult but is distinguished by a partial white collar (as in nonbreeding) as well as retained worn juvenal flight feathers and usually some coverts and outer tail feathers. One-year-old birds begin their complete molt to nonbreeding plumage about a month before adults, starting as early as late May.

**Second life year:** The complete molt to nonbreeding plumage takes place about a month before adults, starting as early as late May (of first life year) and is completed by about August/November. The partial molt to breeding plumage is as adult's, and the resulting plumage is usually indistinguishable from adult.

**Adult:** The partial molt to breeding plumage takes place between about December and April. This molt is limited, involving mostly head and neck feathers. The molt to nonbreeding plumage takes place between about June/July and September/December (timing varies geographically). This molt is usually complete, though outer primaries may be retained for up to 2 years.

#### VOCALIZATIONS

Noisy. Flight call is a shrill, whistled *ku-weet* or a more relaxed, slightly descending *weeup,* similar to American Oystercatcher.

## American Oystercatcher, p. 64

### *Haematopus palliatus*

#### STATUS

Locally common and conspicuous in coastal areas from Massachusetts to the Gulf Coast. Also occurs from the Caribbean south to Argentina and north along the Pacific Coast from Chile to Baja California. Casual stray north to

s. California and Nova Scotia. Gradually expanding range northward along Atlantic Coast and has bred recently in Maine and Nova Scotia. Very rare inland. Largely resident. The world population is estimated at 58,850, with 8,850 in North America.

**TAXONOMY**

Five subspecies are recognized, two in North America: *H. p. palliatus,* of the Atlantic and Gulf coasts, and *H. p. frazari,* which occurs as a stray in s. California. Compared to *palliatus, frazari* is slightly larger, has little or no white in the primaries, less white in the rump, slightly darker upperparts, and often a mottled lower border to breast-band. Hybrids between *frazari* and Black Oystercatcher have been reported.

**BEHAVIOR**

Exclusively coastal. Breeds on sand or shell beaches, dunes, dredge spoil, and salt marshes. Forages on slightly submerged shellfish beds as well as intertidal mud or sand flats. Eats primarily shellfish and other marine invertebrates, particularly mussels, clams, and oysters. Walks steadily while visually searching for prey. Feeds by probing in the substrate or stabbing open bivalves with bill, severing the adductor muscle; also by "hammering" prey item until it breaks open. Gregarious in winter, often forming flocks of up to several hundred.

Nest initiation primarily early April in south to mid-May in north. Egg dates late March to late June in south, early April to early July in north. Incubation 24–27 days. Both adults share parental duties. One brood, but will renest after failure. Peak hatching early May in south to mid-June in north. Young fly after 34–37 days. Juveniles can first fly by early June (Virginia) to early July (Massachusetts) but are still dependent on adults for food for at least a month after that. During courtship, pairs walk together and adopt various postures while giving loud piping calls. These posturing displays frequently continue into tandem flights with continued piping calls, often joined by members of adjoining pairs.

**MIGRATION**

Largely resident, but some individuals migratory, particularly those breeding from Maryland northward. Migration takes place along the coast and over littoral waters, primarily during the day.

**Spring** migration of **adults** and **first-year birds** takes place primarily in March and April. Females arrive on territory before males.

In **fall,** southbound movement of **adults** and **juveniles** begins in August and continues at least through early December. Family groups are maintained throughout the migration period and into early winter.

**MOLT**

**First life year:** The head-and-body molt out of juvenal plumage begins in early September and is usually completed by December. Between about January and March first-year birds undergo a head-and-body molt to breeding plumage, at which point they resemble adults but can be distinguished by worn juvenal flight feathers. One-year-old birds undergo their complete molt to nonbreeding plumage between about April and July. Older subadults probably also molt earlier than adults.

**Adult:** Between about January and March there is a head-and-body molt to breeding plumage. The complete molt to nonbreeding plumage takes place between about June and September.

**VOCALIZATIONS**

Whistled flight call is an easily imitated, loud, clear, downward-arched *wheeu,* also given by standing birds. Alarm call is a similar but sharper *wheep,* given singly or in a variable series. In courtship and posturing displays, gives a piping call comprised of several flight-call-like whistles accelerating into a rapid, descending series of piping notes. Brood contact call is a sharp *peet* similar to Long-billed Dowitcher's call.

## Black Oystercatcher, p. 68

***Haematopus bachmani***

**STATUS**

Uncommon but conspicuous along rocky coastlines from the Aleutian Is. of Alaska to Baja California. More numerous in northern part of range. The world population is estimated at 8,900–11,000.

**TAXONOMY**

No subspecies recognized. Slight clinal variation, with birds in the north being all black and birds toward the south showing increasing amounts of brown on the abdomen. Hybrids with American Oystercatcher have been reported.

**BEHAVIOR**

Strictly tied to rocky intertidal communities. Breeds just above high-tide line on rocky headlands or sand, shell, or gravel beaches. Vegetated habitats are avoided, and unvegetated rock islets are preferred. Forages almost exclusively on intertidal areas, including rocky or gravel shores, slightly submerged shellfish beds, and mud or sand flats. Occasionally uses seaside golf courses. Eats primarily shellfish and other marine invertebrates, particularly mussels, limpets, and chitons. Walks steadily while visually searching for prey. Feeds by probing in the substrate or stabbing open bivalves with bill, severing the adductor muscle. Also by chiseling or prying limpets or chitons off rocks and rarely by "hammering" prey item until it breaks open. Gregarious in winter, though seldom forms flocks of more than 100.

Most egg laying takes place in May and early June. One brood but will re-nest after failure. Both adults share parental duties. Incubation takes 26–28 days; fledging after 38–40 days. Most juveniles can first fly by late July or early August but are still dependent on adults for food for at least a month after that. Family groups are maintained throughout the migration period and into early winter. During territorial display, given year round, pairs walk together and adopt various postures while giving loud piping calls. These displays frequently continue into tandem flights with continued piping calls, often joined by members of adjoining pairs. Pairs remain together year round.

**MIGRATION**

Largely resident. Migration in most individuals is limited to postbreeding flocking. However, a large percentage of Alaskan population withdraws south in the winter, many traveling at least as far as British Columbia. Postbreeding flocks build up in Alaska through July and August and depart in September. Numbers in British Columbia reach their peak in November. Spring migration of **adults** and **first-year** birds takes place primarily in March, and birds return to breeding territories in March and April. Migratory movements take place along the coast and over littoral waters, mostly during the day.

**MOLT**

**First life year:** Some juveniles may undergo a limited partial molt in late fall, while others apparently skip this molt. Between January and March there is a partial molt during which primarily the body feathers are replaced.

**Second life year:** Between May (of first life year) and August, birds undergo a complete molt.

**Adult:** Between April and May there is a partial molt. This molt is more extensive in northern birds, which replace virtually all but the primaries and secondaries. Southern birds replace only scattered body feathers. The complete molt to nonbreeding plumage takes place between July and September.

**VOCALIZATIONS**

Whistled flight call is an easily imitated, loud, clear, downward-arched *wheeu,* also given by standing birds. Alarm call is a similar but sharper *wheep,* given singly or in a variable series. In courtship display, gives a piping call comprised of several flight-call-like whistles accelerating into a rapid, descending series of piping notes.

## STILTS AND AVOCETS (FAMILY RECURVIROSTRIDAE)

### Black-winged Stilt, p. 247

***Himantopus himantopus***

**STATUS**

Eurasian species. Accidental in w. Alaska (May/June). Locally common within range. Breeds locally from France south to South Africa, east to Mongolia, Vietnam, and New Zealand. Winters from Africa east to Taiwan and New Zealand. Regularly overshoots far beyond normal breeding range in spring, sometimes establishing temporary colonies.

**TAXONOMY**

Two subspecies currently recognized. Formerly lumped with Black-necked Stilt. Much debate continues over species limits in stilts. Nominate *himantopus* occurs in s. Eurasia, Africa, and India; *leucocephalus* ("Pied" or "White-headed Stilt") occurs from se. Asia to New Zealand. The nominate race is variable. Some individuals show an entirely white head and neck, while others show varying amounts of black on the crown, face, and nape. The hindneck is either white or at least paler than the crown. *H. h. leucocephalus* has a white head and extensively black nape. Alaskan records refer to the nominate race.

**BEHAVIOR**

Like Black-necked Stilt.

Nests February to July.

**MIGRATION**

Partial to intermediate-distance migrant, primarily through the interior.

**Spring** migration takes place from early March to late May with peak movement from mid-March to late April. Some **first-year** birds remain at wintering areas through the summer.

**Fall** migration takes place from late July to November. Adults begin to depart breeding areas in late July and gather at staging areas from mid-August to October. Southbound **juveniles** begin to depart breeding areas by early August, with peak passage in late August and September.

**MOLT**

**First life year:** The molt out of juvenal plumage begins in July and continues into February. It includes some head and body feathers, but flight feathers, most wing coverts, and sometimes outer tail feathers are retained. Between December and March (sometimes May), there is a limited molt into breeding plumage involving some head and body feathers, some coverts and tail feathers. This plumage resembles adult breeding but may be distinguished by worn juvenal flight feathers, coverts, and tertials as well as more extensive nonbreeding head and body feathers.

**Second life year:** The complete molt to nonbreeding plumage is like adults but may begin in late June. After this molt, second-year birds are usually indistinguishable from adults. Between December and March there is a molt into breeding plumage as in adults.

**Adult:** Between December and March (sometimes into May), there is a molt into breeding plumage involving many head and body feathers, as well as some tail feathers and coverts. The complete molt to nonbreeding plumage takes place from July to December. It is begun on the breeding grounds, suspended during migration, then completed on the wintering grounds.

**VOCALIZATIONS**

Flight and alarm calls are like Black-necked Stilt's.

## Black-necked Stilt, p. 71

### *Himantopus mexicanus*

**STATUS**

Locally common through much of the West and South. Breeds locally from s. Alberta, s. Saskatchewan, and Delaware south through Central America, the West Indies, and South America to s. Chile, s. Argentina, and the Galápagos Is. Regularly overshoots far beyond normal breeding range in spring. The North American population is estimated at 150,000.

**TAXONOMY**

Three subspecies currently recognized. Formerly lumped with Old World Black-winged Stilt *(Himantopus himantopus).* Much debate continues over species limits in stilts. Nominate *mexicanus* breeds in North and Central America, the West Indies, and n. South America from Venezuela to s. Peru. Most northern birds withdraw south in winter. *H. m. melanurus* is resident in South America from e. Peru and se. Brazil to s. Chile and s. Argentina. *H. m. knudseni* is mostly resident on Hawaiian archipelago. Compared to *mexicanus, melanurus* averages larger and has a mostly white crown and a white or grayish collar around the base of the neck. Intergrades between *mexicanus* and *melanurus* occur in Peru and Brazil. *H. m. knudseni* differs from *mexicanus* in larger size and more extensively black face and neck.

**BEHAVIOR**

A noisy bird of quiet waters. Breeds around fresh or salt marshes, shallow grassy ponds or lake margins, and man-made impoundments, often in small colonies. During migration and winter uses similar habitats as well as shallow lagoons, flooded fields, and mangrove swamps. Feeds by walking slowly in moderately deep water, picking food off the surface. Only occasionally submerges head under water or swishes bill from side to side. Feeds by day and night. Eats a variety of aquatic invertebrates, particularly insects and crustaceans, as well as small fish and frogs. During the breeding season often seen in pairs or small groups. In winter gathers in larger flocks, sometimes over 100.

Nest initiation primarily late April to mid-May. Egg dates early April to mid-August. Incubation 18–27 days. Both adults share parental duties. One brood per season. Peak hatching primarily late May to mid-June. Young can fly after about 28–32 days. Age of first breeding 1–2 years. Does not perform aerial courtship display, though aggressively defends territory against predators, often by calling vigorously in flight.

**MIGRATION**

Partial to intermediate-distance migrant, primarily through the interior. Migrates mostly at night and early morning.

**Spring** migration takes place from mid-February to mid-June. Departs wintering areas in Mexico and the West Indies mostly in March and April. Most common across the southern tier of states from mid-April to mid-May. Arrives at breeding areas in the Northwest mostly in mid-April and along the mid-Atlantic Coast mostly in early May. Spring vagrants in the Northeast and Midwest typically appear in May and June.

**Fall** migration takes place from July

to December. **Adults** depart most breeding areas between early July to early September, often gathering in flocks at nearby staging areas in August. Migrants continue into October from s. California to Florida and arrive at wintering sites mostly from August to October (sometimes as late as December). Southbound **juveniles** depart the breeding grounds from mid-July to early September and migrate mostly with adults.

**MOLT**

**First life year:** The molt out of juvenal plumage begins in July and continues into February. It includes many head and body feathers, but flight feathers, most wing coverts, and sometimes outer tail feathers are retained. Between February and April there is a limited molt into breeding plumage, probably involving some head and body feathers. This plumage resembles adult breeding but may be distinguished by worn juvenal flight feathers, coverts, and tertials.

**Second life year:** The complete molt to nonbreeding plumage takes place from about late June to November (mostly July/August). It is begun on the breeding grounds, suspended during migration, then completed on the wintering grounds. After this molt, second-year birds are usually indistinguishable from adults. Between February and May there is a molt into breeding plumage involving many head and body feathers and some tail feathers.

**Adult:** Between February and May there is a molt into breeding plumage involving many head and body feathers and some tail feathers. The complete molt to nonbreeding plumage takes place from July to December (mostly August and September). It is begun on the breeding grounds, suspended during migration, then completed on the wintering grounds.

**VOCALIZATIONS**

Flight and alarm call is a loud, yapping *yip, yip, yip, yip . . .*, often given in a long series, particularly when alarmed. Some variations have a raspy or squeaky quality.

## American Avocet, p. 74

***Recurvirostra americana***

**STATUS**

Locally common through much of the West and South. Scarce in the Northeast. Breeds locally from s. British Columbia and sw. Ontario south to central Mexico. Winters from n. California and North Carolina south to sw. Guatemala, Belize, and Cuba. The world and North American population is estimated at 450,000.

**TAXONOMY**

No subspecies are recognized.

**BEHAVIOR**

A bird of temporary wetlands in arid regions. Breeds around salt ponds, shallow alkaline lakes, and freshwater marshes and impoundments, often in small colonies. During migration and winter uses a variety of shallow-water habitats including ponds, impoundments, fresh and salt marshes, tidal mudflats, and lagoons. Feeds by walking steadily in often belly-deep water, swishing bill from side to side to filter out food items. Also may jab at the water or mud surface or plunge head under water to chase prey. Often forages in small groups. Feeds by day and night. Eats a variety of aquatic invertebrates, particularly insects, as well as small fish and seeds. During the breeding season often seen in pairs or small groups. During migration and winter gathers in large flocks, often numbering in the hundreds or thousands.

Nest initiation primarily mid-May. Egg dates mid-April to early July. Incubation 23–30 days. Both adults share parental duties. One brood per season but will renest after failure. Peak hatching primarily mid-June. Young can fly after about 27 days. Age of first breeding 1–2 years. Does not perform aerial courtship display.

**MIGRATION**

Intermediate-distance migrant, primarily through the interior. Perhaps resident in some areas.

**Spring** migration takes place from mid-February to late June. Departs southernmost wintering areas mostly in March and early April. Peak numbers pass through southern states in April and through northern and interior areas between mid-April and early to mid-May. Some **one-year-old** birds remain on the wintering grounds through the summer, while others return to the breeding grounds.

**Fall** migration takes place from late June to early January. Most **adults** depart breeding areas in July, with peak numbers gathering at staging areas from late July to mid-September. Arrival at wintering areas may begin as early as July and continue into November. Southbound **juveniles** depart the breeding grounds mostly in August and September. Fall migration is generally more coastal than in adults. Peak numbers occur in northern areas in September and southern areas in October/November.

**MOLT**

**First life year:** The molt out of juvenal plumage takes place primarily between July and November. It includes many head and body feathers, but flight feathers, some scapulars, and some wing coverts are retained. Tail remains downy in juvenal plumage and is replaced by feathers during post-juvenal molt. Between about January and April there is a molt into breeding plumage involving most head and body feathers. This plumage resembles adult breeding but may be distinguished by worn juvenal flight feathers, coverts, and often some scapulars.

**Adult:** Between January and April, usually before spring migration, there is a partial molt into breeding plumage involving many head and body feathers. The complete molt to nonbreeding plumage takes place from June to September, mostly at stop-over sites during migration.

**VOCALIZATIONS**

Primary call is a variable loud, whistled *bleet,* higher, more emphatic, and repeated in a long series when alarmed; given as much while foraging as in flight.

## JAÇANAS (FAMILY JAÇANIDAE)

### Northern Jaçana, p. 248

***Jaçana spinosa***

**STATUS**

Mexican and Central American species. Rare and irregular visitor to s. Texas, where it has bred. Recorded from all months of the year, though most sightings fall between October and early May. Juveniles account for about 75 percent of records. Casual visitor to s. Arizona. Resident in lowland areas from central Tamaulipas and s. Sinaloa south to w. Panama and the Greater Antilles.

**TAXONOMY**

No subspecies are currently recognized. Up to four subspecies have been proposed in the past, but differences are slight and overlapping.

**BEHAVIOR**

Breeds in freshwater marshes with abundant floating vegetation. Also feeds in flooded fields, wet meadows, and occasionally dry pastures. Long toes allow birds to walk on sparse floating vegetation, from which they pick invertebrates off root balls and glean insects from emergent vegetation. Also eats seeds, aquatic vegetation, and occasionally small fish. Walks with slow deliberate gait, frequently lifting wings up to regain balance. Makes short low flights across pond, holding wings up briefly after landing. Not typically seen with other shorebirds but often associates with Purple Gallinule and Common Moorhen.

Breeds seasonally or year round, depending on permanence of marsh. In ne. Mexico, egg dates range from late April to mid-August. Incubation 22–24 days. Female polyandrous and mates with up to four males per season. Each male may raise two or more broods. Male incubates eggs and cares for young. Young can fly after about 57 days. No flight display.

#### MIGRATION

Year-round resident wherever appropriate habitat exists. However, when drought or dry season reduces water levels, some birds move locally to more-permanent ponds. Also some dispersal, particularly by **juveniles,** apparently accounts for records north of the Mexican border.

#### MOLT

**First life year:** The molt out of juvenal plumage is complete but very gradual. It begins at about 8 weeks of age and is completed after 10–14 months, after which immature birds are indistinguishable from adults.

**Adult:** Molt patterns of adults are poorly known but appear to be gradual as in first-year birds.

#### VOCALIZATIONS

Primary call, usually given in flight, is a screeching, rail-like chatter, *pit-pit-pit-swee swee swee swee swee . . .*, often interspersed with sharp clicks.

## SANDPIPERS, PHALAROPES, AND ALLIES (FAMILY SCOLOPACIDAE)

### Common Greenshank, p. 249

***Tringa nebularia***

#### STATUS

Old World species. Rare migrant on islands off w. Alaska (May to Sept.). Casual in California (Aug., Oct.) and e. Canada (May and Aug. to Dec.). Common and widespread in the Old World, where it is the ecological equivalent to Greater Yellowlegs. Breeds from Scotland and Norway east to Kamchatka and ne. China. Winters from Ireland south through Africa and the Middle East, east to se. China, Australia, and New Zealand.

#### TAXONOMY

No subspecies recognized.

#### BEHAVIOR

Much like Greater Yellowlegs. Breeds in marshes, wet meadows, clearings within boreal forest, and, at northern edge of range, in boggy tundra. During migration and winter prefers shallow wetlands including flooded fields, fresh and salt marshes, impoundments, and tidal creeks. Active while foraging. Walks quickly with jerky motions and swiftly stabs at the water when prey is found. Like Greater Yellowlegs, often runs frantically with neck extended and bill submerged to chase small fish. Eats a variety of invertebrates, particularly insects, as well as small fish, frogs, and occasionally plant material. Usually solitary or in small flocks, rarely in flocks over 100.

Nests from April to July.

#### MIGRATION

Migrates across both interior and coastal regions, though always more numerous around estuaries along the coast.

**Spring** migration takes place between March and May, with peak movement in April. Southern breeding areas are reached by mid- to late April, northern breeding areas a month later. Some nonbreeders remain at wintering areas through the summer.

**Fall** migration more protracted, from late June to late October. Most **adult females** and failed breeders depart the breeding grounds in late June and early July, followed by **adult** males in late July and August. Peak numbers pass through temperate regions from mid-July to mid-August. Arrives at wintering grounds mostly from late July to late September. Southbound **juveniles** depart the breeding grounds in late July and August, usually accompanied by adult males. Peak numbers pass through temperate regions from mid-August to late September. Arrives at wintering areas mostly in September and October.

**MOLT**

**First life year:** The molt out of juvenal plumage begins from late August to late October and continues into March. It includes most head and body feathers, along with some wing coverts and tail feathers. Some birds, particularly longer-distance migrants, may replace outer primaries between January and May. In April and May there is a highly variable partial molt to breeding plumage. The molt includes some head and body feathers and some wing coverts (often more wing coverts than adult). This plumage may resemble adult breeding but is usually intermediate between breeding and nonbreeding and is further distinguished by the presence of some worn juvenal primaries, coverts, and tail feathers. Birds that remain on the wintering grounds through the summer typically skip this molt.

**Second life year:** For birds that oversummer on the wintering grounds, the complete molt to nonbreeding plumage takes place from June to August. Other birds may be on a more adultlike schedule. After this molt, second-year birds are usually indistinguishable from adults. The partial molt to breeding plumage is as in adults.

**Adult:** The partial molt to breeding plumage takes place primarily from late January to early April. It includes many head and body feathers, plus some coverts and tail feathers. Some feathers acquired early in molt may show a pattern intermediate between breeding and nonbreeding. Timing of the complete molt to nonbreeding plumage is highly variable. Molting may take place on the breeding grounds, wintering grounds, or at stop-over sites, but is suspended during migration. Body molt may begin from late June to mid-October and be completed from September to December. Flight feather molt may begin from late June to mid-November and be completed from mid-September to early January. Most are in heavy molt while at stop-over sites in August and September. Some birds do their entire molt on the wintering grounds.

**VOCALIZATIONS**

Flight call is a loud, ringing *kEeu-teu-teu,* very similar to Greater Yellowlegs but more liquid, less strident.

## Greater Yellowlegs, p. 78

***Tringa melanoleuca***

**STATUS**

Common migrant throughout most of North America at both interior and coastal sites. Breeds in the boreal-forest belt from s. Alaska east to Newfoundland. Winters along the coast from s. British Columbia and Connecticut south and virtually throughout Central and South America and the West Indies. The world and North American population is estimated at 100,000.

**TAXONOMY**

No subspecies recognized.

**BEHAVIOR**

Breeds in open boreal forest with bogs, muskeg, and wet meadows. At northern parts of range, some nest in subarctic tundra. During migration and winter uses a variety of shallow wetlands, including flooded fields, fresh and salt marshes, impoundments, and tidal creeks. Active while foraging. Walks quickly with jerky motions and swiftly stabs at the water when prey is found. In deeper water, often runs frantically to chase small fish (a behavior rarely seen in Lesser Yellowlegs). Also sweeps bill from side to side at water surface, particularly in low light or murky water. Feeds by day and night. Eats a variety of invertebrates, particularly adult and larval insects, as well as small fish, frogs, and occasionally seeds and berries. Often solitary but regularly forages in small groups of 5–10 and sometimes gathers in flocks of up to several hundred, particularly in winter.

Nest initiation primarily mid-May in the West, early June in the East. Egg dates early May to early July. Incubation 23 days. Both adults share parental duties. One brood per season but will renest after failure. Peak hatching early June in West, mid- to late June in East. Young can fly after about 18–20 days. The male's display flight, performed on territory and occasionally during migration, involves a repeated sequence of flapping, gliding, and stooping, resulting in an undulating flight that may last for up to 15 minutes as the bird sings.

#### MIGRATION

Migrates across both interior and coastal regions, though always more numerous along the coast, particularly in fall. Some birds migrate across Atlantic from Northeast to West Indies in fall. Migrates mostly at night and early morning.

**Spring** migration earlier than in most other shorebirds, between early February and early June. Departs southern wintering areas mostly in mid-February and March. Most common across the southern tier of states in April and to the north from late April to mid-May. Somewhat earlier migrant along the West Coast than through the interior and East Coast. Arrival on the breeding grounds takes place mostly late April in the West, early to mid-May in the East. Some **one-year-old** birds remain on the wintering grounds through the summer, while others apparently migrate north to the breeding grounds.

**Fall** migration protracted, from late June to late November. **Adults** begin to depart the breeding grounds in late June (mostly females and failed breeders) and arrive in wetlands throughout North, Central, and n. South America by early July. Peak numbers of adults occur in s. Alaska in mid- to late July and through much of the North American continent in mid- to late August. Southbound **juveniles** depart the breeding grounds in mid- to late July and first arrive in s. Canada in late July, along the central Pacific and mid-Atlantic coasts in early to mid-August, and in n. South America in early October. Peak numbers of juveniles occur through much of North America in late September and October.

#### MOLT

Molts flight feathers during fall migration, unlike Lesser Yellowlegs.

**First life year:** The molt out of juvenal plumage begins in late August or September and continues into February. It includes most head and body feathers, and some wing coverts and central rectrices. Between March and May there is a usually limited molt into breeding plumage involving some head and body feathers and some wing coverts. This plumage resembles adult breeding but usually has less extensive black patterning and is further distinguished by worn juvenal primaries, worn and narrow juvenal outer tail feathers, and often three generations of wing coverts. Birds that remain on the wintering grounds through the summer typically delay or skip this molt and begin their complete molt to nonbreeding plumage as early as late May.

**Second life year:** For birds that over-summer on the wintering grounds, the complete molt to nonbreeding plumage begins between May (of first life year) and July and may be completed as early as September. Other birds may be on a more adultlike schedule. After this molt, second-year birds are usually indistinguishable from adults. Between late February and early May there is a molt into breeding plumage involving many head and body feathers, some wing coverts, and central tail feathers. Some birds wintering in South America may complete this molt before migration.

**Adult:** Between late February and early May there is a molt into breeding plumage involving many head and body feathers, some wing coverts, and central tail feathers. Some birds wintering in South America may complete this molt before migration. The complete molt to nonbreeding plumage begins after breeding in August and is completed in February. The primary molt begins on the breeding grounds and, in many birds, continues through fall migration, mostly August to October. Some birds suspend their primary molt and complete it on the wintering grounds from November to February.

#### VOCALIZATIONS

Flight call is a loud, strident *dee-dee-deer,* usually three or four notes with the last note lower; higher pitched and more strident than Lesser's flight call. When agitated, may give a measured series of sharp *tew* notes. Display song is a rolling, endlessly repeated *kla-wid-kla-wid-kla-wid-kla-wid-kla-wid . . . ,* often lasting up to several minutes.

## Lesser Yellowlegs, p. 83

### *Tringa flavipes*

#### STATUS

Common throughout most of North America. Breeds in the boreal-forest belt from Alaska east to the James Bay region in Quebec. Winters primarily in coastal areas from central California and s. New Jersey south through much of South America. Regular migrant through interior. In spring relatively uncommon along the Pacific Coast and the Northeast. The world and North American population is estimated at 500,000.

#### TAXONOMY

No subspecies recognized.

#### BEHAVIOR

Breeds primarily in open boreal forest and forest/tundra transition habitats. Nest site is typically in open or semi-open forest near a shallow wetland such as a marsh, bog, pond, or lake. Typically nests in drier, more vegetated habitats than Greater Yellowlegs. Breeders forage along lakeshores, pools, and marshes. Migrants commonest in coastal areas but numerous at favorable interior sites through much of North America. In migration and winter uses a wide variety of shallow wetlands, including flooded pastures, rice fields, mangrove swamps, lakeshores, riverbanks, sewage ponds, marshes, tidal creeks, and tidal flats. Very active foraging style, often walking rapidly as it picks at shallow water or substrate. Occasionally swishes bill from side to side in shallow water, much as an avocet does. In typical foraging posture, leans forward with bill held at a 20–30° angle from substrate. Will occasionally swim to feed on surface organisms. Seldom forms tight groups to corral fish as does Greater Yellowlegs. Feeds by day and night. Eats a wide variety of aquatic and terrestrial invertebrates, especially flies and beetles. Occasionally eats seeds or small fish. Typically seen in small flocks, though many hundreds may gather at favorable feeding areas.

Nest initiation primarily mid- to late May in milder areas of Alaska to early to mid-June in central Canada. Egg dates range from early May to late July. Incubation 22–23 days. Both adults share parental duties. One brood per season but will likely renest after egg loss. Peak hatching takes place from mid-June in milder areas to early July in central Canada. Young can fly after 18–20 days. The male's flight display is undulating with bouts of flapping and gliding, tail spread and neck stretched out. The display usually begins and ends at a treetop perch and is usually accompanied by singing throughout.

**MIGRATION**

Migrates during both day and night.

**Spring** migration takes place between late February and late May (rarely early June). Northbound **adults** begin to depart the wintering areas in late February and early March. Spring migrants are most common in the Southeast and interior of the continent, scarcer in the Northeast and along the Pacific Coast. Peak numbers reach the Gulf Coast by mid-April and the northern tier of states by early May. Arrival on the breeding grounds takes place between late April and early June. Some **one-year-old** birds remain on the wintering grounds through their first summer, while others migrate partway or all the way north to the breeding grounds. Spring migration of first-year birds averages later than that of adults.

**Fall** migration takes place between mid-June and November. Most **adult females** depart the breeding grounds by early July and most **adult males** by mid-July. Failed breeders may begin to head south by mid-June. The first fall migrant adults appear in the northern tier of states by late June and peak there between mid-July and early August. The peak along the Gulf Coast is about a week later. Fall migration of adults is somewhat more widespread than spring migration. On the West Coast, adults are somewhat more common in fall than in spring but are still far outnumbered by juveniles. In the Northeast, adults are much more abundant in fall than in spring. Southbound **juveniles** depart the breeding grounds between late July and mid-August and head to both coastal and interior sites. The first juveniles arrive across the northern tier of states in late July, with peak passage in late August and early September. This passage is about 2 weeks later along the Gulf Coast.

**MOLT**

**First life year:** Molt out of juvenal plumage takes place primarily on the wintering grounds and includes the head-and-body feathers as well as outer primaries and inner rectrices. The head-and-body molt takes place between early September and January. The primary and rectrix molt takes place between early November and mid-April. In March and April, first-year birds undergo a partial head-and-body molt. The extent of this molt is variable, but new feathers resemble those of breeding adults. One-year-old birds are most reliably aged by the contrast between fresh outer primaries and tertials and worn inner primaries and secondaries.

**Second life year:** For birds that oversummer on the wintering grounds, the complete molt to nonbreeding plumage may take place a month or more earlier than that of adults. Other birds are on a more adultlike schedule. After this molt, second-year birds are usually indistinguishable from adults.

**Adult:** Between late January and April, adults undergo a head-and-body molt to breeding plumage. The extent of this molt is variable, with some birds replacing virtually all head and body feathers and many wing coverts, while others replace only scattered feathers. Hence there is a great deal of variation in breeding plumage. The complete molt to nonbreeding plumage usually begins in July on the breeding grounds with the replacement of a few head and body feathers. This molt is suspended during fall migration, then completed primarily on the wintering grounds (July to Jan.). Late in the season some migrants begin to show more extensive head-and-body molt, but the flight feather molt takes place entirely on the wintering grounds. Hence, fall migrants invariably show a full set of flight feathers.

**VOCALIZATIONS**

Flight call is a low, whistled *tu,* often doubled. Flight call is given frequently by standing and flying birds. When agitated, may give a measured series of more emphatic *tu* or *kewp* notes. Display song is a rolling, whistled *ta-widia-widia-widia-widia* . . . , often lasting up to several minutes. Fragments of this song are regularly given by migrants, including juveniles. A variety of soft murmuring calls are given by feeding or roosting flocks, and various harsh alarm calls are also given.

## Marsh Sandpiper, p. 251

***Tringa stagnatilis***

**STATUS**

Eurasian species. Accidental in Aleutian Is., Alaska (Aug., Sept.). Common within range. Breeds from e. Ukraine (rarely Finland) east across s. Russia to n. Mongolia. Winters from Africa east to the Philippines and Australia.

**TAXONOMY**

No subspecies recognized.

**BEHAVIOR**

A bird of interior wetlands. Breeds in grassy lowland marshes, flooded meadows, and boreal bogs, usually near open water. During migration and winter prefers marshes, flooded fields, and ponds. Sometimes uses tidal wetlands but always prefers fresh water over salt. Walks steadily, often in belly-deep water, picking delicately at the surface. Often follows waterfowl or herons to seek prey items that get stirred up. Eats a variety of invertebrates, particularly insects, mollusks, and crustaceans. Usually solitary or in small flocks, but groups of over 100 are regular in fall.

Breeds late April to early July.

**MIGRATION**

Long-distance migrant. Migrates across a broad front, mostly inland, though overflies large areas between stop-over sites.

**Spring** migration takes place between early March and late May, with peak movement in April. Departs wintering areas primarily in late March and April and arrives at breeding areas between mid-April and late May. Some nonbreeders remain at wintering areas through the summer, while others migrate north and summer near the breeding grounds.

**Fall** migration takes place from early July to early November. **Adults** depart the breeding grounds between early July and mid-September, but mostly in August after completing extensive molt. Peak numbers occur at temperate staging areas from August to mid-September. Arrives at wintering grounds mostly from late August to October. Southbound **juveniles** depart the breeding grounds from late July to mid-September, about the same time as most adults. Occurrence at stop-over sites and arrival at wintering areas about the same as for adults.

**MOLT**

**First life year:** The molt out of juvenal plumage begins from mid-August to mid-September on the breeding grounds and is mostly completed between November and January on the wintering grounds. It includes most head and body feathers, plus some wing coverts and tail feathers. Many birds are in largely nonbreeding plumage by late September. More advanced birds replace outer primaries in mid-February and March, while other birds replace outer primaries later in spring or not at all. Between early February and late April there is a variable partial molt to breeding plumage. The molt includes a variable number of head and body feathers, wing coverts, tertials, and tail feathers. In more advanced birds this plumage resembles adult breeding but is distinguished by retained juvenal inner primaries, secondaries, and some tail feathers and coverts. Birds that remain on the wintering grounds through the summer typically skip this molt or molt in partial breeding plumage. The latter birds may begin the complete molt to nonbreeding plumage as early as early May.

**Second life year:** For birds that oversummer on the wintering grounds, the complete molt to nonbreeding plumage begins between early May (of first life year) and late July and is completed in August/September. Other birds are on a more adultlike schedule. After this molt, second-year birds are usually indistinguishable from adults. The partial molt to breeding plumage is as in adults.

**Adult:** The partial molt to breeding plumage begins from mid-December to late February and is completed from early February to early April. It includes many head and body feathers, some coverts, tertials, and tail feathers. Timing of the complete molt to nonbreeding plumage is variable. Molting may take place mostly on the breeding or wintering grounds and is suspended during migration. Molt usually begins in early to mid-July and is completed from September (mostly nonbreeders) to early March. Early migrants arrive at wintering grounds with limited molt, late migrants with nearly complete nonbreeding.

#### VOCALIZATIONS

Flight call is a soft, whistled *teu,* often repeated two or more times; much like Lesser Yellowlegs.

## Common Redshank, p. 253

### *Tringa totanus*

#### STATUS

Eurasian species. Casual visitor in Newfoundland. Common within range. Breeds from Iceland, the Faroes, and Spain east to e. Russia and e. China. Winters mostly in coastal areas from Iceland and South Africa east to e. China and Indonesia.

#### TAXONOMY

Six subspecies are recognized, though differences are subtle, and many intermediate populations exist. Nominate *T. t. totanus* breeds from Ireland to w. Russia; *robusta* in Iceland and the Faroes; *ussuriensis* in e. Russia, Mongolia, n. Manchuria; *terrignotae* in e. China; *craggy* in nw. Sinkiang; and *eurhinus* in Kashmir and w. China. Winter ranges are poorly known. Differences in appearance are subtle and overlapping, but eastern birds are generally paler than western. North American records presumably refer to *robusta.*

#### BEHAVIOR

A bird of open wetlands. Breeds in wet meadows, inland freshwater and coastal salt marshes, moorlands, and salt pans. In winter and migration, typically found on coastal mudflats and estuaries, with smaller numbers at interior sites. Walks in a steady, fast-paced fashion, pecking at substrate. Sometimes feeds in shallow water with a sweeping motion of bill and rarely probes. Diet highly variable depending on location and season; mostly crustaceans, mollusks, and marine worms at coastal locations, and earthworms and crane fly larvae inland. In migration and winter, sometimes solitary but more typically in small flocks, occasionally to 1,000 individuals.

Nests March to August.

#### MIGRATION

Short- to medium-distance migrant. Some birds of nominate race in w. Europe and Iceland may be virtually sedentary, moving only short distances to coastal areas for winter.

**Spring** migration takes place primarily February to April, with peak movement in nw. Europe in late April to early May. Some nonbreeders remain at wintering areas through summer months (probably **one-year-olds**), while others migrate north and summer near the breeding grounds. Small numbers of one-year-old birds will nest during first summer.

**Fall** migration takes place from late June to October. **Adults** depart breeding grounds before **juveniles,** with one parent (usually female) leaving before young fledge. Main passage of Fenno-Scandian birds through Europe is July to September, with numbers peaking in Mauritania in October. Wintering grounds primarily reached by late October.

#### MOLT

**First life year:** The molt out of juvenal plumage is partial and takes place from August to January. It includes most body feathers, some tail feathers and tertials, and some lesser and median coverts. Between January and May there is a partial molt to breeding plumage, similar to that of adults, but may involve fewer feathers. Birds in their first breeding plumage differ from adults by more strongly worn primaries and, if present, extremely worn tertials and inner median coverts.

**Second life year:** For birds that oversummer on wintering grounds, the complete molt to nonbreeding plumage is similar to adults', but possibly occurs a little earlier.

**Adult:** The partial molt to breeding plumage takes place from January to May and includes some or most body feathers, scapulars, tertials, wing coverts, and sometimes tail feathers. Northern populations start this molt later. The complete molt to nonbreeding plumage begins in late June (often later in nominate *totanus*) and is usually completed by October but sometimes as late as January. Some birds suspend molt during migration, finishing on wintering grounds. Other birds migrate without arresting molt.

#### VOCALIZATIONS

A social, noisy bird, often first to give alarm calls when danger approaches. Flight call is a quick, musical *teu-dr* or *teu-dr-dr,* intermediate in character between Lesser Yellowlegs and Willet. Alarm call is a drawn-out, piping *teuuuu.*

## Spotted Redshank, p. 254

### *Tringa erythropus*

#### STATUS

Eurasian species. Very rare migrant on w. Alaskan islands. Casual on both coasts and accidental inland. Breeds in low Arctic to subarctic areas from Scandinavia east across n. Russia to e. Siberia. Winters from w. Europe and w. Africa east to Vietnam and se. China. World population estimated in 1983 at about 100,000.

#### TAXONOMY

No subspecies recognized.

#### BEHAVIOR

Breeds on tundra and in open Arctic taiga near bogs and marshes. During migration and winter prefers inland freshwater marshes or brackish lagoons.

Sometimes found in sheltered coastal mudflats. Feeds by picking and probing in muddy substrates, often after a short dash. May probe emphatically with bill held almost vertically, or swish bill side to side like an avocet. Also feeds in deeper water in dense flocks, swimming and upending like puddle-ducks. Diet very diverse but mostly animal material, chiefly adult and larval insects along with fish, crustaceans, mollusks, frogs, tadpoles, and flying insects, which it catches by running in a zigzag manner.

Nests early May to early July.

#### MIGRATION

Mostly a long-distance migrant (w. Africa to Viet Nam and se. China), but some birds now winter in w. Europe, north to Netherlands and Britain. Migration through Europe via long continuous flights to staging areas.

**Spring** migration takes place between March and early May, with peak movements in April. Departs wintering areas in March and moves through Europe in April. First birds reach Finland by early May, with others arriving very soon afterwards. Some nonbreeders remain on wintering grounds through summer months, but most return to Europe and summer just south of breeding areas.

**Fall** migration takes place between early June and October. **Adult females** depart breeding grounds after forming migration flocks before mid-June, while most males are still incubating eggs. **Males** and **juveniles** depart second half of July and August. Adults stage in Europe before prolonged migration to Africa. Peak numbers at European staging areas in mid-September, with numbers dwindling until mid-October. First arrivals at African wintering grounds in August/September, but mainly October.

#### MOLT

**First year life:** The molt out of juvenal plumage typically begins early to mid-September, occasionally mid-August, and is mostly completed by October. It begins with mantle and scapulars, followed by head and breast. Mostly nonbreeding plumage is acquired by early October, but some molt usually continues until April (feathers acquired in March/April are often intermediate in character between breeding and nonbreeding). From April to mid-May there is a variable partial molt to breeding plumage, similar to that of adults but involving a greater retention of nonbreeding feathers on scapulars, underparts, wing coverts, and tertials. Some first-year birds appear mostly nonbreeding, with a scattering of breeding feathers on foreneck, chest, and scapulars.

**Second life year:** For birds that attain a good percentage of breeding plumage in their first summer, a complete molt to nonbreeding plumage is similar to that of adults. Birds that oversummered on wintering grounds possibly start molt a little earlier.

**Adult:** The partial molt to breeding plumage takes place between March and April, and includes head, neck, underparts, mantle, scapulars, some or all tertials, a few central tail feathers, and some wing coverts. Scattered nonbreeding feathers are retained on belly and undertail coverts and lower scapulars. A complete molt to nonbreeding plumage begins in mid-June on breeding grounds and is mostly completed on various staging areas by mid-August. Molt is suspended during earlier migratory movements and is completed in temperate latitudes. All molt is completed before arrival in tropical wintering areas.

#### VOCALIZATIONS

Flight call is an emphatic, whistled *chu-IT,* similar to Semipalmated Plover, but lower and harder. Upon taking flight, sometimes gives a chuckling *chu, chu.*

## Wood Sandpiper, p. 256

***Tringa glareola***

**STATUS**

Old World species. Regular in flocks in spring in w. Aleutians. Accidental elsewhere. Breeds widely across upper-middle and higher latitudes of Palearctic from Britain and Scandinavia to e. Russia and Siberia. Winters mainly in southern part of Old World, often abundant in sub-Saharan Africa and India. Scarce in Australia. World population estimated at slightly more than 1 million birds in 1983.

**TAXONOMY**

No subspecies recognized.

**BEHAVIOR**

A bird of interior freshwater wetlands. Breeds in bogs and marshes in taiga habitat. More rarely uses treeless blanket bogs or alpine zones. During migration and winter prefers inland fresh waters and marshes, sometimes intermittent pools. Generally avoids seashores and other marine habitats. Feeds while walking or wading, picking prey from surface of mud or water or probing with head and neck immersed. Occasionally sweeps bill from side to side under water or captures flying insects after running, jumping, or short flight. Eats a variety of animal prey, primarily insects, including beetles, grasshoppers, crickets, and adult and larval moths, as well as worms and small fish. Sometimes teeters like Spotted Sandpiper and bobs foreparts like yellowlegs. Occurs in large flocks where common, but also found in smaller flocks during migration.

Nests early May to mid-July.

**MIGRATION**

Complete long-distance migrant, although a few individuals have been found as far north as Britain in winter. Winters mainly in tropical and subtropical latitudes of Old World, but scarce as a migrant in w. Europe due to long, nonstop flights from n. European breeding areas.

**Spring** migration takes place from late March/early April to early June, with peak movements in April/May. Departs wintering areas in late March/early April and moves through Europe and Middle East during April and May. Arrives in breeding grounds from late April in milder climates to early June in n. Russia.

**Fall** migration takes place from late June to October. **Adults** begin departure from European breeding grounds in late June, followed by **juveniles** a month later (timing later in Siberia). Some banded birds flew up to 746 mi. (1,200 km) nonstop to staging areas in Italy. Many birds stage for up to a month to molt and accumulate fat. Adults arrive in tropical Africa mostly in August (a few in late July), with juveniles starting in September and increasing into October.

**MOLT**

**First life year:** The molt out of juvenal plumage may begin from early August when still close to breeding area, but typically starts during fall migration at more-temperate staging areas and is completed in winter quarters. Some individuals don't begin molt until wintering areas are reached. Molt includes most of head and body and some wing coverts and tail feathers. First-winter plumage is attained from early October to December. From early March to April there is a partial molt to breeding plumage, though more limited than in adults, with only a few median coverts replaced. Most new wing coverts, tertials, and back and rump feathers are attained late in the molt out of juvenal plumage and may resemble adult breeding feathers or be intermediate between adult breeding and nonbreeding feathers. Birds that acquire only a few breeding feathers usually remain at southern wintering areas throughout the summer months and start the complete molt to nonbreeding plumage as early as late April.

**Second life year:** For birds that oversummer on the wintering grounds, the complete molt to nonbreeding plumage may begin in late April (of first life year) and is completed by midsummer. Other birds are on a more adultlike schedule. After this molt, second-year birds are indistinguishable from adults. The partial molt to breeding plumage is as in adults.

**Adult:** The partial molt to breeding plumage takes place from early February to late March. It involves most of head and body and often some tertials, wing coverts, and tail feathers. Feathers acquired late in the molt to nonbreeding may resemble breeding feathers and are not replaced again in spring. The complete molt to nonbreeding takes place from early August/September to December/January. For most birds, molt begins at southern staging areas and is suspended until wintering grounds are reached. Others begin molt on the wintering grounds. Early migrants arrive at wintering grounds in worn adult plumage, while later arrivals are in suspended molt.

#### VOCALIZATIONS

Flight call is a high, whistled *pseu-hu-hu,* similar in pattern to Greater Yellowlegs but weaker and with the shrillness of Solitary Sandpiper; reminiscent of American Goldfinch's flight call. Alarm call is a persistent *gip, gip, gip.*

## Green Sandpiper, p. 258

### *Tringa ochropus*

#### STATUS

Eurasian species. Casual in spring on w. Alaska islands. Common within range, but often found singly or in small flocks. Breeds across n. Eurasia, from Scandinavia east to e.-central Russia. Winters in s. Europe and Africa, sparingly to w.-central Europe or s. Scandinavia during mild years. World population estimated in 1983 at about 100,000.

#### TAXONOMY

No subspecies recognized.

#### BEHAVIOR

A shy bird of inland fresh waters. Breeds in waterlogged, wooded regions, typically bogs and marshes in open boreal forest. Similar to Solitary Sandpiper in appearance and habits, it prefers standing or gently flowing water. A nervous bird, when flushed it ascends rapidly while giving distinctive sharp whistling calls. During migration and winter typically found at inland fresh waters, including ditches, pond edges, and lakesides. Rarely seen on open mudflats. Usually found singly. Feeds in a deliberate fashion around vegetated margins by picking at substrate or vegetation. Will also probe in shallow water, sometimes submerging head. Sometimes teeters like Spotted Sandpiper. Eats mostly invertebrates, especially aquatic and terrestrial insects, but also crustaceans, small fish, and plant fragments.

Nests mid-April to early July.

#### MIGRATION

Somewhat long-distance migrant. Does not migrate as far south as its close relatives, rarely being found below 10° south in Africa or below equator elsewhere. Migrates overland in a broad front, spanning the full width of w. Palearctic. Rarely found in large concentrations anywhere in range.

**Spring** migration takes place between March/early April and mid-May. Departs wintering areas in late March/early April and virtually completes movement to breeding grounds by mid-May (one week later in n. Russia). No information on summering location of nonbreeders.

**Fall** migration takes place from early June to October. **Adult females** start departing breeding areas in early June, followed by **males** and **juveniles.** First migrants reach nw. and central Europe late June, with peak passage July/August. First arrivals at wintering grounds south of Sahara are in early August, though peak numbers arrive in September.

#### MOLT

**First life year:** The molt out of juvenal plumage takes place from late August/early October to April. Between late August/early October and late September/early November, during migration, most head and body plumage is replaced. Between late November and March, on the wintering grounds, most tertials are replaced with feathers resembling those of breeding adult. Between February and April a few wing coverts and back and rump feathers may be replaced. Between late March and early May there is a variable partial molt to breeding plumage. This molt is more restricted than in adults, typically including head, neck, underparts, and all or some scapulars and mantle feathers. Little information available on whether birds breed in their first year or on timing of molt to nonbreeding plumage.

**Adult:** The partial molt to breeding plumage begins late December to late February and is completed from March to April. It includes head, neck, mantle, underparts, most or all scapulars, some or all tertials, and a small number of wing coverts. The complete molt to nonbreeding plumage takes place between late June and mid-November. Primary molt starts late June/early July on or near breeding grounds and progresses rapidly, sometimes being completed by late August/mid-September, but most birds apparently leave for migration with outer 2–4 primaries old. Occasionally, birds migrate in active wing molt, even during Sahara crossing, but molt is usually suspended, then completed late October to mid-November on wintering grounds. Most head-and-body molt takes place between late June and late August, before arrival at wintering grounds. Remaining body plumage, tail, and a few coverts replaced by mid-November.

#### VOCALIZATIONS

Flight call is a clear, ringing *tuEEt-wit-wit,* similar to Solitary Sandpiper but lower, stronger, and less shrill. Often given by migrants high overhead.

## Solitary Sandpiper, p. 87

### *Tringa solitaria*

#### STATUS

Common throughout most of North America but widespread and never in large flocks. Breeds in the boreal-forest belt from Alaska east to New Brunswick. Winters from s. Texas, s. Florida, and the West Indies south to Peru and Argentina. Migrants equally likely at interior sites as along the coast. The world and North American population is estimated at 25,000, but this number is likely low.

#### TAXONOMY

Two subspecies generally recognized but broad zone of overlap. Nominate *T. s. solitaria* breeds in the southern and eastern portions of the range, from e. British Columbia and southern portions of Prairie Provinces east to Quebec and Labrador. *T. s. cinnamomea* breeds in northern and western portions of range from Alaska and n. British Columbia to ne. Manitoba. Winter ranges of the two subspecies apparently broadly overlap, but *cinnamomea* generally in the West, *solitaria* generally in the East. Differences in appearance subtle and overlapping. *T. s. cinnamomea* averages about 5 percent larger than *solitaria,* has shorter wing point, and is slightly paler above with broader black tail bars, weaker loral stripe, more cinnamon-tinged spots on juvenile upperparts, and white mottling on underside of p10.

#### BEHAVIOR

Breeds primarily in spruce forest with muskeg bogs. Only New World sandpiper to regularly nest in trees, using abandoned passerine nests. Breeders forage in shallow pools, usually in forest opening or at forest edge. During migration and winter prefers stagnant pools or muddy areas near forest or other dense vegetation. Also uses muddy riverbanks, streams, rain pools, and drainage ditches. Seldom found in tidal wetlands. Forages slowly but deliberately, frequently bobbing head and foreparts up and down. Picks food off surface or probes shallowly in mud. Often stirs up substrate with foot to rouse prey items. Feeds by day and night. Eats primarily insects, small crustaceans, small mollusks, and frogs. Typically seen singly but may gather in small groups of up to 10 if prime habitat is limited. When flushed, typically flies up at a steep angle. Upon landing, often holds wings up briefly. Migrates mostly at night and early morning.

Nest initiation primarily late May to mid-June. Egg dates late May to early July. Incubation 23–24 days. Both adults share parental duties. One brood per season but presumably will renest after failure. Peak hatching takes place in late June. Young can fly after about three to four weeks. The flight display, performed by both sexes, is towering with bouts of flapping and gliding while song is given.

#### MIGRATION

Migrates both through the interior and offshore with no significant coastal concentrations. In fall some birds make transoceanic flight from New England to South America, while others take the trans-Gulf route.

**Spring** migration takes place between late March and late May. Most common across the southern tier of states in April and to the north in early to mid-May. Arrival on the breeding grounds takes place mostly in mid- to late May. Some **one-year-old** birds remain on the wintering grounds through the summer, while most apparently migrate north to the breeding grounds.

**Fall** migration takes place between late June and early November. **Adults** begin to depart the breeding grounds in late June, with peak numbers passing through the northern tier of states between mid-July and mid-August and through the south in August. Failed breeders head south before adults. Southbound **juveniles** depart the breeding grounds in late July and early August. Peak numbers pass through the northern tier of states in mid- to late August and through the south in September.

#### MOLT

**First life year:** The molt out of juvenal plumage takes place on the wintering grounds, primarily from November (exceptionally September) to March, and includes head and body feathers along with some wing coverts, outer primaries, and central rectrices. In March and April there is a head-and-body molt into breeding plumage. This plumage resembles adult breeding but is distinguished by contrastingly worn inner primaries.

**Second life year:** The complete molt to nonbreeding plumage probably begins on the breeding grounds and is suspended, then completed after the wintering grounds are reached. After this molt, second-year birds are indistinguishable from adults. In March there is a partial molt to breeding plumage including head, body, tertials, some wing coverts, and central rectrices.

**Adult:** In March there is a partial molt

to breeding plumage including head, body, tertials, and some wing coverts and central rectrices. The complete molt to nonbreeding plumage probably begins on the breeding grounds and is suspended, then completed after the wintering grounds are reached in September/October to December/January.

#### VOCALIZATIONS

Flight call is a shrill, whistled, two- to three-note *pseet-weet-weet,* each note drawn out and rising; higher and more emphatic than Spotted Sandpiper. Given almost exclusively by flying birds and heard regularly in nocturnal migration. Alarm call is a sharp, ringing *pit,* often given with flight call if bird is startled and flushed. Also a husky *chit.* Display song, given from the ground, a treetop perch, or in flight, is a shrill, two-part series of high-pitched, flight-call-like notes, *pit, pit, pit, pit, klee-klee-klee-klee* . . . , increasing in frequency toward the end. Reminiscent of American Kestrel but much higher and shriller.

## Willet, p. 90, p. 95

### *Tringa semipalmata*

#### STATUS

Common and conspicuous in many coastal areas. Locally common breeder and migrant in interior West. Rare migrant in interior East, where it is generally seen only when forced down by stormy weather. Breeds in the prairies from s. Alberta and Manitoba to n. California and Colorado and in salt marshes along the Atlantic and Gulf coasts from Newfoundland to ne. Mexico, the West Indies, and Venezuela. Winters in coastal areas from Washington and New Jersey south to Peru and Argentina. The world population is estimated at 250,000 (160,000 for *inornata,* 90,000 for *semipalmata;* note, however, that subspecies estimates are likely skewed toward *semipalmata* based on misidentification of birds wintering in se. United States and possibly Central America).

#### TAXONOMY

Two distinct subspecies may represent full species and are treated separately here for convenience. Nominate *T. s. semipalmata* ("Eastern Willet") breeds in salt marshes and mangroves along the Atlantic and Gulf coasts from Newfoundland to ne. Mexico with isolated populations in the Bahamas, Greater Antilles, Cayman Is., and the island of Los Roques off Venezuela. Winter range poorly known but apparently centered in e. South America, particularly Brazil; some may also winter in Paraguay, Argentina, n. South America, the West Indies, and possibly Central America. Not documented in winter in the United States. *T. s. inornata* ("Western Willet") breeds in the prairies from s. Alberta and Manitoba to n. California and Colorado and winters in coastal areas from Washington and New Jersey south to Peru, the West Indies, and n. South America. Winter range in South America somewhat conjectural.

Differences in appearance between the two races are moderately distinct in all plumages, though some individuals are problematic, particularly those in nonbreeding plumage. Western averages 10 percent larger than Eastern, but overlap exists. Western is more dimorphic in size than Eastern. Timing and route of migration, timing of molt, and voice differ.

#### BEHAVIOR

**Eastern:** Breeds primarily in salt marshes, barrier islands, barrier beaches, and mangroves. At northern end of range uses pastures and farmland. Nest site is typically in a grassy or shrubby area but may be on bare sand. In winter appears to be closely tied to mangroves and adjacent tidal flats and creeks. Forages in salt marsh pools, tidal creeks and flats, beaches, and oyster beds. Forages primarily by pecking at surface or probing in substrate, walking steadily between foraging attempts. Feeds by day and night. Primary foods are crustaceans, particularly Fiddler Crabs and Mole Crabs, but also takes aquatic insects, marine worms, small mollusks, and fish. Typically seen in pairs or small groups during breeding season and in flocks of 5 to 10, rarely up to 100 or more, at other seasons.

Nest initiation primarily mid-April in the south, late May to early June in the north. Egg dates early March to early July. Incubation 24–26 days. Both adults share parental duties, though female typically abandons young and male after 2 weeks, leaving male to rear chicks alone for next 2 weeks. One brood, but will renest after failure. Peak hatching takes place in May in the south and June in the north. Young can fly after 28 days. The courtship display is a duet, with male rising first, hovering over territory on quivering wings, then joined by female, which hovers and descends to the ground with male. Sometimes displays in small groups with other pairs or unmated birds.

**Western:** Breeds primarily in sparsely vegetated prairie wetlands or adjacent semiarid grasslands. Nest site typically in a sparsely vegetated area adjacent to a windbreak such as a piece of wood, dried cattle dung, or a rock. Interior migrants use marshes, lakeshores, and flooded fields. In coastal areas, migrants and wintering birds prefer rocky coastlines, sod banks, beaches, and shallow bays, where they often associate with Marbled Godwits. Forages primarily by pecking at surface or probing in substrate, walking steadily between foraging attempts. Feeds by day and night. Primary foods include insects, crustaceans, mollusks, polychaete worms, and occasionally small fish. Typically seen in pairs or small groups during breeding season and in flocks of 5 to 10 up to 1,000 or more at other seasons.

Nest initiation primarily early May in the south, early June in the north. Egg dates early April to late June. Peak hatching mid- to late June. Incubation period, fledging period, and parental behavior similar to Eastern's.

**MIGRATION**

Both subspecies migrate primarily at night but also during the day under certain conditions.

**Eastern:** Primarily transoceanic route between South American wintering grounds and North American breeding grounds.

**Spring** migration occurs primarily between early March and early June. Migrants begin to arrive along the Gulf Coast in early March and continue to arrive through mid-April. Peak numbers depart South America in mid-April, arriving along the s. Atlantic Coast in mid- to late April and in New England in late April/early May, but migration continues steadily through May. **Subadult** birds often remain on the wintering grounds through their first summer, and a small percentage apparently remain there at least through their second summer.

**Fall** migration occurs primarily between late June and early September. Most birds depart within a narrow window of time, making the long overwater flight to wintering areas in South America. Peak numbers of **adults** depart the Atlantic Coast in early to mid-July, with stragglers into mid-August (rarely early September). Most **juveniles** fledge by early to mid-July, and peak numbers depart in late July / early August, with stragglers into early-October (possibly later on the Gulf Coast).

**Western:** Primarily overland route through Mississippi Valley and Great Basin between coastal wintering areas and interior breeding areas.

**Spring** migration occurs primarily between late March and mid-June, with peak numbers passing through s. United States in late April and s. Canada in early May. Rare along the n. Atlantic Coast in spring. **Subadult** birds often remain on the wintering grounds through their first summer, and a small percentage apparently remain there at least through their second summer.

**Fall** migration occurs primarily between mid-June and late October. Peak numbers of **adults** pass through the Pacific, Gulf, and s. Atlantic coasts between late July and mid-August. Most **juveniles** fledge by mid-July and begin their southbound migration within a week or two. First juveniles arrive on the Pacific, Gulf, and Atlantic coasts in mid- to late July and peak in late August and early September.

**MOLT**

**Eastern:**

**First life year:** Molt out of juvenal plumage takes place mostly between August and October and includes body plumage and some wing coverts, but many wing coverts are retained. Most birds show only partial scapular molt by the time they leave the United States, and many birds are still completing this molt when they arrive in South America. Between March and May a variable molt takes place, including body feathers but not remiges. This plumage appears to be intermediate between breeding and nonbreeding. **One-year-old** birds may be distinguished from adults by retained juvenal flight feathers and by wing coverts that look comparatively worn.

**Second life year:** For birds oversummering on the wintering grounds (probably including birds in their third life year), the complete molt to nonbreeding plumage presumably begins earlier than in adults.

**Adult:** Between February and April, adults undergo a partial molt to breeding plumage, including most head and body feathers and most or all wing coverts, tertials, inner secondaries, and tail. In fall, adults undergo a complete molt to nonbreeding plumage, mostly between mid-July and October. Flight feather molt apparently takes place mostly in August and September after arrival on the wintering grounds.

**Western:**

**First life year:** Molt out of juvenal plumage takes place between August and October and includes body plumage and most wing coverts. Between March and May a molt takes place including a variable number of body feathers. This plumage may appear as nonbreeding or partial breeding plumage. **One-year-old** birds may be distinguished from adults by retained juvenal flight feathers and (if present) by wing coverts that look comparatively worn.

**Second life year:** For birds oversummering on the wintering grounds (probably including birds in their third life year), the complete molt to nonbreeding plumage may begin a month or more earlier in adults.

**Adult:** Between February and May, adults undergo a molt to breeding plumage, including most head and body feathers and some wing coverts, tertials, inner secondaries, and tail. In fall, adults undergo a complete molt to nonbreeding plumage between mid-July and November. Flight feather molt takes place between late July and November after arrival on the wintering grounds.

**VOCALIZATIONS**

Flight call is a loud, strident *klaay-drr* or *klaay-dr-dr*, typically with a lower, huskier Marbled Godwit–like quality in Western and a higher, Laughing Gull–like quality in Eastern. When flushed, utters a higher, more excited *kli-li-li-li,* often with a trilled quality (more distinctly trilled in Western). Alarm call, given mostly near the nest site, is a sharp, repeated *kleep* or *kalip,* lower and more muffled in Western. Year-round alarm is a more drawn-out, screaming *klaayii* and variations, often with a distinctly curlewlike quality, particularly in Western. All calls of Western average lower pitched and huskier than those of Eastern, but much overlap exists, and many calls are not readily identifiable to subspecies.

Song, given by both sexes in flight display and while perched, may be repeated as many as 200 times on territory but is also given during migration. In Eastern, song is an urgent, rapidly repeated *pidl-will-willit.* In Western, song is a slower, lower-pitched, more clearly annunciated *p'd-weeel-will-wit,* with the second note more drawn out and the last two notes more clearly separated.

## Wandering Tattler, p. 98

***Tringa incanus***

**STATUS**

Locally common but never seen in large numbers. Breeds in montane areas of Alaska and Yukon Territory as well as in extreme n. British Columbia, nw. Northwest Territories, and on the Chukotski Pen. in Siberia. Winters along immediate coastline from n. California to the Galápagos Is., as well as in the Philippines, e. Australia, and throughout Oceania. Very rare inland. The world population is estimated at 10,000.

**TAXONOMY**

No subspecies recognized.

**BEHAVIOR**

Usually seen by itself along rocky coastlines, where it teeters like a Spotted Sandpiper. Breeds in shrubby alpine tundra above treeline near rocky or gravelly streams and lakes. Sometimes also in coastal gravel areas above high-tide line. During migration and winter uses rocky intertidal coastlines and reefs as well as man-made jetties, riprap, and piers. Less often found on tidal flats, coastal ponds, streams, or wet meadows. Active while feeding. Walks quickly over rocks in low horizontal stance with regular teetering motion to rear portion of body. Picks at surface of rocks and probes in sand, among mats of algae, and between rocks. Often runs to chase prey items in receding waves and may jump to catch flying insects. Eats a variety of invertebrates, including insects, worms, crustaceans, and mollusks as well as small fish. When alarmed, often bobs head and foreparts up and down like *Tringa.* Usually found by itself or in small groups of 2 to 3, but flocks of 10 to 15 are seen in some areas. Regularly seen foraging and roosting with Black and Ruddy Turnstones, Surfbird, and Rock Sandpiper.

Nest initiation mostly early June. Egg dates mid-May to late July. Incubation 23–25 days. Both adults share parental duties, but female departs before fledging of brood. One brood per season but will renest after failure. Peak hatching late June and early July. Young can fly after about 18–20 days. Age of first breeding, 2–3 years. Display flights are highly variable. Typically includes an undulating flight at a height of 1,000+ ft. (305+ m), lasting from a few seconds to 6 minutes, followed by a tumbling or nose-dive descent. Song is given during the rise and undulations.

**MIGRATION**

Intermediate- to long-distance migrant. Some birds migrate mostly through coastal areas, while others may cross up to 7,500 mi. (12,000 km) of open ocean.

**Spring** migrants begin to depart southernmost American wintering areas in early March. Peak movement takes place along the Pacific Coast of Mexico and s. California between early April and mid-May, through the central and n. Pacific Coast in early to mid-May, and through s. Alaska in mid- to late May. Arrives at breeding areas mostly in mid- to late May but sometimes as late as mid-June. Departure from oceanic wintering areas slightly later, mostly from mid-April to late May, with peak numbers passing through the Hawaiian Archipelago in early May. Most **one-year-old** and **two-year-old** birds remain on southern wintering areas through the summer.

**Fall** migration takes place between early July and late November. **Adults** begin to depart the breeding grounds in early July (primarily females first). Peak numbers occur from Alaska to California between late July and mid-August and in the Hawaiian Archipelago between mid-July and late September. Wintering sites in Central and South America may not be reached until late September or October. Migration of **juveniles** lasts between mid-August and late November. Peak numbers occur from Alaska to California between late August and late September and in the Hawaiian Archipelago between late September and late November.

**MOLT**

**First life year:** The molt out of juvenal plumage takes place mostly on the wintering grounds from September to July (of second life year). This molt is incomplete, involving most head and body feathers (September to March) and a variable number of primaries, secondaries, and tail feathers (March to July of second life year). Primary molt complex, but most molt outer primaries. This molt may overlap with the beginning of the complete molt to nonbreeding plumage during which all primaries are molted, beginning with inners in about June, so some birds may hold three generations of primaries at once (fresh inner, worn middle, slightly worn outer). Some birds may molt in limited breeding plumage in spring, but most apparently remain in nonbreeding plumage. **One-year-old** birds should be distinguishable from older birds by the presence of some retained juvenal flight feathers and coverts.

**Second life year:** The molt out of juvenal plumage is often completed as late as July of the second life year (see above). The complete molt to nonbreeding plumage may overlap this, beginning as early as June. Molt patterns otherwise probably similar to adult but may attain partial breeding plumage only in spring.

**Adult:** Between about January and May there is a partial molt into breeding plumage involving most head and body feathers. Some birds may be in full breeding plumage as early as March. The complete molt to nonbreeding plumage usually begins in early August on the breeding grounds, but the bulk is completed after the wintering grounds are reached from October to December. Flight feather molt takes place from August to November, but mostly in October. Tail molt often completed by April.

**VOCALIZATIONS**

Flight call is a ringing, trilled whistle, *dididididididi*, all on one pitch. Given regularly by birds on takeoff. Song is a quick, three- to four-note whistle, *treea-treea-treea-tree,* reminiscent of the song of Lesser Yellowlegs.

## Gray-tailed Tattler, p. 260

***Tringa brevipes***

**STATUS**

Eurasian species. Uncommon to rare but regular migrant in w. Alaska, casually to Pt. Barrow, especially in autumn. Accidental in fall to North American Pacific Coast. Breeds in ne. Siberia and winters in se. Asia and sw. Pacific islands south to Australia.

**TAXONOMY**

No subspecies recognized.

**BEHAVIOR**

A bird of rocky, sandy, or muddy coasts. Similar to Wandering Tattler but found more commonly on open mudflats. Breeds along stony riverbeds and lake edges in mountains of ne. Siberia. Perches commonly in trees during breeding season and may nest in old tree nests of other birds. During migration prefers a variety of coastal habitats, but also occurs inland in flooded fields. Found in winter on coastal mudflats, sandflats, and estuaries, where it feeds on exposed areas at low tide; also on rocky coasts. Feeds singly or in small groups, picking or probing actively at substrate or surface of rocky habitats. Eats a variety of invertebrates, including insects, worms, crustaceans, and mollusks, as well as small fish. Occasionally teeters like Spotted Sandpiper, and bobs foreparts like yellowlegs when alarmed. Forms communal roost flocks where commonly found, unlike Wandering Tattler. Often found roosting with Eurasian Curlew, Terek Sandpiper, and Lesser Sand-Plover.

Nests late May to July.

**MIGRATION**

Long-distance migrant. Migrates southward from breeding areas in ne. Siberia to continental coastlines and larger islands in sw. Pacific.

In **spring,** most of population migrates northward in April and May. Departs wintering areas in April and arrives at breeding areas in late May/early June. Many nonbreeders (mostly **one-year-olds**) remain at wintering areas through the summer.

In **fall, adults** begin to depart the breeding grounds in July, with most birds departing in August. **Juveniles** depart the Arctic from mid-August to late September. Peak passage of adults and juveniles takes place in September/October.

**MOLT**

Little information available on timing and region of molts.

**First life year:** The molt out of juvenal plumage is protracted. Molt of head, body, and some wing coverts and tail feathers takes place between September and December. The outer 4 or 5 primaries are replaced during January to August (of second life year). Birds that remain south during their first summer attain only partial breeding plumage, and many appear as full nonbreeding. Birds returning north to the breeding grounds may attain a nearly adultlike plumage.

**Adult:** In spring, adults undergo a partial head-and-body molt to breeding plumage. The complete molt to nonbreeding takes place between July and January.

**VOCALIZATIONS**

Flight call is an up-slurred whistle, *tuu-eet* or *tuwi-di-di,* much like Common Ringed-Plover; very different from Wandering Tattler.

## Common Sandpiper, p. 262

***Actitis hypoleucos***

**STATUS**

Eurasian species. Rare but regular migrant in w. Alaska, particularly in spring. Common within range. Breeds widely in Britain and Ireland and across n.-central portion of entire Eurasian continent, with the exception of snow-covered or very hot areas. Winters from s.-central Africa east to India, se. Asia, islands of sw. Pacific, and Australia. World population unknown, but incomplete data from early 1980s estimates almost 2 million birds.

**TAXONOMY**

No subspecies recognized.

**BEHAVIOR**

A widespread bird found in virtually all wetland habitats. Breeds near water in forested areas, preferring gravel or stony shores. Habitats include rivers, small pools, lakes, and sheltered seacoasts. During migration and winter found in virtually all wetland habitats where shorebirds are found, including ocean beaches, rocky shores, estuaries, inland ponds and rivers, and roadside ditches. Feeds in a slow and deliberate fashion, teetering and bobbing constantly like a Spotted Sandpiper, especially immediately after landing. Restless and almost always in motion, walking and running from point to point. Feeds visually, picking prey from surface, often between stones or in cracks or low vegetation. Adept at creeping up on and snatching stationary or flying insects, with head and body at low horizontal angle. Diet very diverse, but eats mostly stationary or flying invertebrates, especially insects. Usually solitary or in small flocks, but may form large flocks of up to 200 during migration.

Nests mid-April to mid-July.

**MIGRATION**

Short- to long-distance migrant. Migrates frequently at night in small to medium-sized groups. Most breeders migrate long distances to wintering areas in Southern Hemisphere, while smaller numbers winter well to the north in maritime climates of w. Europe and se. Asia. Some birds breed in s. Europe, and a few have bred in Africa, resulting in a much shorter migration.

**Spring** migration takes place between March and early June, depending upon latitude and climate. Departs wintering grounds in late March, after accumulation of large fat deposits, and arrives at temperate breeding grounds in second half of April, at more northerly breeding areas in late May/early June. Many nonbreeders summer in wintering areas, while others migrate north and summer near breeding grounds.

**Fall** migration takes place from late June (probably failed breeders) to October. Some **adults** of either sex depart breeding grounds starting in late June/early July (later in northern breeding areas), while brooding males and **juveniles** follow several weeks later. Peak passage of adults through central Europe is in late July, with juveniles late July to late September, peaking in August. Arrives at wintering grounds mostly from late August through October, with most birds arriving in September.

**MOLT**

**First life year:** The molt out of juvenal plumage is variable, sometimes complete, and takes place primarily on the wintering grounds from late September/December to January/March. This molt includes most head and body feathers, some wing coverts, tertials, and tail feathers, and a variable number of primaries. Many juveniles replace all flight feathers from late October to January, with no juvenal plumage remaining. Others replace only outer primaries and some secondaries from January to March, retaining some juvenal flight feathers through summer months. From March to April (somewhat later than adults), there is a variable partial molt to breeding plumage. This molt is similar to that in adults but involves fewer feathers.

**Second life year:** No information on birds that oversummer on wintering grounds, but the complete molt to nonbreeding plumage likely occurs earlier than in adults or nonbreeders that summered near breeding grounds. After this molt, second-year birds are usually indistinguishable from adults. The partial molt to breeding plumage is probably as adults.

**Adult:** The partial molt to breeding plumage occurs from February to May. It includes most head and body feathers, most tertials and tail feathers, and some wing coverts. Some nonbreeding feathers on back and rump are retained. A complete molt to nonbreeding plumage typically takes place on wintering grounds, although some birds molt some feathers of body, tail, and tertials while on breeding grounds in late June and July. Heavy molt begins upon arrival on the wintering grounds in August/September and is largely complete by late September. Flight feather molt takes place from late August/October to January/February. Some birds suspend molt of flight feathers during winter and complete molt by April.

#### VOCALIZATIONS

Flight call is a shrill, drawn-out *swEE-dee-dee-deu,* falling slightly in pitch; similar in quality to Solitary Sandpiper but faster.

### Spotted Sandpiper, p. 101

***Actitis macularia***

#### STATUS

Fairly common at inland and coastal sites throughout North America, though usually solitary. Our most widespread breeding shorebird, it nests from Alaska to California and Labrador to North Carolina. Winters from British Columbia and South Carolina south through Central America and the West Indies to n. Chile and n. Argentina. The world and North American population is estimated at 150,000, but this number is likely low.

#### TAXONOMY

No subspecies recognized.

#### BEHAVIOR

Teeters along sandy or rocky shorelines of almost any watery habitat, usually by itself. When flushed, flies low to the water on shallow vibrant wingbeats mixed with stiff-winged glides. Breeds in a wide variety of grassy, brushy, or forested habitats near water, particularly in areas where ground-dwelling predators are few. During migration and winter uses almost any habitat near water, from sandy borders of lakes and ponds to exposed rocks along fast-moving streams and rivers to wrack lines on beaches. Regularly found in man-made habitats such as concrete drainage ditches, sewage ponds, and seawalls. Forages visually on ground, seldom in water. Walks steadily in low horizontal stance with continuous teetering motion to rear portion of body. Picks at surface and occasionally runs to chase flies and other insects. Feeds only during the day. Eats almost any animal small enough to be captured, but bulk of diet consists of adult and larval insects. Usually found by itself or in small groups, sometimes near other shorebirds.

Nest initiation mid-May in South to mid-June in North. Egg dates mid-May to early August. Incubation 19–22 days. Female often polyandrous, mating with up to four males, each of which hatches and raises a brood. Female assists with incubation in monogamous pairs but does little brood rearing. One brood per male per season but will renest after failure. Peak hatching mid-June to early July. Young can fly after about 16–18 days. The display flight, performed by both sexes, is an alternating series of flapping and gliding with wings pointed down, head held high, and back arched upward as the song is delivered.

#### MIGRATION

Intermediate- to long-distance migrant. Broad-front migrant across entire continent. Somewhat less common inland in fall than spring. Some birds make transoceanic flight from New England to South America. Females winter farther north than males. Migrates during both day and night.

**Spring** migrants depart s. South American wintering areas between late February and late April and pass through n. South America from late April to late May. Peak numbers pass through the southern states from mid-April to mid-May and through northern states and s. Canada in mid- to late May. Arrives at northernmost breeding areas by early June. In the Rocky Mountains arrives in lowland areas in mid- to late April and at high elevations by late May or early June. Females arrive at breeding sites before males. Most **one-year-old** birds breed in their first season but migrate slightly later than adults.

In **fall, adults** begin to depart the breeding grounds in early July (primarily females first). Peak numbers pass through the northern tier of states from mid-July to mid-August and through the South in August. First arrivals in South America take place in early July, suggesting that some birds fly there nonstop. Southbound **juveniles** depart the breeding grounds between early July and early September and peak about 2 to 3 weeks after adults.

#### MOLT

Unlike most shorebirds, the molt out of juvenal plumage is usually complete.

**First life year:** The molt out of juvenal plumage takes place on the wintering grounds from September/October to April/May. This molt is usually complete, but it is variable, with some birds retaining inner primaries along with some secondaries and coverts. The body molt is usually completed by early December. Between about February and early May there is a partial molt into breeding plumage involving most head and body feathers. This molt overlaps the completion of flight feather molt. First-year birds may be distinguished from adults only if retained worn juvenal inner primaries or inner wing coverts are present.

**Adult:** Between about February and April there is a partial molt into breeding plumage involving most head and body feathers. The complete molt to nonbreeding plumage may begin as early as July and is completed by February/March. Only limited molt takes place before the wintering grounds are reached. Body molt is completed by late October or November.

#### VOCALIZATIONS

Flight call is a clear, whistled, two- to four-note *peet-weet-weet;* lower pitched and often slower and more wavering than Solitary Sandpiper's flight call. Given regularly by birds as they flush and also by nocturnal migrants. Also gives a variety of similar calls, including a single sharp *peet* when alarmed. Flight song, given by both sexes, is a monotonously repeated *cree-cree-cree-cree . . .* or a more complex, stuttering *bbbrEEt bbbrEEt bbbrEEt . . .*

## Terek Sandpiper, p. 264

### *Xenus cinereus*

#### STATUS

Eurasian species. Rare migrant on w. Alaska islands, casual elsewhere in Alaska and along Pacific Coast; accidental in Manitoba and Massachusetts. Fairly common within range. Breeds from Finland eastward to Siberia. Winters along coastlines of Africa, India, s. Asia, and Australasia.

#### TAXONOMY

No subspecies recognized.

#### BEHAVIOR

A bird of coastal mudflats away from breeding grounds. Breeds in a variety of habitats, but primarily in lowland boreal taiga by rivers and lakes. During migration and winter found mostly on coastal mudflats and estuaries, or on lake edges and streams during overland migration. Almost frantic feeding behavior, running back and forth while making frequent stops and changes of direction. Charges around with tail up and head down as it picks up insects and crabs from surface of mud. Also probes in mud for food. Occasionally wags rear body like Spotted Sandpiper. Feeds mostly on invertebrates, especially crustaceans, but also eats seeds on breeding grounds. Typically found singly or in small numbers among large feeding flocks with Curlew Sandpiper and Red-necked and Little Stints. Communal roosts with other species also common. Often perches on mangrove branches and pilings when resting in winter.

Nests mid-May to early July.

#### MIGRATION

Long-distance migrant. Migrates mostly overland through Eurasia and coastally in Southern Hemisphere.

**Spring** migration takes place between March and June. Departs wintering areas in Africa in late March and April and arrives at breeding areas in Finland and southern locations in early May, n. Russia in early June. Some nonbreeders remain at wintering areas through the summer, while others migrate north and summer near breeding grounds.

**Fall** migration takes place from early July to November. **Adults** depart breeding grounds from early July to August. Arrives at wintering grounds from late August to November. **Juveniles** depart breeding areas from August to September.

#### MOLT

**First life year:** The molt out of juvenal plumage is variable in timing. Some birds molt most head and body feathers at stop-over sites from late August to early October. Others arrive at wintering sites in full juvenal plumage and molt from October to February. Retained juvenal feathers include flight feathers, some wing coverts, outer tail feathers, some tertials, and part of back and rump. More advanced birds molt outer primaries starting in late December. Between mid-February and late April some birds undergo a partial molt to breeding plumage, less extensive than that of adults. Birds remaining on the wintering grounds through the summer generally lack this molt and start to molt inner primaries by March.

**Second life year:** For birds that oversummer on the wintering grounds, a complete molt to nonbreeding plumage takes place from March (of first life year) to August.

**Adult:** The partial molt to breeding plumage begins in late February and is completed by late April. It includes head, body, some wing coverts, and some or all central tail feathers and tertials. Timing of the complete molt to nonbreeding plumage is variable. Molting may take place while on breeding grounds, during migration stop-overs, or not until wintering grounds are reached. Birds that start this molt in July usually complete it by late November to January. Birds that arrive on wintering grounds in breeding plumage may start the molt in November and complete it by March.

#### VOCALIZATIONS

Flight call is a quick series of plaintive whistles, *pri-pi-pi*, similar to Whimbrel but softer.

## Upland Sandpiper, p. 104

***Bartramia longicauda***

**STATUS**

Locally fairly common in the core of the Great Plains. Uncommon and declining in the remainder of its range. Breeds from se. Alaska, sw. Manitoba, and Maine south to n. Oklahoma and Delaware. Winters in South America east of the Andes Mts., from Suriname and n. Brazil to n. Argentina and Uruguay. Uncommon migrant down Atlantic Coast in fall. The world and North American population is estimated at 350,000.

**TAXONOMY**

No subspecies recognized.

**BEHAVIOR**

A bird of the prairies, this species is seldom seen near water. Breeds most commonly in native tall-grass or mixed-grass prairie, particularly where there are scattered rocks or fenceposts for perches. To a lesser degree also uses upland heath tundra, high-altitude meadows, blueberry barrens, airports, and hayfields. During migration and winter prefers areas with shorter vegetation such as well-grazed pastures, sod farms, and cultivated fields. Forages visually, walking steadily with head-bobbing motion, periodically picking insects off grass. Horizontal stance when walking but very upright when alarmed. Eats primarily adult and larval insects and other small invertebrates. Occasionally eats seeds. Usually found by itself or in small groups. Often seen in the same habitat as Killdeer, American Golden-Plover, and Buff-breasted Sandpiper, but does not particularly associate with them.

Nest initiation mostly early May in South to late May in North. Egg dates late April to early July. Incubation 21 days. Both adults share parental duties. One brood per season but on rare occasion will renest after failure. Peak hatching late May to mid-June. Young can fly after about 30 days. The flight display is performed by both sexes over territory and occasionally during migration. The display is at a height of 100–1,300 ft. (30–400 m) on shallow fluttering wingstrokes, interrupted by periods of gliding when whistled song is given. As the bird lands on a prominent perch, it momentarily holds its wings high in the air and sings again.

**MIGRATION**

Long-distance migrant. Primary route is through the Great Plains, with small numbers passing east of the Mississippi R., mostly in fall. Migrates primarily at night and early morning.

**Spring** migrants depart South American wintering areas mostly from mid-March to early April. Peak numbers pass through the w. Gulf Coast in late March and early April and through the n. Plains and New England in early May. Alaskan breeders arrive by mid-May. Most **one-year-old** birds apparently breed in their first season.

In **fall,** most **adults** depart the breeding grounds in early to mid-July (failed breeders in late June). Peak numbers pass through the n. Great Plains in mid- to late July, the central Plains in early August, and the Gulf Coast in late August. Smaller numbers of adults pass down the East Coast, peaking in the mid-Atlantic in late July and early August. First arrivals in South America take place in mid- to late August. Southbound **juveniles** depart the breeding grounds mostly between late July and mid-August. Peak numbers pass through the n. Great Plains in early August, the central Plains and mid-Atlantic Coast in mid- to late August, and the Gulf Coast in September. Most birds are gone from North America by mid-October.

**MOLT**

Unlike with most shorebirds, the molt

out of juvenal plumage is usually nearly complete, and a spring molt is often lacking.

**First life year:** The molt out of juvenal plumage takes place from August to March, mostly on the wintering grounds. This molt is usually nearly complete, including all feathers except some wing coverts and occasionally some flight and tail feathers. Usually no spring molt. Those birds that did not undergo a complete molt in the fall may be distinguished from adults by retained worn juvenal wing coverts, flight feathers, or tail feathers.

**Adult:** A spring molt is lacking in many individuals, though some undergo a very limited head and body molt. The complete molt to nonbreeding plumage may begin as early as July and is completed by January. Only limited molt takes place before the wintering grounds are reached, and primaries are molted entirely on the wintering grounds.

#### VOCALIZATIONS

Flight call is a low, whistled *qui-pi-pi-pi*, with a mellow yet explosive quality. Alarmed birds on the breeding grounds may extend this call into a long chatter. Song, given in flight or while perched, is an extraordinary series of gurgling notes that slowly rise to an abrupt end, followed by a long, descending "wolf whistle," *prrrrrrrEEP-hlEEEoooooo.* The whole song lasts about 3 seconds.

## Little Curlew, p. 266

### *Numenius minutus*

#### STATUS

Asian species. Casual in fall along Pacific Coast; accidental in spring in w. Alaska. Uncommon throughout range. Breeds in n.-central and ne. Siberia, winters primarily in s. New Guinea and n. Australia.

#### TAXONOMY

No subspecies recognized.

#### BEHAVIOR

A bird of open fields and grasslands. Breeds in loose colonies in partial clearings created by fire or storms in dry taiga zones, usually in river valleys. Only small percentage of population located in limited breeding areas. During migration and winter prefers interior grasslands, cultivated fields, and freshwater margins. Rarely found along coast. Feeds mostly by picking and shallow probing, similar to *Pluvialis* plovers. Walks deliberately with occasional upright stances. Eats mostly invertebrates, especially beetles, earthworms, and insects; similar to foods preferred by golden-plovers. Also eats berries. At stop-over sites sometimes seen in dense flocks of up to several hundred birds. Wintering birds on the open plains of Australia may gather in groups of up to several thousand.

Nesting season probably late May to mid- to late July.

#### MIGRATION

Long-distance migrant. Southbound migration mostly overland through Asia. Apparent nonstop flight from e. China to wintering areas in Australia. Little information on migration timing, but probably similar to other Siberian shorebirds that winter in Australasia.

**Spring** migration takes place between April and early June. Probably departs wintering areas in April and arrives at breeding areas in May. Limited data suggests first-year birds return to breeding areas, but breeding data incomplete.

**Fall** migration takes place from late July to October. Postbreeding flocks of adults and juveniles gather from late July and depart in August to early September. Arrives in Australia from September to October, with no information on movements of **adults** versus **juveniles.**

#### MOLT

**First life year:** The molt out of juvenal plumage takes place primarily on wintering grounds, but some feathers are replaced before fall migration begins. These include some feathers of head, mantle, scapulars, chest, and flanks. The remainder of this partial molt to nonbreeding plumage takes place during winter stay and includes rest of head and body feathers and many upper wing coverts. Juvenal flight feathers and some outer tail feathers are retained until nonbreeding molt in the following summer, though some outer primaries may be replaced by spring. In early spring (probably March/April), a partial molt to breeding plumage takes place, but to a much lesser extent than in adults. Timing is similar to adult.

**Second life year:** Probably similar to failed breeding adults, which molt more extensively before migration than successful breeders.

**Adult:** The partial molt to breeding plumage occurs in early spring, shortly before departure in April, and is sometimes completed during migration (April to May). Extent of molt is variable, including some feathers of head, mantle, scapulars, tertials, and often parts of chest and flanks. Occasionally, molt includes all feathers of head, neck, mantle, scapulars, tertials, and underparts. The complete molt to nonbreeding plumage starts before southbound migration (late July/August) and is completed on wintering grounds after suspension during movements. Premigratory molt includes some feathers of head, mantle, scapulars, tertials, and wing coverts. Molt is more extensive in failed nesters. Molt of flight feathers starts in September/October, shortly after arrival in wintering grounds, and is completed by following March.

#### VOCALIZATIONS

Flight call is a rapid, whistled *twi-ti-ti-ti*, each note rising slightly in pitch; higher and weaker than Whimbrel's call.

## Eskimo Curlew, p. 268

***Numenius borealis***

#### STATUS

Presumed extinct. Once tremendously abundant, darkening the skies of the Great Plains in spring and the Northeast in fall. Bred in high Arctic regions from w. Alaska to nw. Canada and wintered in the pampas of central Argentina and s. Brazil. Population decimated by market and sport hunting between 1850 and 1875 and nearly extinct by early 20th century. Suppression of wildfires and extinction of a principal spring food source, the Rocky Mountain Grasshopper, may also have contributed to its demise. Last confirmed sightings were of one photographed near Galveston, Texas, in March and April 1962, and one shot in Barbados in September 1963. Some subsequent sightings are compelling but lack documentation.

#### TAXONOMY

No subspecies recognized.

#### BEHAVIOR

Bred in treeless high Arctic tundra and wintered on pampas grasslands. At fall staging areas foraged in heath barrens, drier portions of salt marshes, and tidal pools, where it fed almost exclusively on berries (especially crowberry) and snails. In spring used upland habitats, including coastal pastures, plowed fields, tall-grass prairie, and shallow grassy sloughs. It particularly favored burn areas. In these habitats fed on a variety of invertebrates, particularly grasshoppers and their eggs, crickets, grubs, and ants. Highly gregarious, it was often found in dense flocks numbering in the thousands, though small flocks were also seen. Strongly associated with American Golden-Plover, with which it was almost invariably seen during migration.

Little known of its breeding biology. Nest initiation probably as early as late

May. Egg dates from June 8 to July 12, with a peak June 18–25. Presumably single-brooded. Peak hatching probably early July. Young probably fledged by early to mid-August.

**MIGRATION**

Long-distance, partially transoceanic migrant. Followed an elliptical migration route.

**Spring** migrants departed wintering areas in late February and early March and passed through the prairies, primarily west of the Mississippi R. and east of the Rockies. Peak numbers passed through the Texas coast between late March and mid-April and through the n. Plains states in late April and early May. Arrivals on the breeding grounds took place in late May and early June.

**Fall** migrants began to depart breeding areas in mid- to late July, moving eastward to staging areas from coastal Labrador to Newfoundland and Nova Scotia. Staging areas were occupied between late July and mid-September, with peak numbers occurring between early August and early September. The first wave in late July and early August was surely **adults,** probably augmented by **juveniles** in late August and September. Irregular occurrence in New England, mostly storm-driven, was somewhat later, with birds arriving in mid-August and departing as late as early October. After putting on adequate fat reserves, southbound migrants made the long transoceanic flight to s. South American wintering grounds, arriving mostly in September. Some birds may have stopped over in the West Indies or Bermuda, while others probably flew nonstop to South America. Occasional birds would appear along the Atlantic Coast south of New England after being forced down by stormy weather, but the species was apparently always rare south of Long Island.

**MOLT**

Little information. Presumably did a complete molt in fall, probably taking place largely on the wintering grounds. Body molt likely began on the breeding grounds and was suspended during migration. In spring there was at least a limited head-and-body molt. This was visible on spring migrants, with new feathers appearing more pinkish-buff.

**VOCALIZATIONS**

The Inuit name for Eskimo Curlew was "pi-pi-pi-uk," an onomatopoetic reference to its flight call. The flight call was a soft series of low whistled notes, said to resemble that of Upland Sandpiper though possibly closer in quality to that of Eastern Bluebird. Shorter two-syllable calls have also been described. Also reported to have given a sharp squeak, similar to some notes from Common Tern but weaker. No information on other vocalizations but probably had repertoire similar to Little Curlew's.

## Whimbrel, p. 107

***Numenius phaeopus***

**STATUS**

Common and widespread, occurring nearly around the globe, but populations quite localized at all seasons. Breeds in several disjunct populations in boreal, subarctic, and low Arctic regions of Alaska and nw. Canada to Hudson Bay; Iceland to central Siberia; and scattered sites in e. Siberia. Winters in coastal areas from s. United States to s. South America, sw. Europe, Africa, se. Asia, Australia, and New Zealand. Generally rare migrant inland except at a few sites. The world population is estimated at 797,000, with 57,000 in North America.

**TAXONOMY**

Four subspecies recognized, three occurring in North America. *N. p. hudsonicus* breeds in Alaska and Canada and winters from s. United States to s. South

America. Nominate *N. p. phaeopus* breeds from Iceland to w. Siberia, winters from sw. Europe to w. Indian Ocean, and is a casual vagrant to the Atlantic Coast. *N. p. variegatus* breeds in e. Siberia, winters in se. Asia, Australia and New Zealand, and is a rare but regular migrant in w. Alaska. All Old World races differ from *hudsonicus* in having a variably whitish rump, whiter ground color to underparts and underwing, more coarsely marked face, neck, and breast, and colder overall coloration. Vocalizations differ slightly (see below). Additionally, *phaeopus* differs from *variegatus* in a whiter underwing and tail.

**BEHAVIOR**

Breeds primarily in open subarctic and alpine tundra. Habitat ranges from dry heath uplands to soggy lowlands, often with abundant dwarf shrubs. Breeding birds generally form loose aggregations. Breeders forage in heath uplands, shallow pools, and intertidal flats. During migration and winter forages mostly in marshes, tidal creeks, tidal flats, and mangroves. May also use upland meadows. Forages mostly visually, walking steadily, then picking or probing when prey items are found. May pick food from surface or insert bill partway or all the way into a burrow of a Fiddler Crab or other prey item. Bill curvature matches Fiddler Crab burrows. Feeds primarily during the day but sometimes also at night. Eats primarily marine invertebrates, particularly crabs, but also marine worms, mollusks, and fish. On the breeding grounds also eats insects and berries. Typically seen in small flocks of up to 40 or 50 birds.

Nest initiation primarily late May in most of Alaska, early June in Canada and Alaska's North Slope. Egg dates range from late May to mid-July. Incubation 27–28 days. Both adults share parental duties. One brood per season but will renest after failure. Peak hatching late June in Alaska, early July in Canada. Young can fly after 5–6 weeks. The male's display flight begins with a steep ascent (45° angle) up to 500–1,000 ft. (150–300 m), followed by wide circles over territory, alternately flapping and gliding as it gives song.

**MIGRATION**

Migrates in coastal areas as well as offshore, mostly at night. Separate migration routes for western and eastern breeding populations. Western birds migrate along and off the Pacific Coast to wintering areas primarily along the Pacific Coast from the United States to South America. Eastern population winters primarily in se. United States and ne. South America. In fall many birds make nonstop flight of 2,500 mi. (4,000 km) over the Atlantic from staging areas in s. Canada and New England to wintering areas in South America. Spring route passes primarily through the Middle Atlantic and Great Lakes. Fall migration more protracted than spring.

**Spring** migration takes place between early March and early June. In the West, peak numbers pass through s. California in late March and April and through the Pacific Northwest between mid-April and mid-May. From there, birds fly directly to the Gulf of Alaska, where large numbers gather in mid-May. In the East, peak numbers pass through the Gulf and s. Atlantic coasts between April and early May and through the Middle Atlantic in early to mid-May. Scarce in New England in spring. Seen regularly in the eastern Great Lakes region in mid- to late May. Arrives at breeding sites in Alaska in early to mid-May and in Canada in late May. Most **one-year-old** and some **two-year-old** birds remain on the wintering grounds through the summer.

**Fall** migration takes place between late June and late October. Failed breeding **adults** begin to depart the breeding grounds in late June, adult females by mid-July. Numbers build up along the s. coast of Alaska and the Pacific Northwest in late June and July and through s. California in late July and early August. In the East, peak numbers of adults gather from Hudson Bay and Labrador to the Middle Atlantic from mid-July to early August. Arrival on the Gulf Coast is protracted. Southbound **juveniles** depart the breeding grounds mostly in early August, about a month after the adults. Peak numbers of juveniles pass through Alaska and the Pacific Northwest in late August and through s. California and the Northeast in late August and early September.

**MOLT**

**First life year:** Molt out of juvenal plumage takes place on the wintering grounds, starting October/December and ending January/April. Molt includes all but flight feathers, some tertials, most coverts, and some tail feathers. The complete molt to nonbreeding plumage may begin as early as May (see below). In Old World races some birds begin primary molt as early as January.

**Second life year:** The complete molt to nonbreeding plumage takes place on the wintering grounds, beginning mid-May (of first life year) to mid-July and ending in December. In March some birds undergo a limited body molt, after which birds are on the same molt schedule as adults.

**Adult:** In March some birds undergo a limited body molt. A complete molt takes place between August/September and November/January. Body molt and occasionally primary molt may begin on migration but is completed after the wintering grounds are reached.

**VOCALIZATIONS**

Flight call is a far-carrying staccato whistle, *pi-pi-pi-pi-pi-pi,* all on one pitch. Given almost exclusively by flying birds and heard regularly in nocturnal migration. During aggressive encounters gives a weak, trilled *cooo-reee* and various short whistles. Flight song is a series of long haunting whistles, increasing in pace, pitch, and volume, culminating in a long, flight-call-like trill. In flight display *phaeopus* gives whistle call during climb, unlike *hudsonicus.*

## Bristle-thighed Curlew, p. 269

***Numenius tahitiensis***

**STATUS**

Rare and local endemic breeder in remote tundra of w. Alaska's Seward Peninsula and Yukon-Kuskokwim Delta region. Winters on small islands and atolls in central and s. Oceania. Stages in coastal Alaska (Yukon-Kuskokwim Delta) in fall. Casual along Pacific Coast. The world population is estimated at 10,000.

**TAXONOMY**

No subspecies recognized.

**BEHAVIOR**

Breeds in rolling hilly tundra with short vegetation and scattered shrubs. Staging birds in late summer use coastal dwarf-shrub habitat with abundant black crowberry. On wintering grounds uses a wide variety of open areas, including saltpans, lagoons, beaches, reefs, and grassy areas. May occasionally perch in trees. Forages mostly visually, walking steadily, then picking, probing, or chasing when prey items are found. May "slam" selected prey items against hard surface to kill or dismember them. Also may slam piece of stone or coral onto large seabird eggs to break them open. Such tool use is unique among shorebirds. On breeding grounds eats primarily insects and spiders as well as fruit and other vegetation. Black crowberry and other fruits are primary food at late summer staging areas. On wintering grounds, eats a wide variety of intertidal and terrestrial invertebrates, seabird eggs and hatchlings, small lizards and rodents, carrion, and fruit. Typically seen in pairs or in small groups. Migrating flocks typically number 15–20.

Nest initiation primarily late May and early June. Egg dates range from mid-May to early July. Incubation about 26 days. Both adults share all parental duties. One brood per season. Peak hatching mid- to late June. Young can fly 21–24 days after hatching. The display flight, mostly by male, begins with a steep ascent on rapid shallow wingbeats to a height of 15–500 ft. (5–152 m), followed by a slow linear or circular glide as the song is delivered. Several variations of the flight display include paired flights, undulating flights, and a fluttering hover.

#### MIGRATION

Transoceanic migrant. Southern breeders from the Yukon-Kuskokwim Delta area make 2,500–3,700 mi. (4,000–6,000 km) nonstop flight to wintering areas in n. Oceania. Northern breeders from the Seward Pen. winter in more southerly areas and apparently use stop-over sites in central Pacific. Migrates by day and night.

Peak numbers of **spring** migrants gather in northern parts of wintering range in April, with departure in the first half of May. Spring migrants fly directly to breeding grounds, seldom stopping over elsewhere in Alaska. Arrival on breeding grounds takes place in early to mid-May, southern populations arriving before northern populations. Males arrive 1–6 days before females. Most subadults remain on the wintering grounds until 3–4 years old.

**Fall** migration takes place between June and October. Failed breeding **adults** depart the breeding grounds in mid-June, successful breeders by late July. Stages along coastal Yukon-Kuskokwim Delta between mid-June and August. Scarce elsewhere in w. Alaska during this period. Arrival at wintering areas takes place from late July to early September. Southbound **juveniles** depart the breeding grounds mostly in early August and depart Alaska from mid-August to early September. Arrival on the wintering grounds takes place mostly in late August and early September.

#### MOLT

All molting takes place on the wintering grounds. Only shorebird known to become flightless during molt, leaving them highly vulnerable to predators.

**First life year:** Head-and-body molt out of juvenal plumage begins from late September to mid-October and continues slowly throughout the year. Flight feather molt begins March/June and is completed in second life year.

**Second life year:** Flight feather molt is completed by July/October, though some birds retain juvenal outer primaries and their coverts. Some third-generation flight feathers are also molted from July/August to October/November, so second-year birds may have up to three generations of flight feathers simultaneously. Body molt continues virtually year round.

**Third life year:** Complete primary molt takes place July/August to October/November, one month ahead of adults. Body molt continues virtually year round.

**Adult:** In January/April some birds undergo a limited body molt involving body feathers, some coverts, and some tail feathers. In late July to late August, soon after arrival at wintering grounds, adults undergo a complete molt. Head-and-body molt begins first, followed by primary molt between early September and early December. Primaries are dropped rapidly, resulting in a flightless period of about 2 weeks in many individuals.

**VOCALIZATIONS**

Flight call is a clear, easily imitated whistle, *ee-o-weet;* similar to Black-bellied Plover but faster, higher, and repeated more frequently. Flight song is a whistled phrase, *wiwiwi-chyooo,* repeated several times with a variable number of low haunting whistles before and after. Other calls include variations on flight call and song elements.

## Far Eastern Curlew, p. 271

### *Numenius madagascariensis*

**STATUS**

Asian species. Casual in spring and early summer on w. Alaska islands. Accidental in British Columbia (one fall record). Uncommon within range and possibly declining. Breeds in ne. Asia, including Siberia, Kamchatka, Mongolia, and n. Manchuria. Winters primarily in Australia and less commonly in New Zealand. World population estimated at 38,000.

**TAXONOMY**

No subspecies recognized.

**BEHAVIOR**

A bird of coastal beaches, estuaries, and inland wetlands in winter. Breeds in peat bogs or swampy moorland with scattered bushes. During migration and winter frequents sandy beaches, muddy estuaries, or large inland wetlands. Feeds deliberately by probing deeply in substrate with its long bill (longest of any shorebird). Also picks from surface of mud or wet marshlands. Eats a variety of invertebrates, particularly burrowing insects and crustaceans. Typically feeds singly or in loose flocks, but known to number in the hundreds when moving to roosts or new feeding areas. Sometimes roosts communally in tall trees on islands or mangroves in winter.

Nests mid-May to mid-July.

**MIGRATION**

Long-distance migrant. Large postbreeding flocks migrate through the Philippines and New Guinea en route to wintering grounds in Australia and New Zealand. Limited records of vagrancy, with the exception of Aleutian Is. in spring.

**Spring** migration takes place between March and May, with peak movements through China in early April. Large numbers use Yellow Sea locations during spring migration, staging in early May at north end of Yellow Sea prior to departure for breeding areas. Departs wintering areas in Australasia late March/May, and arrives at breeding areas in May/early June. Many nonbreeders (probably **one-year-old** birds and older subadults) remain at wintering areas during northern summer months, while large numbers of nonbreeders spend summers at migratory stop-over sites, such as north end of Yellow Sea in China.

In **fall,** postbreeding flocks of **adults** start to gather in early July and depart for wintering areas by late July/August. Peak passage at Yellow Sea in China during August/September. Arrives at wintering areas in Australasia in September, with **juveniles** continuing into October.

**MOLT**

No information available.

**VOCALIZATIONS**

Flight call is a clear mournful whistle, *cuu-ree* or *cuu-reer,* often repeated several times; flatter and less melodic than Long-billed Curlew and more often repeated. Also gives a whistled *cui-ui-ui-ui,* more musical than Whimbrel.

## Slender-billed Curlew, p. 273

### *Numenius tenuirostris*

**STATUS**

Eurasian species. Accidental in Ontario. Extremely rare throughout range or possibly extinct. Breeds in sw. Siberia. Winters in Mediterranean Basin, especially

Morocco and Tunisia. World population believed to be no more than 50–200 birds.

**TAXONOMY**

No subspecies recognized.

**BEHAVIOR**

Breeds in Siberian taiga zone. Habitats include marshlands and peat bogs within steppe woodlands. During migration and winter found in a variety of habitats, including coastal mudflats, marshes, estuaries, fresh or brackish inland waters, and moist to dry grasslands. Often associates with flocks of other shorebirds, especially Eurasian Curlew and Black-tailed Godwit. Often feeds while wading in shallow water, probing into mud with longish decurved bill. Also feeds on mudflats and grasslands, picking food off surface or probing into soft ground. Food is chiefly invertebrates, including insects, crustaceans, mollusks, and worms. Historically seen in large flocks during migration and winter, but most recent sightings involve from 1 to 10 individuals.

Nesting season likely between late May and July.

**MIGRATION**

Fairly long-distance migrant. Movements overland in spring and fall from sw. Siberian breeding grounds to wintering areas in Mediterranean Basin.

**Spring** migration takes place between March and May, with peak movements in April. Departs wintering areas in March and probably arrives at breeding grounds in late May, based on sight records from Balkan states (especially Hungary) in May.

**Fall** migration takes place between August and November, with peak movements in September/October. Departure from breeding grounds begins in August, and passage to wintering areas is probably completed by November.

**MOLT**

**First life year:** The molt out of juvenal plumage probably begins soon after fledging and is mostly completed by October/December. This partial molt includes most head and body feathers and some wing coverts and tertials. Between late December and March there is a partial molt to breeding plumage, similar to adult's but more restricted, involving scattered feathers on crown, scapulars, chest, and flanks only.

**Second life year:** Information incomplete, but suggests that between June and August juvenal flight feathers, wing coverts, tertials, and outer tail feathers are replaced. After this molt, second-year birds are like adults. The partial molt to breeding plumage is as in adult.

**Adult:** The partial molt to breeding plumage is variable. It begins in December and is completed by March and includes head, mantle, most scapulars, breast-sides, and some flank feathers. Some birds molt all underparts and some tertials, upperwing coverts, and central tail feathers. Other birds replace only a scattering of feathers on crown, upperparts, breast-sides, and flanks. The complete molt to nonbreeding plumage is believed to take place from about June to January/February. Most flight feathers, some wing coverts, and some body and tail feathers are replaced rapidly on the breeding grounds from June to August. Some flight feather molt is suspended during migration. The rest of the molt to nonbreeding plumage is completed on wintering grounds before January/February.

**VOCALIZATIONS**

Flight call is shrill *kew-EE*, higher and shorter than Eurasian Curlew. Also a repeated *pi-pi-pi*, slightly higher and faster than Eurasian.

## Eurasian Curlew, p. 274

### *Numenius arquata*

**STATUS**

Old World species. Casual in fall and winter along Atlantic seaboard from New York to Greenland. Common within range. Breeds widely across Eurasia from Britain to central Siberia in temperate climates. Winters in coastal areas from Iceland through Africa to s. Japan and Indonesia.

**TAXONOMY**

Two subspecies are recognized. Nominate *N. a. arquata* breeds from Europe east to Urals and winters from Britain and w. Europe south to Mediterranean and w. Africa. Small numbers winter in Iceland and the Faroes. *N. a. orientalis* breeds from the Urals east to central Russia and winters from e. Mediterranean and Red Sea coasts east across Asia to Japan and Philippines. Compared to nominate birds, *orientalis* is slightly larger with more heavily marked rump and underwing and more streaked rather than spotted flanks. However, much variation exists, and many individuals are not identifiable to subspecies. North American records presumably pertain to *arquata.*

**BEHAVIOR**

A bird of coastal sands and mudflats, estuaries, and wet fields. Breeds in a wide variety of temperate habitats, including open grassy or boggy moorland and moist meadows in steppe zones. During migration and winter prefers mostly coastal habitats such as tidal mudflats, estuaries, sand beaches, and salt marshes. Winters inland at wetland habitats in some areas. Walks deliberately, like most large curlews. Feeds by deep probing of long bill in mud or wet soil. Also picks prey from surface of substrate or vegetation. Omnivorous, though prefers mostly invertebrates, including worms, crabs, mollusks, and insects. Also eats berries and sometimes eggs. Often seen singly, but may form large flocks of up to several thousand, particularly at roosts.

Nests late April to early July in southern areas, up to a month later in the north.

**MIGRATION**

Short- to medium-distance migrant. Some European breeders move very short distances, while eastern birds generally move farther south to locations in Africa, Philippines, and Japan. Both races winter in Mediterranean region.

**Spring** migration takes place between February and March for southern and western breeders; mainly April/early May for far northern breeders. Arrival at breeding areas mainly in March for southern breeders and early May for northern birds. Some nonbreeders remain at wintering areas through the summer, while others migrate north and summer near breeding grounds.

**Fall** migration begins in late June in the south and July/August in Russia, extending to October/November. **Adults** depart w. European breeding grounds in late June and journey to North Sea molting areas, where many birds remain to winter. Adults from Britain, France, and w. Germany depart by mid-July, and from Russia by July/August. Movement through e. Europe continues into October/November. Arrival at southern wintering grounds takes place from late July to November, with peak probably in September. **Juveniles** move to molt or winter locations in w. Europe much later, from mid- to late September.

**MOLT**

**First life year:** The partial molt out of juvenal plumage begins in w. Europe in mid-September and is mostly completed by late November, with little or no molt through the winter. This molt involves most head and body feathers. Some early-fledged birds may complete this molt as early as August. In eastern populations of *orientalis,* timing is similar to *arquata,* but some birds do not start molt until October. In central and se. Europe and w. Asia, molt to nonbreeding plumage is delayed until December and not completed until April. A partial molt to breeding plumage takes place from mid-February to April in w. Europe and from January to April in Asia. This molt is similar to adult's but is more restricted, involving only head, breast-sides, mantle, and some scapulars.

**Second year life:** For birds that oversummer on the wintering grounds the complete molt to nonbreeding plumage starts from late February/May (of first life year) and is completed after June. Other birds that summered near breeding areas are on a more adultlike schedule, though slightly earlier. After this molt, second-year birds appear similar to adults. The partial molt to breeding plumage is as in adults.

**Adult:** Between January/February and April there is a variable partial molt to breeding plumage. This molt may involve only a few head and body feathers or may include all of head and body, plus some tail feathers, tertials, and wing coverts. The complete molt to nonbreeding plumage is variable among populations. In *arquata* it takes place from early July/mid-August to early October/mid-November. Some primary molt may begin on the breeding grounds, but molt takes place largely at staging/wintering areas. In *orientalis,* which generally has a longer migration, this molt takes place later, from July/mid-September to December, also mostly at wintering areas. Birds that undertake longer migration may start body molt on the breeding grounds in July/August, but primary molt takes place entirely on the wintering grounds (September–December).

#### VOCALIZATIONS

Flight call is a rich, whistled *courrr-E,* slightly lower and flatter than Long-billed with a more drawn-out first syllable. Also a liquid, melodic, two- to four-note *coy-coy-coy,* of a similar cadence to Spotted Sandpiper but much lower and richer.

## Long-billed Curlew, p. 111

***Numenius americanus***

#### STATUS

Fairly common but declining. Breeds in the Great Plains, Great Basin, and intermountain valley regions from s. British Columbia and sw. Saskatchewan south to e.-central. California and n. Texas. Winters from Washington and Florida (rarely to North Carolina) south through Mexico, rarely to Costa Rica. Now very rare along the Atlantic Coast where formerly a common migrant as recently as the middle of the nineteenth century. The world population is estimated at 20,000.

#### TAXONOMY

Two poorly defined subspecies are recognized. Nominate *N. a. americanus* breeds primarily in southern and eastern portions of range; *N. a. parvus,* in northern and western portions. On average, *parvus* is smaller and shorter billed, but much overlap exists, and some authorities question the validity of these subspecies.

#### BEHAVIOR

Breeds in short-grass and mixed-grass prairie habitats with flat or rolling topography and little or no tall vegetation. During migration and winter uses mostly wet pastures, marshes, and tidal mudflats. Less common on sandy beaches than Marbled Godwit or Willet. Forages by walking steadily, picking at surface or probing deeply in soil or mud. Eats a variety of invertebrates, including insects, crustaceans, mollusks, and worms, as well as fish and other small vertebrates. Bill length and curvature adapted for foraging on earthworms in pastures and on burrow-dwelling crabs and shrimp on tidal mudflats. Typically seen in small groups of 5–10 individuals but may form flocks of 50 or more.

Nest initiation primarily mid-April to mid-May. Egg dates range from late March to early July. Incubation 27–31 days. Both adults share parental duties. Female abandons mate and young after 2–3 weeks. One brood per season. Peak hatching mid-May to mid-June. Young can fly after 38–45 days. Age of first breeding 2–4 years. The male's undulating display flight begins with a steep ascent (45° angle) on rapid fluttering wingbeats up to 30–50 ft. (9–15 m), followed by a long descending glide as it gives song. This flight path is repeated several times before the bird touches down.

#### MIGRATION

Short-distance migrant. Family groups may sometimes migrate together.

**Spring** migration takes place between early February and mid-May. Peak numbers pass through the Southwest from mid-March to mid-April, with arrival at most northern breeding areas by mid- to late April. Some nonbreeders, presumably subadults, remain at wintering sites through the summer.

**Fall** migration takes place between early June and mid-December. Females and failed breeders first arrive at wintering sites along the Pacific Coast in mid-June and the Gulf Coast in early July. Peak numbers pass through the n. Plains from mid-July to late August and the s. Plains and Southwest in September. Arrival at southernmost wintering areas in Central America may be as late as mid-December. Differential timing between **adults** and **juveniles** poorly known.

#### MOLT

**First life year:** The partial molt out of juvenal plumage takes place from September/October to about December/January. Molt is usually limited, including a variable number of head and body feathers and innermost secondaries, but precise extent and timing poorly known. In spring there is another limited partial molt of head and body feathers, but extent and timing poorly known; this molt may be a continuation of the post-juvenal molt. First-year birds in spring are best distinguished from adults by retained worn juvenal flight feathers.

**Second life year:** Molts probably similar to adult's, but poorly known. The complete molt to nonbreeding plumage probably begins earlier than adult's.

**Adult:** Between February and June there is a limited molt involving head and body feathers along with many wing coverts and tail feathers. The complete molt to nonbreeding plumage takes place between August and November. The molt begins during migration and is completed on the wintering grounds.

#### VOCALIZATIONS

Flight call is a clear, whistled *cuur-lEE,* often combined with a rapid *qui-pi-pi-pi-pi,* very similar to Whimbrel but huskier. Given almost exclusively by flying birds and heard regularly in nocturnal migration, particularly in spring. Flight song involves several low haunting whistles and trills, culminating in a repeated *werr-EEEer,* similar to flight call but more drawn out and more pure toned.

### Black-tailed Godwit, p. 276

***Limosa limosa***

#### STATUS

Eurasian species. Rare but regular spring migrant on w. Alaska islands; casual elsewhere in w. Alaska and along

Atlantic Coast; accidental in Ontario and Louisiana. Common within range. Breeds widely across Europe from Iceland and France east to w. Asia and more sparsely to e. Asia. Winters from w. Europe and Africa east to Indonesia and Australia.

#### TAXONOMY

Three subspecies recognized. Nominate *L. l. limosa* breeds from w. Europe eastward across Russia and winters in sub-Saharan Africa and n. India. *L. l. islandica* breeds in Iceland, Norway, and Scotland and winters in Ireland, Britain, and w. France. *L. l. melanuroides* breeds from e. Siberia south to Mongolia and at scattered locations in e. Asia; winters from e. India south to Australia.

Nominate *limosa* is largest in size (becomes gradually larger from west to east) and has intermediate brightness in breeding plumage. *L. l. islandica* is slightly smaller and shorter billed and has the most extensive rufous below in breeding plumage. *L. l. melanuroides* is smallest with the dullest breeding plumage and the darkest nonbreeding plumage. Their bills are finer-tipped than other subspecies' and often subtly drooped.

#### BEHAVIOR

Breeds in a variety of habitats in lowland temperate and boreal zones, including damp hummocky moorlands and wet meadows. During migration and winter frequents a variety of wet habitats, including coastal estuaries, mudflats, inland lakeshores, flooded grasslands, and inland freshwater marshes. Walks slowly, often in deep water, with bill vertical while probing slightly to find food. If prey is located, will probe deeply to secure food. Also probes soft soil and picks food from surface of ground and vegetation. Eats a variety of invertebrates, especially insects and their larvae, worms, mollusks, crustaceans, and tadpoles. Also eats vegetation during migration and winter. Often forms large wintering flocks involving many thousands of birds.

Nests early April to late June.

#### MIGRATION

Moderate- to long-distance migrant. *L. l. islandica* moves moderate distance to wintering areas in n. Europe, while other two races migrate farther over a broad front, mainly overland. Migration is characterized by long flights between scattered staging areas to wintering grounds.

**Spring** migration takes place primarily between February and April. Departs wintering areas during February and arrives at breeding areas in Iceland in late April; in w. Europe mid-March to mid-April; and in n. areas of Asia in April to early May. Most **one-year-old** birds remain at wintering areas through the summer.

**Fall** migration takes place from late June to October. **Adults** depart breeding grounds late June to September, with e. Asian birds slightly later. Peak passage of *islandica* to n. Europe from mid-August to mid-September. Nominate *limosa* begins departure from European breeding grounds in late June, with concentrated exodus in July. Migration through Europe occurs mid-July to September, with trans-Saharan crossing taking place from mid-August onward, peaking during second half of September. Arrives at wintering areas mainly in October. E. Asian *melanuroides* moves slightly later than nominate. Arrival of first **juveniles** in African wintering grounds from 1 to 2 weeks later than adults.

#### MOLT

**First life year:** The molt out of juvenal plumage begins in August and is mostly completed by November on the wintering grounds. It includes most head and body feathers, inner tertials, some wing coverts, and some or all tail feathers. General nonbreeding appearance is mostly acquired by September. Between March and June there is a variable head-and-body molt to breeding plumage for birds that migrate to locations near the breeding grounds. This molt is typically more extensive in males than females and includes some to most of head and body. Molt typically less extensive than adults', with most tertials, wing coverts, and tail feathers probably retained. Birds that remain on wintering grounds typically skip this molt or molt into partial breeding plumage. These birds may start the molt to nonbreeding plumage from early May onward.

**Second life year:** For birds that over-summer on wintering grounds, the complete molt to nonbreeding plumage begins in early May (of first life year) and is probably completed by July. Other birds begin molt about one month earlier than breeding adults (typically by late May). Molt is variable, as in adult, with some birds attaining nonbreeding appearance before migration and others waiting until wintering areas are reached before beginning substantial portion of molt. After this molt, birds are on same schedule as adults.

**Adult:** Between January/February and April there is a partial molt to breeding plumage involving most head and body feathers as well as some tertials, coverts, and tail feathers. This molt is more extensive in males than females. The complete molt to nonbreeding plumage takes place between May and December. Limited body molt begins on the breeding grounds as early as May, but most molt takes place from late June onward at stop-over sites or from late July/August to December at wintering areas. Birds from center of continent tend to molt later and mostly on the wintering grounds.

#### VOCALIZATIONS

Noisy on breeding grounds, but mostly silent in migration and winter. Flight call is a nasal, creaking *ke-weeku,* often repeated.

## Hudsonian Godwit, p. 115

***Limosa haemastica***

#### STATUS

Uncommon along upper Texas/Louisiana coast and along narrow central U.S. flyway during spring migration; locally scarce along upper Atlantic Coast in fall. Rare elsewhere. Surprisingly little is known about this large, striking sandpiper due in part to its remote breeding and wintering areas. Breeds in small numbers in scattered tundra locations from Alaska to Hudson Bay, with many breeding sites still unaccounted for. Winters in concentrated numbers at a handful of locations on the Atlantic side of s. South America, which subjects population to potentially dangerous crashes. North American and world population estimated at 50,000.

#### TAXONOMY

No subspecies are recognized.

#### BEHAVIOR

Breeds in sedge meadows and muskeg habitats at widely scattered locations near treeline in Canada and Alaska. During migration and winter prefers various freshwater and saline habitats. During winter in s. South America substantial numbers concentrate at sites with large expanses of coastal mudflats. Feeds by probing long bill into mud, locating food by touch. When in deeper vegetated ponds, submerges head while probing underwater substrates, similar to dowitchers. Occasionally feeds by sight, picking food from mudflat or base of vegetation. Diet on breeding grounds consists mainly of insect larvae, beetles, and various other invertebrates, while plant tubers make up major portion of diet in fall migration. During winter, diet includes worms, insects, crustaceans, and various other items. Typically seen in small numbers in fall, but may occur in flocks of up to 100 or more in spring rice fields of sw. Louisiana and e. Texas. Numbers on East Coast have declined over last 30 years. During migration, occurs most often with typical "flooded field" shorebirds such as Long-billed Dowitcher, Stilt Sandpiper, Lesser Yellowlegs, and Wilson's Phalarope.

Nest initiation presumably occurs mid-May in s. Alaska, later at inland locations. Egg dates mid-May to mid-July. Incubation 23.5 days. Both adults share parental duties of incubation and brood rearing. Single brooded. Peak hatching in mid-June in s. Alaska and early July in Churchill, Manitoba. Young can fly after approximately 26 days. Courtship involves pursuit flight by female followed by male, and a quivering flight that involves both sexes, with birds hovering above nest area.

**MIGRATION**

Long-distance migrant, undertaking spectacular nonstop flights in fall from Canada to n. South America, and possibly farther. Absence of known staging area in n. South America suggests possible nonstop flights of up to 5,000 mi. (8,000 km), similar to those of Bar-tailed Godwit.

**Spring** migration occurs from March to early June. Northbound movements through South America are incompletely known; numbers peak in March in s. Brazil, with most gone by late April. Uncommon in Mexico from mid-April to May along Pacific and Gulf coasts. Texas coast arrival dates late March to late May, with peak concentrations in mid-May. Large concentrations at small number of staging areas from n. Texas to Canada in central flyway from early April to late May, with peak numbers varying from late April to late May, depending upon latitude. Arrival on breeding grounds in s. Alaska in last week of April, peaking in second week of May. Earliest arrival on Seward Pen. is mid-May. Churchill arrivals average late May/early June, peaking in early to mid-June. Some **one-year-old** birds travel to breeding grounds, while others remain in Southern Hemisphere.

**Fall** migration takes place from late June to November. Southbound **adults** depart breeding areas from late June (mostly males) to mid-July (mostly females) for a number of critical staging areas in Alaska and Canada. **Juveniles** and most females arrive at these sites from late July to mid-August. Alaskan adults start departing for South America by late July, with most juveniles leaving by late August. Relatively few birds are seen away from staging areas in migration, with flight path to South America unknown. Hudson Bay nesters stage at the south end of James Bay in large numbers, with adults arriving in July and departing by late August. Juveniles start to gather in early August and depart mid-September to mid-October. Nonstop flight to South America takes birds over Maritime Provinces of Canada and New England and out over w. Atlantic. Low numbers are regularly seen in these areas, with fairly large numbers occurring after northeasterly storms or winds. Sightings are infrequent to rare inland, with the exception of a few locations. East Coast sightings occur from mid-July to November, with most records mid-August to mid-September.

**MOLT**

**First life year:** Molt out of juvenal plumage begins in October during migration and is completed after arrival on wintering grounds. Molt involves head and body feathers, with retention of flight feathers and some coverts, tertials, scapulars, and mantle feathers. Molt to first breeding plumage is partial but poorly known. Individuals that migrate north to the breeding grounds seem to be on a similar molt schedule as adults, though some attain only partial breeding plumage. Others that remain in South America retain much of nonbreeding plumage into first summer. One-year-old birds are distinguished from adults by retained worn juvenal primaries and inner wing coverts.

**Second life year:** For birds remaining on the wintering grounds, the complete molt to nonbreeding plumage likely occurs earlier than in adults, but timing poorly known. For birds that summered on the breeding grounds, this molt is as adult's.

**Adult:** Between February/early March and mid-April, adults quickly undergo a partial molt to breeding plumage, including most head and body feathers, tertials, and some wing coverts. Molt begins on wintering grounds and is mostly completed before arrival in North America in mid-April. The complete molt to nonbreeding plumage takes place from July to November. Body molt starts on breeding grounds and is often mostly complete by August or September (sometimes not complete until November). Fall migrants typically show extensive head-and-body molt but fully retained wing coverts, tertials, and flight feathers. Primaries are not replaced until southbound migration is completed.

#### VOCALIZATIONS

Flight call an emphatic, piping *pEEd-wid,* with each syllable rising. Also a high *peet* or *kwee,* similar to Black-necked Stilt's but softer. Display song begins with several drawn-out, "worried" whistles followed by a rapidly repeated two- to three-syllable phrase, *to-WIda to-WIda to-WIda* . . . or *to-WIT to-WIT to-WIT to-WIT* . . . , with a desperately emphatic quality.

### Bar-tailed Godwit, p. 278

***Limosa lapponica***

#### STATUS

Primarily Old World species. Uncommon to locally common breeder in w. Alaska. Rare to casual migrant in Pacific Northwest. Casual along Atlantic Coast. Widespread in Old World. Breeds at high latitudes from Norway east to Siberia and Alaska. Winters in coastal areas from w. Europe and Africa east to se. China, Australia, and New Zealand. World population estimated at about 1.2 million, with 120,000 in Alaska.

#### TAXONOMY

Two subspecies are recognized. Nominate *L. l. lapponica* breeds in Scandinavia and w. Siberia and winters in Europe, Africa, and w. India. *L. l. baueri* breeds in e. Siberia and w. Alaska and winters from s. China south to Indonesia, Australia, and New Zealand. Nominate race shows mostly white underwing, rump, and lower back, whereas these areas are heavily barred brownish in *baueri.*

#### BEHAVIOR

Breeds on lowland tundra, coastal wetlands, foothills, and uplands of Arctic and subarctic regions, occasionally in areas of scattered trees. During migration and winter prefers mudflats, coastal estuaries, ocean shorelines, and interior wetlands. Feeds by steadily walking in shallow water or wet mud, probing substrate with long bill. Often submerges head under water, burying long bill to the hilt in muddy substrate. Diet very diverse. On Alaskan breeding grounds includes berries from previous summer and mostly insects, especially beetles, along with adult crane flies and their larvae. Some birds from n. Alaska riparian zone consume numerous land snails. During critical Alaska staging area, diet is almost exclusively marine bivalves and worms. In nonbreeding grounds in Australasia diet includes seeds, marine worms, crustaceans, insect larvae, grains, and small fish. Seen singly, in small flocks, and often in large flocks. Numbers may be substantial in favored nonbreeding locations, with sightings of up to a half million birds in Mauritania. Associates with numerous other coastal shorebirds in migration, especially Hudsonian and Marbled Godwits, Black-bellied Plover, Red Knot, Dunlin, and Whimbrel.

Nest initiation in Alaska varies from mid-May on Yukon-Kuskokwim Delta to late May on Seward Pen. to mid- to late June in n. Alaska. Early egg dates follow same timetable and range occasionally to mid-July in n. Alaska. Incubation period 20–21 days. Both parents share parental duties, with either parent taking major share of incubation. Little known about brood rearing, but both parents accompany young. Single-brooded. Peak hatching varies in Alaska from mid-June to mid-July. Young can fly at about 30 days. A great variety of displays exhibited during courtship, both aerial and terrestrial. Displays similar to those of other godwit species.

**MIGRATION**

Complete long-distance migrant, with Alaskan population undertaking most spectacular nonstop migration of any animal, a 6,000+ mi. (9,700+ km) nonstop flight over open ocean during fall migration to wintering grounds in Australia and New Zealand.

**Spring** migration complex and incompletely known. *L. l. baueri* begins to depart wintering areas in New Zealand and Australia as early as February, with peak departure in March and early April. Data indicates two long flights from New Zealand/e. Australia, first to the central e. Asian coast, then on to breeding grounds. First arrivals at Alaskan staging areas (Alaska Pen. estuaries) take place during last week of April. Arrives on w. Alaska breeding areas in early to mid-May, n. Alaska by early June. The few spring records along the U.S. Pacific Coast span from mid-February to early June. Most **one-year-old** birds remain at wintering areas through the summer and apparently do not breed in their first year. Nominate *lapponica* starts to return to staging areas in March/April for prebreeding molt and departs for breeding grounds in May. Spring records along the U.S. Atlantic Coast span from late April into June.

**Fall** migration also needs more study. For Alaskan *baueri,* movements to major staging areas on the Yukon-Kuskokwim Delta take place July/August, with **adults** preceding **juveniles.** Numbers peak to over 100,000 in September (sometimes early October), after which mass departures occur. Arrival via nonstop flight over Pacific in Australia/New Zealand takes place in August to November, with peak arrival of adults in late September and juveniles in October. Fall records along the U.S. Pacific Coast span from early July to late November, with a clear peak of mostly juveniles in September. For nominate *lapponica,* migration occurs mostly from July to October. In w. Europe, peak passage of adults takes place in July/August, juveniles from August to October. Fall records along the U.S. Atlantic Coast span from June to late November, with slight peaks of adults in July/August and juveniles in September and early October.

**MOLT**

**First life year:** Molt out of juvenal plumage takes place from mid-September to February and includes most head and body feathers and some wing coverts and tertials. During April and May there is a partial molt to breeding plumage, involving some head and body feathers. This molt may be as extensive as in adults but is usually more restricted and is sometimes skipped altogether. Juvenal underwing coverts, tertials, and flight feathers are retained, which distinguish one-year-old birds from adults. Data from Australia shows that most birds attain about 25 percent adult breeding plumage, but none migrate to breed.

**Adult:** Between late February and May adults undergo a partial molt to breeding plumage including much of head and body. *L. l. baueri* begins molt on wintering grounds and completes at e. Asian staging areas. Females typically show much less extensive molt than males. The complete molt to nonbreeding plumage occurs from July to February, beginning with molt of body feathers on Alaska breeding and staging grounds. Flight feathers are molted from late August to early February on the wintering grounds.

#### VOCALIZATIONS

Flight call is an emphatic, slightly nasal *klEY-did* or *klEY-dey-did;* lower and less piping than Hudsonian. Display song begins with several drawn-out whistles followed by a rapidly repeated two- to three-syllable phrase, *ta-WEA ta-WEA ta-WEA . . .*; much like Hudsonian's song but lower and less shrill.

## Marbled Godwit, p. 119

### *Limosa fedoa*

#### STATUS

Fairly common in the West and South, particularly at coastal wintering areas. Generally scarce and local along Atlantic Coast. Breeds in n. Great Plains from Alberta and sw. Ontario south to Montana and South Dakota. Small isolated populations breed along coast of sw. James Bay in Canada and on Alaska Pen. Winters along Pacific Coast from s. Washington south to Costa Rica, and in smaller numbers along Gulf and Atlantic coastlines, from New Jersey south to the Yucatán Pen. North American and world population estimates vary between 140,000 and 200,000.

#### TAXONOMY

Two subspecies recognized. Nominate *L. f. fedoa* breeds in Great Plains and around s. James Bay and winters along most temperate coastlines in North and Central America. *L. f. beringiae* breeds on Alaska Pen. and probably winters south along Pacific Coast to San Francisco. Little known about migration routes and extent of winter range. The recently described *beringiae* averages larger with proportionally shorter wings and legs, but these differences are unlikely to be recognizable in the field.

#### BEHAVIOR

Breeds in short to sparsely vegetated grasslands, interspersed with wetland complexes. During migration and winter prefers sandy beaches, coastal mudflats with adjoining meadows or savannas, and inland wetlands and lake edges. Surprisingly tame while sharing s. California beaches with bathers. Walks deliberately while probing substrate for food. Also picks insects from ground or vegetation as well as fish from shallow water. Sometimes feeds in shallow water, burying bill to the hilt in mud and submerging head underwater. Diet is diverse. In breeding season feeds mostly on insects, but also aquatic plant tubers, leeches, and small fish. During the remainder of the year eats marine worms, small mollusks, fish, and crabs in coastal habitats, as well as earthworms in flooded fields. Typically seen in small flocks, with larger numbers occurring at favored wintering areas along the California and Baja coastlines. Often associates with similar-sized Western Willet, Whimbrel, and Long-billed Curlew, but also with most coastal shorebirds on mudflats and estuaries.

Nest initiation primarily mid-May to early June. Egg dates range from early May to early July. Incubation 23–24 days. Both parents share incubation, and either parent may brood young. Single brood per season. Most nests hatch early June to early July, with peak around mid-June. Young can fly at about 30 days. Displays include high flights that result in circling territory on slow wingbeats while calling, then gliding on stiff wings to ground. Also undulating joint flight of both partners prior to incubation. Normally both parents reunite on breeding grounds in subsequent years, but small numbers change mates.

**MIGRATION**

Relatively short- to long-distance migrant, with movements from primary interior breeding grounds to temperate North American and Central American coastlines.

**Spring** migration takes place between late March and early May. Northbound adults start to leave wintering grounds in late March in southern areas, peaking along coastlines in April and in interior mid-April to mid-May. Arrival on prairie breeding grounds takes place mid- to late April and possibly several weeks later in James Bay and Alaska. Some birds (presumably one-year-olds) remain at coastal wintering sites during summer months. Limited data documents that some **one-year-old** birds return to breeding grounds, but none apparently breed.

**Fall** migration takes place between early July and December. Central interior **adults** form large postbreeding flocks in early July and remain through August, with numbers of birds migrating from mid-July through September. **Juveniles** migrate several weeks after adults, arriving on California coast into December. Atlantic Coast migrants peak late July to late September, with some birds arriving in Florida into early December. Most James Bay birds presumed to winter along se. coast of United States, with direct flight to that area, but data incomplete. Most birds from Alaska Pen. typically remain until late September, after which they move to wintering areas from British Columbia south to central California coast.

**MOLT**

**First life year:** Molt out of juvenal plumage takes place rapidly, mostly between August and October, and includes head and body plumage and most wing coverts and tertials. Molt to first breeding plumage undescribed but likely more limited and variable than in adult. One-year-old birds may be distinguished from adults by retained worn juvenal primaries.

**Second life year:** The complete molt to nonbreeding plumage is poorly known but likely takes place earlier than in adults. After this molt, birds are on the same schedule as adults.

**Adult:** Between February and March adults undergo a partial molt to breeding plumage, including most head and body feathers and probably some wing coverts and tail feathers as well. The complete molt to nonbreeding plumage occurs from July to November. Flight feather molt takes place on the wintering grounds.

**VOCALIZATIONS**

Flight call a strong, nasal, laughing *kwEH-wed* or a single cracking *kwEH!;* lower pitched than other godwits. On breeding grounds, gives a rapidly repeated, nasal *ga-WIda ga-WIda ga-WIda ga-WIda* . . . , particularly during territorial chases.

### Ruddy Turnstone, p. 123

***Arenaria interpres***

**STATUS**

Common visitor along sand or gravel seacoasts throughout North America. Generally rare inland except through the prairies and Great Lakes, where sightings are uncommon. One of the most widespread shorebirds in the world, occurring on every continent except Antarctica. Breeds in high Arctic regions throughout the Northern Hemisphere. Winters along almost all temperate and tropical coastlines from Washington and Newfoundland south to central Chile and central Argentina. Also in the Old World from the British Isles and Denmark south through Africa and east to Taiwan, Australia, and Oceania. The world population is estimated at about 450,000, with 235,000–267,000 in North America.

**TAXONOMY**

Two subspecies recognized. Nominate *A. i. interpres* breeds in high Arctic regions from northern parts of the Canadian Arctic Arch. and n. Greenland east across Arctic Eurasia to w. Alaska. Winters in the Old World as well as along the North American Pacific Coast. *A. i. morinella* breeds from ne. Alaska (Beaufort Sea coast) east to Baffin I. and probably w.-central Greenland. Winters along Pacific and Atlantic coasts, south through the Caribbean and West Indies and along both coasts of South America. Differences in appearance are slight but often distinct. Compared to nominate birds, *morinella* shows brighter upperparts (particularly wing coverts) and less streaked crown as well as slightly shorter wings, longer bill, and longer tarsus. Plumage differences are most noticeable in breeding adults.

**BEHAVIOR**

Breeds in Arctic regions on rocky coasts and tundra, usually near water. During migration and winter prefers gravel or sand beaches and rocky shorelines, where it is typically found investigating the high-tide line. Walks quickly, stopping to overturn seaweed and debris or dig in sand or rock crevices in search of food. Diet extremely diverse. During the breeding season eats primarily insects as well as some plant material. During the remainder of the year eats crustaceans, mollusks, snails, worms, fish, carrion, as well as human garbage. Typically seen singly or in small flocks. In spring may gather in groups of over 1,000 in some areas. Associates with a wide variety of other shorebirds. Along rocky coastlines associates with other "rockpipers" such as Purple and Rock Sandpipers, Black Turnstone, and Surfbird. In spring forms large mixed flocks with Sanderling, Semipalmated Sandpiper, and Red Knot.

Nest initiation primarily early June in much of Alaska, mid-June in Canada and on Alaska's North Slope. Egg dates range from late May to early August. Incubation 21–23 days. Both adults share parental duties, with females doing most incubation and males most brood-rearing. Single-brooded. Peak hatching late June in most of Alaska, mid-July in Canada and Alaska's North Slope. Young can fly after 24–26 days. The display flight, performed by male at a height of 30–150 ft. (9–46 m), involves short bursts of slow, deep wingbeats interrupted by short glides with wings bowed down. On the ground both sexes may perform aggressive displays with back hunched, feathers ruffled, tail lowered and vibrated, and continuous calling.

**MIGRATION**

Long-distance migrant, many populations making long transoceanic flights.

**Spring** migration takes place between early March and mid-June. Northbound adults depart wintering grounds mostly in late March. Peak migration takes place along the Gulf Coast in April and along the Pacific Coast between mid-April and mid-May. Atlantic Coast population stages in large numbers at Delaware Bay between late April and early June, with peak numbers there in late May. Small concentrations around the Great Lakes take place mostly in the second half of May. Migrants around Hudson Bay peak in early June. Arrival at the breeding grounds takes place in late May and early June (w. Alaska) and early to mid-June (Canada). Most **one-year-old** birds remain at wintering areas through the summer and apparently do not breed in their first year.

**Fall** migration takes place between early July and early December. Southbound **adults** depart breeding areas mostly between late July and mid-August. Failed breeders may begin to head south by early July. The first fall migrants appear in early July on the West Coast and mid-July on the East Coast. Adults pass through much of the continent in August, peaking slightly earlier on the West Coast. Most **juveniles** depart the tundra between mid-August and early September. Juveniles first arrive along the Pacific Coast in early August, peaking there from mid-August to late September. Elsewhere in the continent juveniles arrive in mid- to late August and peak in September and early October.

### MOLT

**First life year:** Molt out of juvenal plumage takes place primarily between late September and November and includes most head and body feathers, some tertials, wing coverts, and tail feathers, and possibly some primaries. Between about April and early June there is a partial molt involving mostly head, neck, and body feathers and some wing coverts. This molt is highly variable resulting in a range of plumages from one like breeding adult (mostly birds that return to the breeding grounds) to one much like nonbreeding (mostly birds that remain on the wintering grounds). One-year-old birds may be distinguished from adults by retained juvenal wing coverts and flight feathers.

**Second life year:** For birds remaining on the wintering grounds the complete molt to nonbreeding takes place mostly between June and September. After this molt, birds are on the same schedule as adults. For birds that summered on the breeding grounds the fall molt is later, like that of adult.

**Adult:** Between late February and June (mostly mid-March to mid-May), adults undergo a partial molt to breeding plumage, including most head and body feathers and some wing coverts (more on males than females). The complete molt to nonbreeding plumage takes place between July/October and November/February. Head, neck, and some body molt begins on the breeding grounds or at stop-over sites. The molt may be completed either at stop-over sites or after arrival on the wintering grounds. Longer-distance migrants of both subspecies tend to molt later (from 2–12 weeks later) and mostly on the wintering grounds.

### VOCALIZATIONS

Usually highly vocal. Flight call is a low, rapid *cut-a-cut,* sometimes extended into a long chatter; many variations including higher, scratchy chattering notes. Alarm call is a harsh, squeaky *keeu.* Also, especially on the breeding grounds, a chattering *ti-woy-ti-woy-ti-woy-tititititititititi.*

## Black Turnstone, p. 128

### *Arenaria melanocephala*

**STATUS**

Locally common along Pacific Coast. Breeds only in Alaska, mostly within a little over 1 mi. (2 km) of coast. Winters in coastal areas and nearshore islands from s. Alaska (west to Kodiak Is.) south to s. Baja California, and along shorelines of the Gulf of California to s. Sonora, Mexico. Numbers of birds determined by suitable rocky habitat within winter range. World and North American population estimated between 80,000 and 95,000, with over 80 percent of total breeding population on Yukon-Kuskokwim Delta in Alaska.

**TAXONOMY**

No subspecies recognized.

**BEHAVIOR**

Most breed in coastal sedge meadows throughout w. Alaska, although some found well inland along rivers or lakeshores. Northernmost birds nest away from the coast along alpine gravel streambeds up to 2,000 ft. (610 m) above sea level. Strictly coastal during migration and winter, preferring tidal zone of rocky shorelines, sand and gravel beaches, intertidal mudflats, and man-made jetties. Forages by using uniquely shaped bill to flip or turn over stones, mud, algae, and other detritus found on wave-washed locations to expose prey living on or beneath these objects. May also "hammer" crustaceans or barnacles to loosen them or extract soft tissue. Typically in small flocks of several to a few dozen, with possible migrant concentrations of up to 1,000 birds.

Nest initiation primarily late May/early June. Egg dates range from late May to early July throughout range. Incubation averages 22–24 days, with both parents attending. Brooding is shared equally, with small percentage of females deserting brood after hatching. Typically one brood but may renest if failure occurs in early stages of incubation. Young attain flight at 25–34 days after hatching. Breeding behavior includes elaborate ground and aerial displays as well as aggressive nest protection, with aerial predators subjected to shrieking vocalizations and strong, relentless pursuit, often resulting in bodily contact or pulling of feathers.

**MIGRATION**

Distances covered range from short to intermediate from breeding sites in w. Alaska to coastal Pacific wintering sites from Alaska to Mexico. Migrates in flocks of a few birds to several hundred, probably diurnal at temperate latitudes.

In **spring,** northbound migration occurs along the Pacific Coast between mid-March and mid-May, with more southerly wintering birds departing earlier. Exact timing of northern birds difficult to determine due to addition of southern migrants. Principal staging area for large portion of population before dispersal to breeding areas is Prince William Sound, Alaska. Birds arrive in last week of April and depart after first week of May. Breeding areas on Yukon-Kuskokwim Delta are reached in early May, with more northern areas from mid-May to early June. Some one-year-old birds return to breeding grounds but unclear as to what proportion breeds. Others remain on wintering grounds or along northern migration route.

In **fall,** southbound **adults** depart breeding grounds from mid-June to early September, with failed breeders departing in mid-June followed by successful breeders. Females depart before males. Movements determined by breeding success. Migration from w. Alaska is south along the coast and across base of Alaska Pen. to Gulf of Alaska. Large numbers using Prince William Sound may bypass se. Alaska en route to temperate wintering areas. Southbound movements in British Columbia begin in mid-June, with migrants common along U.S. Pacific Coast by mid-July. Southern migration continues into mid-October, with some birds moving later. Most **juveniles** fledge by mid-July and begin migration in mid-August through early September, with some continuing into November.

**MOLT**

**First life year:** Molt out of juvenal plumage takes place between August/September and November/December. Much of this molt is apparently completed rapidly upon arrival on the wintering grounds. Molt includes some head and body feathers and possibly a few wing coverts and tertials. Flight feathers and some wing coverts are retained. In late winter there is a partial molt involving most head and body feathers but not flight feathers. One-year-old birds may be distinguished from adults by retained juvenal flight feathers.

**Second life year:** The complete molt to nonbreeding plumage takes place between late June to late November, but most actively August/October.

**Adult:** Between January and May adults undergo a partial molt to breeding plumage, including most head and body feathers, tertials, and some wing coverts. The complete molt to nonbreeding plumage takes place between June (rarely May) and November. Limited head-and-body molt takes place on the breeding grounds, but heavy molt does not begin until birds are away from the breeding grounds, mostly south of Alaska. Primary molt begins in July and finishes in 2–3 months.

**VOCALIZATIONS**

Flight call is a shrill, rattled *breerp*, often extended into a chatter; much higher pitched than Ruddy Turnstone. Also gives a variety of purring sounds, trills, and a *tu-whit* call given by both sexes. On breeding grounds male gives a long, repeated series of staccato calls.

## Surfbird, p. 131

### *Aphriza virgata*

**STATUS**

Locally fairly common along the Pacific Coast. Breeds in mountainous regions of Alaska and Yukon Territory. Winters strictly along the Pacific Coast from s. Alaska to n. Mexico and locally south to Chile. Very rare inland. The world population is estimated at 70,000–100,000.

**TAXONOMY**

No subspecies recognized.

**BEHAVIOR**

Except during the brief breeding season, this species is strictly coastal. Breeds on dry stony alpine tundra, often on steep slopes. During migration and winter closely tied to rocky, wave-pounded coastlines with mussel beds and marine algae. Less often uses sheltered rocky shores or tidal flats near rocks. Usually very active while foraging. Walks continuously over rocks, stopping occasionally to peck or tug at prey items. Short stout bill well adapted for gripping firmly attached prey items. On the breeding grounds runs quickly to chase insects. Eats primarily insects on the breeding grounds and intertidal invertebrates, particularly mollusks and barnacles, in the nonbreeding season. Apparently does not eat fruit as most tundra-nesting shorebirds do. Typically found in small groups, often with Black Turnstones. At spring concentration points may gather in flocks of a few hundred to a few thousand.

Nest initiation primarily late May and early June. Egg dates range from mid-May to early July. Incubation probably about 22–24 days by male. Unknown if female incubates. Brood tended by both adults but mostly male. Peak hatching late June. Fledging period unknown. The display flight covers a large area, apparently not a discrete territory, and occurs 100–150 ft. (30–46 m) above ridgetops or hundreds of meters above adjacent valleys. Flights involve alternating bouts of fluttering and gliding, with song given during glides.

**MIGRATION**

Medium-distance migrant, mostly along the coast but probably also offshore. Peak concentrations occur in spring in the Gulf of California and Gulf of Alaska.

**Spring** migration takes place between mid-February and early June. Northbound adults depart southernmost wintering areas by early March. Peak numbers pass through the Gulf of California and s. California from mid-March to late April and through British Columbia from mid-April to early May. Largest concentrations occur at Prince William Sound, Alaska, in early May. Arrival at the breeding grounds takes place between early May and early June. Nonbreeders, presumably mostly **one-year-old** birds, arrive in s. Alaska in late May and early June and remain through the summer. Small numbers remain at more southerly wintering areas through the breeding season.

**Fall** migration takes place between late June and early December. Southbound **adults** depart breeding areas mostly between mid-July and early August. Failed breeders and nonbreeders may begin to head south by late June. Peak numbers occur in s. Alaska in late July and August and in California in August and September. Most **juveniles** depart the tundra by late July or early August and gather in small flocks at nearby coastal areas. The first juveniles appear south of Canada in late July and early August. Passage of juveniles occurs about a month later than that of adults.

**MOLT**

**First life year:** Molt out of juvenal plumage takes place primarily between September and March and includes many head and body feathers and some wing coverts and tail feathers. Between about March and June there is a partial molt to breeding plumage, involving a variable number of head and body feathers and wing coverts. The resulting plumage is duller than that of adults on birds summering at nonbreeding sites. Others may attain a more adultlike plumage. One-year-old birds may be distinguished from adults by retained juvenal feathers, particularly flight feathers and some wing coverts.

**Adult:** Between about February and May adults undergo a partial molt to breeding plumage, including most head and body feathers and a variable number of wing coverts, tertials, and tail feathers. The complete molt to nonbreeding plumage takes place mostly between mid-July and November. Body molt may begin on the breeding grounds but is suspended during migration and completed on the wintering grounds with heaviest molt taking place in August and September. Flight feather molt takes place primarily at wintering areas between mid-July and early September.

**VOCALIZATIONS**

Relatively quiet away from breeding grounds. Flight call is a soft, mellow *whif-if-if.* Feeding flocks give a constant chatter of soft, high *whik* notes. Display song is a series of scratchy, squeaky notes, *kree kree kree . . . ki-drrr ki-drrr ki-drrr . . .* and a more laughing *quoy quoy quoy . . .*

## Great Knot, p. 281

### *Calidris tenuirostris*

**STATUS**

Asian species. Casual in w. Alaska, usually in spring. Accidental south to Oregon.

Common within range. Breeds in alpine and Arctic zones of ne. Siberia. Winters in se. Asia and Australia. Winter numbers from Australia in mid-1990s suggest population of at least 250,000.

**TAXONOMY**

No subspecies recognized.

**BEHAVIOR**

Breeds on upland slopes and ridges, primarily in barren, lichencovered gravelly habitats, generally up to 3,500 ft. (1,067 m) in elevation. During migration and winter prefers sandy beaches and tidal mudflats, often in large flocks. Less frequent at inland fresh and salt lakes. Walks in a steady, straightforward fashion while foraging, similar to Red Knot. Feeds by picking food from surface or probing in soft mud or sand, sometimes below shallow water. Eats mostly intertidal invertebrates, including mollusks and other crustaceans. Also eats plant material in breeding season. Typically found in large flocks away from breeding grounds, but may appear in smaller numbers during migration. Often found in the company of Red Knots during migration and winter.

Nests late May to early July.

**MIGRATION**

Long-distance migrant. Movements mainly along upper Asian coastlines en route to wintering areas in se. Asia and Australia. Migration data incomplete.

**Spring** migration takes place between early April and early June. Departs Australian wintering areas from early to late April. Spring movements noted along Chinese coasts from early April to early June, but mainly first half of May through Japan. Arrives at breeding areas in Siberia in second half of May. Virtually all **one-year-old** birds remain at wintering areas through the summer.

**Fall** migration takes place from early July to probably November. **Adults** begin departure from breeding grounds in July, with juveniles and brooding adults probably following a month or more later. Migrants recorded from se. Siberia and e. China from late July to late October, which reflects typical later passage of **juveniles** for Arctic breeders. Peak numbers occur in these areas August to September. Arrives at wintering areas from September to early November, with juveniles usually making up bulk of later arrivals.

**MOLT**

**First life year:** The partial molt out of juvenal plumage begins during fall migration in September/October and is completed by following March, including most head and body plumage and some wing coverts, tertials, and tail feathers. Nonbreeding appearance is generally attained by November/December. Between April and May there is a partial molt to breeding plumage, including scattered head and body feathers. This molt is typically less extensive than adult's, but some individuals attain well developed breeding appearance. Between April and June some outer primaries are replaced.

**Second life year:** For birds that oversummer on wintering grounds (presumably all first-summer birds), the complete molt to nonbreeding plumage is variable, with the loss of p1 taking place any time from late April (of first life year) to August. Flight feather molt is completed between September and December. Molt of head, body, tail, and wing coverts takes place primarily during August to September. After this molt, second-year birds appear as adults. The partial molt to breeding plumage is as in adults.

**Adult:** The partial molt to breeding plumage begins by late February and is completed before spring migration in mid-April. It includes most head and body feathers as well as some inner tertials and wing coverts. The complete molt to nonbreeding plumage takes place between July and February. Limited molt may occur before migration, including scattered feathers of crown, mantle, scapulars, and upper chest, but most fall migrants are in largely worn breeding plumage. The majority of molt takes place on wintering grounds from late September to early November and is completed with regrowth of p10 from late December to February.

**VOCALIZATIONS**

Mostly silent. Flight call, heard occasionally, is a low, muffled *nyut-nyut,* similar to Red Knot.

## Red Knot, p. 134

### *Calidris canutus*

**STATUS**

Common but quite localized and declining. Breeds at widely scattered locations in the high Arctic from Siberia east to Canada and Greenland. Small numbers winter along the southern coasts of the United States, particularly in Florida. The bulk of the population winters in large concentrations at widely scattered locations, principally Argentina, w. Europe, Africa, Australia, and New Zealand. Rare inland. The world population is estimated at 1,029,000, with 400,000 breeding in North America.

**TAXONOMY**

Five subspecies are recognized, three of which are known to occur in North America. *C. c. roselaari* breeds in w. Alaska, principally Seward Pen., west to Russia's Chukchi Pen. Winter range poorly known but believed to include southern coasts of the United States south to tropical coasts of South America north of the equator. Most migrants along the Pacific Coast are believed to be *roselaari. C. c. rufa* breeds in Nearctic Canada (primarily Victoria I., Melville Pen., and Southampton I.) and winters primarily in s. Argentina though possibly in numbers north to equatorial Brazil and possibly se. United States. In migration, the bulk of the *rufa* population stages along Delaware Bay shore in spring and New England coast in fall. Birds wintering in s. United States may be either *rufa* or *roselaari. C. c. islandica* breeds in ne. Canada (principally the Parry and Queen Elizabeth Is.) and Greenland and migrates southeast to winter in w. Europe.

Differences in appearance between the three North American subspecies are minor and largely outweighed by individual variation. Definitive field recognition is seldom possible. In breeding plumage, *rufa* averages palest overall with relatively limited rufous and black in upperparts, extensive white from lower belly to undertail coverts, and contrasting silvery gray crown and nape. Underparts on *roselaari* slightly darker rufous than *rufa* but with similar extent of white. Upperparts on *roselaari* brighter rufous and more extensively black, with more golden cast to crown and nape. Wing point averages longer than *rufa,* but much variation. *C. c. islandica* has upperparts similar to *roselaari* but often more gold-spangled. Underparts more extensively rufous than either *rufa* or *roselaari.* Differences in juvenal and nonbreeding plumages need more study.

**BEHAVIOR**

Breeds on flat barren tundra in high Arctic islands and peninsulas. The nest is placed on dry rocky ground, often among dense, low vegetation. Highly coastal during migration and winter, preferring sandbars, beaches, and tidal flats. During high tides roosts on salt marsh pools and adjacent fields. Occasionally found on rocky coastlines or jetties. Feeds by probing in the mud or picking at the surface. On the breeding grounds eats primarily insects and plant material. During migration and winter prefers small mollusks, crustaceans, worms, and other invertebrates. In spring migration the *rufa* subspecies feeds almost entirely on Horseshoe Crab eggs gleaned from beaches in the Delaware Bay region. Sometimes seen singly or in small groups, but regularly gathers in large flocks of up to several thousand at favored stop-over sites. Often associates with Dunlin, Sanderling, Ruddy Turnstone, and Black-bellied Plover. Rarely occurs inland except at a few preferred stop-over sites.

Nest initiation primarily early to mid-June. Egg dates early June to late July. Incubation 21–22 days. Both adults share parental duties, but male does most incubation and most or all brood-rearing. Single-brooded. Young can fly after 18–20 days. The male's display flight involves a steep ascent to 100 ft. (30 m) or more, followed by a slow downward spiral with wings set and tail spread. Both parents tend the young initially, but the female leaves the group before the young fledge.

**MIGRATION**

Long-distance migrant, many populations making long transoceanic flights.

**Spring** migration takes place between mid-February and mid-June. Birds wintering in South America gradually move northward from mid-February, with peak numbers in Brazil in late April and early May. Some of these birds (*rufa*) fly nonstop to staging areas along the mid-Atlantic Coast, particularly along the shores of Delaware Bay. Peak numbers occur in the Delaware Bay region in mid- to late May, after which these birds make another nonstop flight to their breeding grounds in Canada. Arrival on the breeding grounds takes place primarily in early June. Birds wintering along s. coasts of the United States and Central America, and perhaps some wintering in n. South America (presumed to be *roselaari*), apparently fly nonstop to a staging area along the Pacific Coast, where peak numbers occur between late April and mid-May. After feeding for several weeks to build up fat reserves, these birds fly nonstop to their Alaskan breeding grounds, arriving by late May. Some **one-year-old** birds remain on the wintering grounds through the summer, while others migrate partway north toward the breeding grounds.

**Fall** migration takes place between early to mid-July and mid-November. Southbound migration of **adults** begins by early to mid-July. As in spring, fall migrants make their whole movement in one or two nonstop flights. The population breeding in Nearctic Canada *(rufa)* arrives in New England and the Delaware Bay region by late July and remains there for a few weeks to build fat reserves. By mid- to late August these birds make the flight to their wintering grounds in Argentina. The population breeding in Alaska *(roselaari)* apparently disperses to a number of sites along s. coasts of the United States, Central America, and perhaps n. South America, arriving by early to mid-July. Some of these birds arrive on the mid-Atlantic Coast and apparently remain well into September, after which most gradually move south. Generally less common along the Pacific Coast than elsewhere in fall. One-year-olds probably return to the wintering grounds earlier than adults. After spending several weeks feeding near the breeding grounds, **juveniles** head straight to the coast, usually in one long flight, so are rarely seen inland. Juveniles first arrive along the mid-Atlantic and mid-Pacific coasts by mid- to late August, with peak passage there in September and early October. Arrival on the wintering grounds takes place between October and December.

**MOLT**

**First life year:** Molt out of juvenal plumage takes place between September/October and October/December and includes most head and body feathers and usually some wing coverts and rectrices. Birds that will oversummer at South American wintering sites may also molt some or all primaries between about January and July. Between about April and June, while still on the wintering grounds, first-year birds undergo a highly variable partial molt to first summer plumage. This plumage varies from a gray, nonbreeding-like plumage (mostly birds that remain on the wintering grounds) to a nearly full, adultlike breeding plumage (mostly birds that migrate partway north).

**Second life year:** For most one-year-old birds (and possibly older subadults), the complete molt to nonbreeding plumage takes place earlier than in adults, generally between late May and late July, though molt may be protracted to December in some individuals. After this molt, second-year birds are usually indistinguishable from adults.

**Adult:** Between February and early June (mostly March and April), adults undergo a partial molt to breeding plumage involving most head and body feathers and some wing coverts. This molt takes place largely on the wintering grounds. In most *rufa* the complete molt to nonbreeding plumage takes place between early July and early March. Some head-and-body molt takes place while birds are staging in New England, but flight feather molt takes place mostly at South American wintering grounds between early October and early March. In populations wintering in North America (*roselaari* or *rufa*), the complete molt to nonbreeding plumage takes place in coastal areas of the United States mostly between late July and November. Heavy flight feather molt on these birds takes place between August and October.

**VOCALIZATIONS**

Generally quiet. Flight call, given infrequently, is a godwit-like or Sora-like *kuEEt* or *kawit-wit-wit*. Also gives a soft, muffled *kuup*. Display song is an eerie whistle, *por-meeee por-meeee por-meeee* . . . , the second note in each phrase drawn out and slightly rising. This is usually followed by a second element, *por-por por-por por-por* . . .

## Sanderling, p. 139

### *Calidris alba*

**STATUS**

Common visitor along ocean beaches throughout North America. Also common at stop-over sites in the Great Plains and around the Great Lakes, but otherwise uncommon to rare inland. Probably the most widespread shorebird in the world, occurring on every continent except Antarctica. Breeds in high Arctic regions from n. Northwest Territories to ne. Greenland and disjunctly on the island of Spitsbergen, Norway, and the Taymyr Pen. and New Siberian Is. in Russia. Winters along almost all temperate and tropical beaches from British Columbia and Nova Scotia south to s. South America. Also from the British Isles and Denmark south through Africa and east to Taiwan, Australia, and Oceania. The world population is estimated at 643,000, with 300,000 in North America.

**TAXONOMY**

No subspecies recognized.

**BEHAVIOR**

Probably our most familiar shorebird, with small flocks commonly seen running along the surf line "chasing" waves. Breeds on coastal tundra in the high

Arctic, particularly on gravel ridges or slopes near water. During migration and winter uses wave-washed sandy beaches and, to a lesser degree, gravel or rocky coastlines and tidal mudflats. Interior migrants use sandy margins to lakes, ponds, and streams. Foraging behavior often very active with much running. On ocean beaches Sanderlings chase receding waves, picking up stranded mollusks or crustaceans, or rapidly probe and dig before the next wave chases them back. On higher ground they often dig in one spot, particularly while searching for Horseshoe Crab eggs. Less active on mudflats or quiet shorelines, where they pick and probe at a slower pace. Some individuals territorial in nonbreeding season, so chases and fights are frequent. Territorial birds often exhibit threat posture in which back is hunched and head is held low. Feeds mostly during the day, occasionally at night. Eats a variety of aquatic and terrestrial invertebrates, especially insects on the breeding grounds and small crustaceans and mollusks in the nonbreeding season. Typically seen in flocks of 5–30 individuals but may gather in groups of over 1,000 in some areas. Associates with a wide variety of shorebirds, particularly Dunlin and Ruddy Turnstone.

Nest initiation primarily mid-June in Greenland, late June or early July in Canada. Egg dates range from late May to early August. Incubation 23–32 days. Both adults share parental duties. Some birds are single-brooded, while others will lay a second or third clutch. In such cases male incubates and raises the first clutch, female the second clutch, and a different male the third clutch. Peak hatching early to mid-July in Greenland, late July in Canada. Young can fly after 17 days. The display flight, usually performed by male away from nest site, involves rapid fluttering wingbeats interspersed with bow-winged glides as the bird flies and hovers 6–30 ft. (2–9 m) above ground. The display lasts from 30 seconds to 2 minutes.

**MIGRATION**

Long-distance migrant. Different populations travel between 3,000 and 10,000 mi. (4,828 and 16,090 km) between breeding and wintering grounds. In spring most common along Atlantic Coast, particularly in the Delaware Bay region, and through Great Plains, particularly in s. Saskatchewan. Less common through Great Plains in fall than spring. Most common in winter along the Pacific Coast and in South America.

**Spring** migration takes place between early March and mid-June. Northbound adults depart South American wintering grounds mostly in late March and early April. Peak numbers pass through the s. Pacific and Gulf coasts from late April to mid-May and through the n. Pacific Coast, n. Plains, and New England from mid-May to early June. Arrival at the breeding grounds takes place between late May and mid-June. Most **one-year-old** birds remain at wintering areas in South America through the summer, while only a few oversummer at North American wintering sites. Most apparently do not breed in their first year.

**Fall** migration takes place between late June and mid-November. Southbound **adults** depart breeding areas mostly between mid-July and mid-August. Failed breeders may begin to head south by late June. The first fall migrants appear from the n. Pacific Coast to New England in early July, with peak numbers passing through much of the continent in August. Adults arrive in South America by late August or early September. Most **juveniles** depart the tundra by early to mid-August and gather in flocks at nearby coastal areas. They begin to head south by late August and early September. The first juveniles arrive across s. Canada and n. United States in mid- to late August, with peak passage there from mid-September to early October. The last juveniles linger at interior sites into early November. Juveniles arrive across the southern states in early September, with peak passage from mid-September to late October. Arrivals in South America take place from early October to mid-November.

#### MOLT

**First life year:** Molt out of juvenal plumage is highly variable geographically. For birds wintering in temperate North America it takes place primarily between mid-September and mid-October and includes many head and body feathers, inner tertials, and rarely a few wing coverts and tail feathers. Birds wintering in the Tropics molt somewhat later, from about October to December, and usually undergo a more extensive molt, including many wing coverts. For birds wintering in s. South America this molt is often complete, with flight feathers and other remaining juvenal plumage being replaced from January to April. In some individuals the juvenal secondaries and inner primaries are retained. In May and June there is a partial molt to breeding plumage involving a highly variable number of head and body feathers resulting in a range of plumages from one like breeding adult to one much like nonbreeding. Many one-year-old birds may be distinguished from adults by retained juvenal feathers, particularly rump, wing coverts, and flight feathers. Individuals that underwent a complete molt over the winter are indistinguishable from adults.

**Adult:** Between mid-March and June, adults undergo a partial molt to breeding plumage, including most head and body feathers and a variable number of median wing coverts, tertials, and tail feathers. Some feathers grown in late March/early April or in early July show characters intermediate between breeding and nonbreeding plumage, adding to the highly variable appearance of this plumage. The complete molt to nonbreeding plumage takes place between mid-July (on the breeding grounds) and late October or early November. Most body molt takes place at stop-over sites in August. Flight feather molt takes place at stop-over sites in birds wintering in temperate regions, and near the wintering grounds in birds wintering in the Tropics. Primary molt is rapid, with some birds becoming nearly flightless.

#### VOCALIZATIONS

Flight call is a soft, squeaky *pweet,* given by standing and flying birds. This call produces a playful chatter from flocks. Threat call is a high, whistled *sew-sew-sew-sew* . . . Display song, given mostly by male, is a repeated series of harsh, buzzy, froglike notes and a harsh, squeaky, chattering *cher-cher-cher-cher* . . .

## Semipalmated Sandpiper, p. 145

***Calidris pusilla***

#### STATUS

Common to locally abundant in e. North America, particularly along the coast, where huge concentrations gather at favored staging areas. By far the most common "peep" at many migration sites along the Atlantic Coast. Less common inland and uncommon to rare throughout much of the West south of Alaska. Breeds in subarctic regions from extreme ne. Siberia to Newfoundland. Winters in coastal areas from s. Mexico and extreme s. Florida to n. Chile and Uruguay. The world population is estimated at 3.5 million.

#### TAXONOMY
No subspecies recognized.

#### BEHAVIOR
Breeds in subarctic to Arctic tundra near water. Nest site is in wet to dry tundra, shrubby areas, or sand. Migrants use areas of shallow fresh or salt water and open muddy or sandy areas with little vegetation, such as intertidal flats or lake shores. Forages by pecking at surface or probing shallowly in mud, walking quickly between foraging attempts. Pecking rates may average faster than with other peeps. More aggressive than other peeps while foraging. Often chases or fights with other birds and threatens them with neck feathers fluffed up, wings drooped, and tail held high in the air. Feeds by day and night. Eats a variety of aquatic and terrestrial invertebrates, especially arthropods, mollusks, marine worms, insects, and spiders. Typically seen in flocks of a few dozen up to several thousand.

Egg laying may begin within a week of arrival on the breeding grounds but may be delayed for several weeks in inclement weather. Egg dates range from early June to early July in Alaska and mid-June to late July in Canada. Incubation 20–22 days. Both adults share parental duties, though female typically abandons young and male within 11 days. One brood but will renest after failure. Peak hatching late June in Alaska, second week of July in Canada. Young can fly after 16–19 days. The male's flight display is performed while hovering over its territory on alternating bursts of fluttering and gliding.

#### MIGRATION
Western Arctic populations migrate primarily through the prairies to wintering areas in South America. Eastern Arctic populations migrate down the Atlantic Coast, many making the long transoceanic migration from New England and the Maritime Provinces to wintering areas in more eastern portions of South America, a flight of up to 2,000 mi. (3,200 km). Migrates primarily at night but also by day during long transoceanic flights.

In **spring,** northbound **adults** depart the wintering grounds mostly in April. Migrants first arrive along the Gulf Coast in late March, with peak numbers there in late April and early May. Arrivals in the Northeast take place in late April, with peak numbers passing in mid- to late May and into early June. Most breeding areas are reached by late May and early June (males precede females by several days). About two-thirds of **one-year-old** birds remain at wintering areas through the summer, while others migrate partway or all the way north to breeding grounds.

In **fall,** southbound **adults** begin to depart breeding areas by mid-July (females before males by about 5 days). Failed breeders may begin to head south by late June. The first fall migrants appear from the Pacific Northwest to New England in late June with peak numbers passing through between mid-July and early August. More common in New England and the Maritime Provinces in fall than spring. Migration through the Gulf Coast peaks about two weeks later. Arrivals begin in South America by mid-July, and most are gone from North America by early October. Most **juveniles** fledge in mid- to late July and begin their southbound migration within a week or two. The first juveniles arrive across s. Canada and n. United States in the first days of August, with peak passage there in late August and early September. The first juveniles arrive along the Gulf Coast in mid-August, peaking there in September. Most are gone from North America by mid-November.

**MOLT**

**First life year:** Molt out of juvenal plumage takes place between September and May. Fall migrants usually hold extensive juvenal plumage well into the fall, at best replacing a scattering of scapulars. Most of the molt, including all of the flight feather molt, takes place on the wintering grounds. All but some inner primaries, secondaries, and underwing coverts are replaced. Primary molt takes place between November and May. Some birds undergo a partial molt in spring, replacing a variable number of head and body feathers. One-year-old birds may attain partial breeding plumage or nearly full nonbreeding. One-year-old birds are distinguishable by retained worn juvenal inner primaries and secondaries.

**Second life year:** Between late June and January, second-year birds undergo a complete molt to nonbreeding plumage, after which they are usually indistinguishable from adults.

**Adult:** Between January and May, before spring migration, adults undergo a partial molt to breeding plumage, including most head and body feathers and some wing coverts. In fall, adults undergo a complete molt to nonbreeding plumage between late June (during late incubation) and January. Virtually all southbound adults show a mixture of breeding and nonbreeding scapulars while in North America, and only rare individuals exhibit full nonbreeding plumage before the wintering grounds are reached. Primary molt takes place between mid-August and mid-January and takes 2½–3½ months. Birds passing through the prairies begin primary molt at stop-over sites, while Atlantic Coast migrants, which make a long overwater flight, do not begin primary molt until the South American wintering grounds are reached.

**VOCALIZATIONS**

Flight call is a variable rough *chrrk* or higher, sharper *chit,* sometimes with several notes rolled together in a twitter; generally lower pitched than calls of Western Sandpiper, but variable. Given mostly in flight but also while standing. Threat call is a playful, chattering *toy-toy-toy-toy* . . . or a quicker, descending *ti-d-d-do.* Display song, given by male, is a long, pulsating, liquid trill (like an idling engine) mixed with variable scratchy notes.

## Western Sandpiper, p. 151

***Calidris mauri***

**STATUS**

Common to locally abundant in coastal areas, particularly in the West, where huge concentrations gather at favored staging areas (for example, 2 million in a single day at Copper River Delta, Alaska). Progressively scarcer toward the East but still locally common north to New York in fall and Georgia in spring. Locally common migrant inland in West; rare inland in East. Breeds in subarctic and low Arctic regions of Alaska and extreme northeastern Siberia. Winters in coastal areas from s. British Columbia and New Jersey south to n. Peru and Suriname. Winters locally at interior sites in Mexico and Salton Sea, California. The world population is estimated at 3.5–4 million, though numbers are declining (up to 6.5 million were estimated at Copper River alone in 1973).

**TAXONOMY**

No subspecies recognized.

**BEHAVIOR**

Breeds primarily in low Arctic and subarctic tundra near water. Nest site typically in a small patch of dry heath tundra among large expanses of wet grass-sedge tundra. Migrants use areas of shallow fresh or salt water and open muddy or sandy areas with little vegetation, such as intertidal flats or lakeshores. Tends to prefer sandier substrate than Semipalmated Sandpiper. Forages by pecking at surface or probing shallowly in mud. Usually walks with slower, more hesitant steps than Semipalmated Sandpiper. Feeds by day and night. Primary foods include insect larvae, crustaceans, and marine worms. May be seen in small groups or in large flocks of 1,000 or more. At spring staging areas in Alaska, flocks of up to 2 million or more may gather at a single location.

Egg laying usually begins in late May/early June, soon after snowmelt. Egg dates range from late May to early July. Incubation 21 days. Both adults share incubation, but the male does most or all of the chick rearing. One brood but will renest after failure. Peak hatching late June. Young can fly after 17–18 days. The male's flight display involves short or long flights over territory. Unlike other "peeps," does not usually hover during display flight.

#### MIGRATION

The bulk of the population migrates along the Pacific Coast, with smaller numbers through the interior, especially in fall. Males winter farther north than females.

In **spring,** northbound **adults** depart the wintering grounds in late March and April. Peak numbers pass through the Gulf Coast in early April and through the central and n. Pacific Coast in late April. Huge masses, accounting for the bulk of the population, gather at the Copper River Delta, Alaska, in early to mid-May. Smaller numbers pass through the interior, mostly between late April and mid-May. Most breeding areas are reached by mid- to late May (males precede females by a few days). A small number of **one-year-old** birds remain at wintering areas through summer, while most apparently migrate partway or all the way north to breeding grounds.

In **fall,** southbound **adults** begin to depart breeding areas by late June (females before males). Failed breeders may begin to head south slightly earlier. The first fall migrants appear in the Pacific Northwest, interior West, and mid-Atlantic Coast in late June, with peak numbers passing through these areas in mid-August. Much more common along the Atlantic Coast in fall than spring. Passage through the Gulf Coast about two weeks later, with many of those birds apparently stopping over in the interior West. The first **juveniles** arrive in the Pacific Northwest by late July, with peak numbers passing there between late August and mid-September. Along the Atlantic Coast juveniles arrive two weeks later but peak at the same time as on the West Coast. At interior sites juveniles arrive in early to mid-August, peak in late August/early September, and often linger into November. The first juveniles arrive along the Gulf Coast in mid-August, peaking there in September.

#### MOLT

**First life year:** Molt out of juvenal plumage takes place between late August and late November and includes head and body plumage as well as tertials. Some wing coverts and uppertail coverts may be replaced later in the winter. Between February and April there is a variable partial molt during which some birds (those migrating north to the Arctic) replace most head and body feathers and attain nearly full breeding plumage, while others (those remaining on or near the wintering grounds) undergo a more limited molt and acquire new nonbreeding feathers.

**Second life year:** As early as May (of first life year), one-year-old birds begin their complete molt to nonbreeding plumage. This molt is completed earlier than in adults, likely by August to October. After this point, second-year birds are on the same molt schedule as adults.

**Adult:** Between February and April adults undergo a partial molt to breeding plumage, including most head and body feathers and some wing coverts. The complete molt to nonbreeding plumage takes place between late June and February. Birds wintering in the United States undergo this molt between late June and late October. Limited head-and-body molt may begin on the breeding grounds, but most molt, including that of the flight feathers, takes place on the wintering grounds or at nearby coastal stop-over sites. Birds wintering in South America do most of this molt, including flight feather molt, between October and February after the wintering grounds are reached.

#### VOCALIZATIONS

Flight call is a husky *jrrk* or higher, sharper *jeek*, sometimes with several notes rolled together in a twitter; generally higher and weaker than calls of Semipalmated Sandpiper, but variable. Given mostly in flight but also while standing. Threat call is a high, twittering *twi-di-di-di-di . . .*, higher and faster than Semipalmated. Display song, given by male, is a short series of husky, scratching calls accelerating into a trill, *jrrrr-jrrrr-jrrrr-j-j-j-jee*, repeated up to 20 times per minute.

## Red-necked Stint, p. 283

### *Calidris ruficollis*

#### STATUS

Asian species. Regular migrant and rare breeder in w. Alaska. Rare but nearly annual migrant along Atlantic and Pacific coasts, particularly in July/August. Accidental inland. Common within its core range. Breeds primarily in Arctic regions of Siberia from Laptev Sea eastward. Winters from e. India (possibly farther west) and Taiwan south through Australia and New Zealand.

#### TAXONOMY

No subspecies recognized.

#### BEHAVIOR

Breeds in low dry tundra or cotton-grass bogs or in high tundra near water. Migrants prefer open muddy or sandy areas such as tidal flats and lakeshores. Less often in shallow water. Forages by pecking rapidly at surface or sometimes probing shallowly in mud, walking quickly between foraging attempts. Eats a variety of aquatic and terrestrial invertebrates, especially insects and crustaceans. In North America typically found among flocks of Semipalmated Sandpipers.

Nests late May to July. Incubation probably about three weeks. Both adults share parental duties, though young apparently tended by female alone. One brood. The male's flight display is performed while hovering over its territory on alternating bursts of fluttering and gliding. This display is terminated by a descent to the ground with wings held up in a deep "V".

#### MIGRATION

In **spring,** northbound **adults** depart wintering areas primarily in March and April and reach ne. Siberia and extreme w. Alaska in late May and early June. The few spring records along the Atlantic and Pacific coasts are from May. Many **one-year-old** birds remain at wintering areas through the summer, while others migrate partway or all the way north to breeding grounds.

In **fall,** southbound **adults** depart the breeding areas mostly in July, with failed breeders departing as early as mid- to late June. Migrant adults occur in w. Alaska mostly between early July and early August. Along the Atlantic and Pacific coasts adults have occurred between mid-June and late August, with most records from mid- to late July. The first adults arrive in Australia by late August. Migrant **juveniles** are seen in w. Alaska primarily from mid-August to mid-September. Although there are very few records, juveniles along the Atlantic and Pacific coasts are most likely to occur in September and October. Juveniles arrive at wintering areas in Australia through November.

**MOLT**

**First life year:** Molt out of juvenal plumage poorly known but apparently takes place largely on or near the wintering grounds. At least inner primaries and secondaries are retained. In spring there is a variable partial molt during which some birds attain partial breeding plumage while others remain in nearly full nonbreeding.

**Second life year:** For birds that oversummered on the wintering grounds, the complete molt to nonbreeding plumage probably takes place earlier than in adults. After this molt, second-year birds are on the same schedule as adults.

**Adult:** Between late March and April adults undergo a partial molt to breeding plumage, including most head and body feathers but very few wing coverts or tertials. In fall, adults undergo a complete molt to nonbreeding plumage between about July and February. Many southbound adults show nearly full breeding plumage, but molt progresses rapidly upon arrival on the wintering grounds, and some birds appear as nearly full nonbreeding by mid-August. Primary molt takes place after the wintering grounds are reached and is completed by February.

**VOCALIZATIONS**

Flight call is a slightly rough *kiirp*, similar to Semipalmated Sandpiper but usually higher; lower pitched and longer than calls of Little Stint. Also gives a high, descending chatter, *ti-d-d-do*, much like that of Semipalmated. Display song is a repeated series of deep rising notes, *rrooa rrooa rrooa rrooa* . . . , sometimes alternated with a series of higher, harder notes, *ek ek ek ek ek* . . .

## Little Stint, p. 285

### *Calidris minuta*

**STATUS**

Old World species. Rare but nearly annual visitor to North America, mostly along the Atlantic and Pacific coasts and in Alaska. Common within its core range. Breeds primarily in Arctic regions from Norway east to the New Siberian Is. Winters from s. Europe and Africa east to Myanmar.

**TAXONOMY**

No subspecies recognized.

**BEHAVIOR**

Breeds primarily in dry, high Arctic tundra at low elevation but also in marsh ground at higher elevations. Migrants prefer muddy or sandy areas such as tidal flats and lakeshores. Less often in shallow water. Tends to prefer more sheltered sites than Red-necked Stint. Forages by pecking rapidly at surface or sometimes probing shallowly in mud, walking quickly between foraging attempts. Diet consists mainly of small invertebrates such as insects, beetles, worms, mollusks, and crustaceans. In North America typically found among flocks of Semipalmated Sandpipers.

Nests June to August.

**MIGRATION**

Migrates on a broad front, though route is generally shifted more easterly in spring, more westerly in fall.

In **spring**, northbound **adults** depart wintering areas primarily in April and early May and reach breeding grounds by late May and early June. Spring records in North America come mostly from Alaska in late May and early June. Most **one-year-old** birds are believed to return to the breeding grounds.

In **fall,** southbound **adults** depart the breeding areas by July, with failed breeders leaving as early as late June. Peak numbers of adults pass through Europe in July and early August and reach Africa between late July and September. Records along the Atlantic Coast fall between mid-June and early September, with the bulk of these in July. Many of these are likely unpaired birds or failed breeders that headed south early. Peak numbers of **juveniles** pass through Europe in late August and September, with the latest migrants continuing into early November. The few records of juveniles in North America are mostly from the Pacific Coast in September.

**MOLT**

**First life year:** Molt out of juvenal plumage takes place between September/October and April, primarily at wintering areas. This molt is nearly complete, usually including all but inner primaries and some coverts. Head and body molt takes place primarily between late September/October and January. Primary molt takes place between December and April and is usually restricted to outers. In spring there is a partial molt to breeding plumage as in adult, though it probably takes place slightly later. One-year-old birds may be distinguished from adults by relatively more worn retained juvenal inner primaries and some coverts.

**Adult:** Between February and May adults undergo a partial molt to breeding plumage, including most head and body feathers and many wing coverts, tertials, and central tail feathers. The complete molt to nonbreeding plumage takes place between August (rarely late July) and March. Primary molt takes place on or near the wintering grounds between August/September and January/March.

**VOCALIZATIONS**

Flight call is a very short, squeaky, high-pitched *chit* or *tik,* much shorter and harder than Semipalmated. Threat call is a short, trilled *di-di-di-dut.*

## Temminck's Stint, p. 288

### *Calidris temminckii*

**STATUS**

Eurasian species; rare migrant on w. Alaska islands. Accidental on Pacific Coast in fall. Fairly common within range. Breeds in subarctic zone across Eurasia from Scandinavia east to Chokotski Pen., with small population in n. Scotland. Winters primarily from w. Africa east to se. China and Indonesia. Small numbers winter in Europe, rarely as far north as Britain.

**TAXONOMY**

No subspecies recognized.

**BEHAVIOR**

A bird of freshwater habitats. Breeds in bogs, marshes, and elevated open tundra with woodland fringes. During migration and winter prefers freshwater wetlands and estuaries. Typically avoids open coastal habitats. Feeds in a relatively slow fashion, unlike faster-paced Semipalmated Sandpiper and Little Stint. Often creeps along on flexed legs in vegetated margins of marshes or estuaries, picking deliberately at surface of mud. Rarely probes for food. Eats a variety of invertebrates, including insects and their larvae (especially beetles and flies), worms, crustaceans, and mollusks. Usually seen singly or in small groups, but may form concentrated flocks from 50 to several hundred during migration. When flushed, flies off high with erratic jerky movements, giving its distinctive trilled call.

Nests from late May to late July.

#### MIGRATION

Mostly long-distance migrant. Majority of birds migrate largely overland across a broad front to tropical wintering areas, with smaller numbers spending winter in Europe or Arabia.

**Spring** migration takes place between March/April and mid-May/early June, with peak movements mid-April to mid-May. Departs wintering areas in March/April and arrives at European breeding areas in second half of May. Siberian arrivals take place in late May/early June. Some nonbreeders presumably remain at wintering areas through the summer, while others migrate north and summer near the breeding grounds.

**Fall** migration takes place between early July and early November. **Adults** depart breeding grounds between early July and August. Peak movements occur through temperate Europe from late July through August. Arrives at wintering grounds from August through September, with most birds appearing in September. Southbound **juveniles** leave natal areas in first half of August and pass through Europe and Turkey through late September/early October. Arrival on wintering grounds slightly later than for adults, extending into early November.

#### MOLT

**First life year:** The molt out of juvenal plumage begins in August to late September and is mostly completed by November (occasionally December). This partial molt includes head and body, some wing coverts, and some central tail feathers. Outer primaries (p6–10) and inner secondaries are replaced from December through May. Between February/April and June, a variable partial molt to breeding plumage takes place. The molt includes most of the body and wing coverts and may occur slightly later than adults.

**Second life year:** For birds that do not breed in their first summer, the complete molt to nonbreeding plumage probably starts in late June/July and is completed by late August (a similar timetable to failed breeders). After this molt, second-year birds are usually indistinguishable from adults. The partial molt to breeding plumage is the same as in adults.

**Adult:** The partial molt to breeding plumage begins from February to April and is completed by June. It includes variable amounts of body feathers and wing coverts as well as some tail feathers and tertials. The complete molt to nonbreeding plumage takes place from July to November/April. Head and body molt begins in July and is completed by October/November. Primary molt begins in September and is completed by November or suspended, then completed in April. Failed breeders may begin primary molt as early as July while still on the breeding grounds.

#### VOCALIZATIONS

Flight call is a crisp, ringing trill, *tidididup,* often repeated; recalls Smith's Longspur's rattle but is higher and more rapid.

## Long-toed Stint, p. 290

***Calidris subminuta***

#### STATUS

Primarily Asian species. Fairly common migrant on w. Aleutians; casual elsewhere in w. Alaska, primarily in spring; accidental on Pacific Coast s. of Alaska. Fairly common within limited range. Breeding area incompletely known, but includes from w.-central Siberia east to Kamchatka Pen. Winters primarily from se. India to se. China, the Philippines, and Indonesia, with smaller numbers in Australia.

**TAXONOMY**
No subspecies recognized.

**BEHAVIOR**
A bird of primarily inland freshwater habitats. Occasionally found on tidal mudflats. Breeds in a wide variety of Arctic and boreal habitats, often in boggy and tundralike openings where mosses, sedges, and dwarf willows are present. During migration and winter prefers inland freshwater habitats. Less common on tidal mudflats. Often associates with Little, Red-necked, and Temminck's Stints. Typically feeds among vegetation at water's edge or on floating mats of algae or aquatic weeds. Long toes, adapted for walking on floating vegetation, are lifted awkwardly while walking. Less mouselike in stance and gait than Least Sandpiper due to proportionally longer legs. Walks slowly and deliberately while picking at surface of mud for food. May probe in shallow water as well. Food source not described, but invertebrates probably make up major portion of diet.

Nests from late May to early July.

**MIGRATION**
Long-distance migrant. Migrates across a broad front, both overland and coastally. Timing of migration movements incompletely known.

**Spring** migration takes place between March and late May/early June, with peak movements April/May. Departs southern wintering areas in March and is common during passage in China during April/May, with few birds found in Korea or Japan during this period. Arrival at known breeding areas late May/early June.

**Fall** migration poorly known but apparently takes place from mid-July to November. **Adults** begin departure from breeding areas in mid-July, with peak passage through Mongolia, Japan, and Korea in August/September. Arrives at wintering areas from August to November, with southbound movements through e. Siberia continuing through October. Later migrants mostly **juveniles.**

**MOLT**
**First life year:** The partial molt out of juvenal plumage begins in September and is mostly completed by October. It includes most of body, some back, rump, and tail feathers, and shorter tertials. Between April and May there is a partial molt to breeding plumage, similar to adult's but with juvenal flight feathers and many wing coverts retained.

**Second life year:** For birds that do not breed, the complete molt to nonbreeding plumage most likely takes place sooner than adults, starting in June and July and being completed by late summer/early fall. Second-year birds that breed are on a more adultlike schedule. After this molt, second-year birds are on the same schedule as adults. The partial molt to breeding plumage is as in adults.

**Adult:** The partial molt to breeding plumage begins in March and is completed by May, with many birds in almost full breeding plumage by late April. It includes most head and body feathers, central tail feathers, tertials, and most wing coverts. The complete molt to nonbreeding plumage takes place from August to December. Molt of body feathers begins in August and is mostly complete by late September, with some birds showing a few breeding feathers into October. Tail molt takes place from September to October. Primary molt begins in late August and is mostly completed in December.

**VOCALIZATIONS**
Flight call is a rich, trilled *churrd,* similar to Pectoral Sandpiper and much lower pitched than Least Sandpiper's call.

## Least Sandpiper, p. 156

***Calidris minutilla***

**STATUS**

Common throughout most of North America. Usually the most numerous "peep" at interior sites during migration and winter but often outnumbered by Semipalmated and Western Sandpipers along the coast. Breeds primarily in subarctic tundra and northern boreal forest from w. Alaska to Newfoundland. Winters at both coastal and interior sites from Washington and Virginia south to Chile and Brazil. The world and North American population is estimated at 600,000.

**TAXONOMY**

No subspecies recognized.

**BEHAVIOR**

Breeds primarily in subarctic and northern boreal-forest regions near or above treeline. Nest site is typically in tussock-heath or boggy areas near water and mudflats. Migrants use a wide variety of muddy areas such as lakeshores, riverbanks, and rain pools. In coastal areas uses sheltered tidal pools and margins of tidal flats, usually near vegetation or debris. Less numerous on broad tidal flats preferred by Semipalmated and Western Sandpipers. Forages by walking slowly, pecking at surface or probing shallowly in mud. Often assumes crouched posture while feeding and tends to stay in a smaller area than Semipalmated or Western Sandpipers. Feeds by day and night. Eats a variety of aquatic and terrestrial invertebrates, especially amphipods, gastropods, and flies. In flight, tends to take a more erratic path than other peeps. Typically seen in small flocks of 10–20, though larger flocks may gather in prime habitat.

Nest initiation primarily late May in south, early to mid-June in north. Egg dates range from mid-May to early August. Incubation 19–23 days. Both sexes incubate, though male takes primary role. Female abandons young and male after about 1 week. One brood but will renest after failure. Peak hatching takes place in late June in south, mid-July in north. Young can fly after about 15 days. The male's flight display is performed while circling over its territory on alternating bursts of fluttering and gliding.

**MIGRATION**

Medium- to long-distance migrant. Some populations make long transoceanic flight in fall. Migrates during both day and night.

**Spring** migration takes place from mid-March to early June. Northbound **adults** depart the wintering grounds mostly in April. Migrants first arrive through the interior West and Northeast in mid-April, with peak movement in May. Migration on the West Coast about 1–2 weeks earlier. More common through interior than along Atlantic Coast in spring. Arrival on southern breeding areas begins as early as early May, while northernmost breeding areas are reached by late May and early June. Egg laying begins shortly after arrival. Small numbers of **one-year-old** birds remain at wintering areas through summer, while most migrate partway or all the way north to breeding grounds.

**Fall** migration takes place between mid-June and early November. **Adults** begin to depart breeding areas by late June (females before males). Failed breeders may begin to head south by mid-June. The first fall migrants appear from the Pacific Northwest to New England in late June. Peak numbers of adults pass through the Pacific Northwest in early to mid-July and through much of the remainder of the continent in late July/early August. More common in New England and the Maritime

Provinces in fall than in spring. Eastern population makes transoceanic migration from New England and the Maritime Provinces to wintering areas in South America and Lesser Antilles. Other populations move in shorter hops. The first adults arrive in South America by mid-July. **Juveniles** fledge between late June and early August and begin their southbound migration within a week or two. The first juveniles arrive across s. Canada and n. United States in mid-July, with peak passage there in mid- to late August. The first juveniles arrive along the Gulf Coast in early August and in South America by mid-August, with arrivals continuing through October. Eastern population makes transoceanic migration from New England and the Maritime Provinces to wintering areas in South America and Lesser Antilles. Other populations move in shorter hops.

**MOLT**

**First life year:** The molt out of juvenal plumage takes place between August and December, primarily at wintering or staging areas, and includes head and body feathers and a variable number of tertials and rectrices. Between January and June birds undergo a partial molt to first summer plumage, including most head and body feathers, outer primaries, and inner tertials. This plumage is similar to breeding adult's but may not be as fully developed and is distinguishable by retained worn juvenal inner primaries, secondaries, and wing coverts.

**Second life year:** Birds that summer south of the breeding grounds undergo the complete molt to nonbreeding plumage earlier than adults, sometimes beginning as early as June. After this molt, second-year birds are on the same molt schedule as adults.

**Adult:** Between January and June adults undergo a partial molt to breeding plumage including most head and body feathers, some wing coverts, and some rectrices. Feathers replaced early in the season appear intermediate between breeding and nonbreeding plumage. The complete molt to nonbreeding plumage takes place between mid-July and January. Longer-distance migrants wintering in South America molt on the wintering grounds (August to January), while shorter-distance migrants wintering in the United States undergo much of this molt at staging areas (July to November).

**VOCALIZATIONS**

Flight call is a variable, drawn-out *kreeet* with a trilled musical quality; often rising and often repeated or given as a two-syllable *kree-eet.* Given mostly in flight and regularly given by birds flying high overhead, including nocturnal migrants. Threat call is a chattering *chi-di-di-di.* Display song, given by male and occasionally by female, is a repeated, trilled *b-b-reeee, b-b-reeee, b-b-reeee . . .* ; each note rising but longer, harsher, less musical than flight call; sometimes interspersed with higher, buzzier notes and chattering.

## White-rumped Sandpiper, p. 161

### *Calidris fuscicollis*

**STATUS**

Fairly common migrant mostly through Great Plains in spring and Atlantic Coast in fall. Rare in the West. Usually seen in relatively small numbers among larger flocks of "peeps." Breeds in high Arctic regions from the North Slope of Alaska east to Baffin I. Winters in temperate regions of South America from s. Brazil and Chile to s. Argentina. The world population is estimated at 400,000.

**TAXONOMY**

No subspecies recognized.

**BEHAVIOR**

Breeds on wet lowland or upland tundra with abundant grassy or mossy tussocks, preferably near ponds, lakes, or streams. Most numerous near the seacoast. During migration and winter shows a decided preference for wet grassy areas, such as flooded ricefields, grassy margins of pools, and grassy margins of tidal flats. Less often on open mudflats or beaches. Forages by walking slowly, pecking sparingly at surface and shallowly probing. Diet consists of a variety of invertebrates, including insects, spiders, earthworms, and marine worms. Also takes some plant material. Usually seen singly or in small flocks of 20–30, but occasionally forms groups of up to several hundred. Often associates with peeps, particularly Semipalmated Sandpiper, but tends to be in the perimeter of the flock in slightly grassier habitats. Flight swift and swerving, flashing white rump-patch as the bird banks from one side to the other.

Nest initiation primarily mid-June. Egg dates range from early June to early August. Incubation 21–22 days. Mating is polygynous, with males attempting to attract and mate with more than one female. Males usually depart the breeding territory shortly after egg laying, leaving the care of eggs and young entirely to the female. One brood per season per female. Peak hatching late July. Young can fly after about 16–17 days. During display flight, male ascends to a height of about 30–80 ft. (9–24 m), then hovers on shallow wingbeats while singing before gliding back to the ground. Male also performs a grouselike dance with tail raised and white rump exposed.

**MIGRATION**

White-rump's annual movements between wintering areas in s. South America and breeding areas in n. Alaska and the upper Canadian Arctic represent one of the longest migrations in the Western Hemisphere. A good portion of the 8,000+ mi. (12,870+ km) journey is covered in several nonstop flights that can last up to 60 hours and cover 2,500 mi. (4,000 km). These marathon flights are enabled by the species' long pointed wings and fueled by extensive body-fat reserves acquired during critical migratory stop-overs.

**Spring** migration takes place between mid-February and mid-June. Departure from the wintering grounds take place as early as mid-February, with peak departure from late March to late April. Some birds apparently head to staging areas in n. South America, while others head straight to se. United States, arriving along the Gulf Coast by late April. Most spring migrants pass through the interior of the United States east of the Rocky Mts., with relatively small numbers along the Atlantic Coast and only rare strays showing up in the West. Major staging areas in the Great Plains hold peak numbers between late May and early June, where birds need to build enough fat reserves to make their flight to the Arctic and begin breeding activities. Arrival on the breeding grounds takes place primarily in mid-June, with males and females arriving at nearly the same time. Some **one-year-old** birds oversummer on the wintering grounds, while others migrate partway north to the breeding grounds.

**Fall** migration takes place between mid-July and December. **Adult** males depart the breeding grounds in early to mid-August, followed by females in late August. Failed breeders may depart as early as mid-July. Most birds head to staging area at James Bay and along the Atlantic Coast before heading to wintering grounds in South America. Peak numbers of adults pass through the mid-Atlantic states in late August and the first half of September. Adults reach the n. coast of South America between late August and early October, then arrive at their wintering areas a month or so later. **Juveniles** depart the breeding area by about mid- to late September and head primarily toward staging areas at James Bay and along the n. Atlantic Coast, with relatively few passing through the s. Atlantic Coast and the interior of the continent. Peak numbers of juveniles pass through e. Canada from mid-October to mid-November, with progressively smaller numbers farther south at the same time. Juveniles arrive on the n. coast of South America mostly in October/November, then arrive at their wintering areas a month or so later.

#### MOLT

**First life year:** Molt out of juvenal plumage takes place almost entirely on the wintering grounds between late September/late November and late winter/early spring. The molt includes most head and body feathers and a variable number of tail feathers, tertials, coverts, and primaries. Some individuals undergo a nearly complete molt. Between about March and April, while still on the wintering grounds, there is a partial molt to breeding plumage, involving a variable number of head and body feathers, tertials, coverts, and sometimes outer primaries. This plumage may resemble breeding or nonbreeding or be intermediate.

**Second life year:** Birds that remain south of the breeding grounds probably return to the wintering grounds earlier than adults, after which they molt to nonbreeding and become essentially indistinguishable from adults.

**Adult:** During March and April, while at South American staging areas, adults undergo a partial molt to breeding plumage. This molt includes most head and body feathers and a few tertials and wing coverts. The complete molt to nonbreeding plumage takes place between June and early February. Extensive head-and-body molt takes place on the breeding grounds beginning in June, then is suspended during fall migration. Extent of molt varies, possibly in relation to breeding success or sex of the bird. Flight feather molt and remainder of head-and-body molt take place after arrival on the wintering grounds from mid-October/late December to early December/early February.

#### VOCALIZATIONS

Call, given frequently day and night by perched and flying birds, is a very high pitched, insectlike *tzeet,* much higher pitched than calls of other shorebirds. Flight song high pitched and insectlike, consisting of mechanical, "typewriter-like" buzzing and bubbling noises interspersed with piglike grunts.

## Baird's Sandpiper, p. 165

### *Calidris bairdii*

#### STATUS

Fairly common migrant through the w. Great Plains and Rocky Mts., where it is sometimes the most numerous small shorebird. Scarce elsewhere in fall. Breeds in the high Arctic from the Chukotski Pen. in e. Russia east to nw. Greenland and Baffin I. Winters in s. South America from Ecuador and s. Brazil south to s. Chile and Argentina. The world population is estimated at 300,000.

#### TAXONOMY

No subspecies recognized.

#### BEHAVIOR

Breeds in exposed, dry to moist coastal and upland tundra with sparse vegetation. Typically uses the first snow-free sites available. During migration and winter uses primarily inland habitats such as lake and river shorelines, rain-soaked pastures, and rice fields. At high elevations, above treeline, often stops at snowmelt pools. In coastal areas sometimes uses tidal sand flats. Forages by walking steadily, pecking sparingly at surface. Sometimes shallowly probes and often runs briefly between foraging attempts. Eats almost exclusively insects but also spiders and crustaceans. Less gregarious than most other shorebirds. Usually seen singly or in small flocks of 20–30, but occasionally forms groups of up to several hundred, rarely over 1,000. Often associates with Least Sandpipers on lakeshores and, loosely, with Pectoral and Buff-breasted Sandpipers in pasture habitats. Flight swift and steady, with much less twisting and banking, more level gliding than other similar shorebirds.

Nest initiation early to late June. Egg dates early June to mid-August. Incubation 21 days. Both adults share parental duties, but male deserts female and young after 5–7 days. One brood per season. Rarely renests. Peak hatching early July. Young can fly after about 20 days. During display flight male ascends steeply to about 30–150 ft. (9–46 m), then sings while alternating between deep, butterfly-like wingstrokes and shallow, quivering wingstrokes along with an occasional stall with wings up in a deep "V". After a 1–3 minute display, the bird coasts to the ground with wings up in a deep "V", then holds one wing up for about a minute after landing.

**MIGRATION**

Long-distance migrant, traveling up to 9,300 mi. (15,000 km) in as little as 5 weeks. Primary route for adults is through the Great Plains in spring and through the Great Plains and Rocky Mts. in fall. Adults are scarce on the West Coast, very rare on the East Coast. Juveniles move south at a slower pace and on a broader front, occurring uncommonly along both coasts, more commonly inland.

**Spring** migration takes place between mid-February and early June. Most birds depart wintering grounds in early March and arrive along the Gulf Coast in mid- to late March, with peak passage there in mid- to late April. Passage through the n. Plains peaks in early May, with arrival at the breeding grounds between mid-May and early June.

**Fall** migration takes place between late June and mid-November. **Adults** depart the breeding grounds between late June (failed breeders) and early August (females before males) and head directly to staging areas in the n. Great Plains and Rocky Mts. Peak numbers occur in these regions in August. Less common in the w. and s. Rocky Mts. and along the Gulf Coast. Arrival at wintering sites mostly by late August. Southbound **juveniles** depart the breeding grounds mostly from late July to mid-August and arrive in the Pacific Northwest and n. Plains by late July and in New England and the Middle Atlantic by mid-August. Peak numbers pass through the northern tier of states from mid-August to mid-September and through the south from late August to early October. Arrival at wintering sites mostly late September and early October.

**MOLT**

**First life year:** The molt out of juvenal plumage takes place entirely on the wintering grounds from about late October to late March. This variable molt may be complete or may involve most head and body feathers and a few outer primaries. Between February and May there is a partial molt into breeding plumage involving most head and body feathers and inner coverts. This molt begins on the wintering grounds and is completed on spring migration. It resembles adult breeding, but some individuals retain worn juvenal inner primaries.

**Adult:** Between February and May

there is a partial molt into breeding plumage involving most head and body feathers and inner coverts. This molt begins on the wintering grounds and is completed on spring migration. The complete molt to nonbreeding plumage begins on the breeding grounds in late June or early July with a variable number of head and body feathers. The molt is then suspended during migration and completed after the wintering grounds are reached from about October to late March or early April.

**VOCALIZATIONS**

Flight call is a low, trilled *preep,* similar to Pectoral Sandpiper's but higher pitched. Song begins with a buzzy, frog-like series of gently rising notes, *drray drray drray . . . ,* followed by a repeated, buzzy *hee-aaw, hee-aaw, hee-aww . . . ,* not unlike a distant braying donkey. The song is usually finished with a chatter.

## Pectoral Sandpiper, p. 169

### *Calidris melanotos*

**STATUS**

Fairly common migrant through much of North America, being most numerous in the Great Plains and scarce in the interior West. Breeds in the Arctic from the Yamal Pen. in w. Russia east to w. Hudson Bay. Winters in s. South America from s. Peru, Bolivia, and s. Brazil south to central Chile and s. Argentina. Also in Australia, New Zealand, and Polynesia. The North American population is estimated at 400,000.

**TAXONOMY**

No subspecies recognized.

**BEHAVIOR**

Breeds in moist to wet grassy tundra with scattered raised ridges or hummocks, particularly on the coastal plain. During migration and winter prefers grassy habitats such as rain-soaked pastures, grassy margins to ponds or lakes, and salt marshes. Rarely ventures onto open mudflats or beaches. Forages by walking slowly with head down, pecking and probing at surface. Usually stays in a relatively small area and seldom runs. When alert, may stand upright with neck extended. Eats a variety of invertebrates, particularly adult and larval insects, as well as seeds and algae. Less gregarious than many other shorebirds. Usually seen singly or in small flocks of 20–30, but sometimes forms groups of up to 1,000 or more at prime stop-over sites in the Great Plains. Often loosely associates with American Golden-Plover and Buff-breasted Sandpiper in pasture habitats. On the breeding grounds male has fatty inflatable throat sac that hangs down like a dewlap while the bird is standing.

Nest initiation primarily early to late June. Egg dates late May to late July. Incubation 21–23 days. Female does all incubation and brood rearing. One brood per season but will likely renest after failure. Peak hatching early July. Young can fly after about 23–25 days. During display flight male flies in low (3–6 ft.; 1–2 m), often undulating circles around territory on slow, high wingbeats interspersed with glides. As the bird flies, its head bobs rhythmically and the throat sac expands and contracts, producing hooting call.

**MIGRATION**

Longer-distance migrant than any other North American shorebird. Majority pass through the e. Great Plains, particularly in spring, with smaller numbers elsewhere. Many Siberian breeders apparently cross Arctic Ocean and North America to wintering grounds in South America. Unlike most other shorebirds, males migrate before females in fall.

**Spring** migration takes place between late February and late June. Most birds depart wintering grounds in March and arrive along the Gulf Coast in early to mid-March, with peak passage there between late March and late April. Passage through the n. Plains takes place from mid-April to late May, with a peak in early to mid-May. Less common east of the Mississippi R. with passage through the Midwest and Northeast from early March to late May, peaking in mid- to late April. Rare along the s. Pacific Coast and scarce along the n. Pacific Coast in April and May. Arrival at breeding grounds takes place in late May to mid-June. First-year birds may arrive as late as mid- to late June. Rarely oversummers at wintering or migration sites.

**Fall** migration lasts between late June and early December. **Adult** males depart the breeding grounds in early to mid-July, females in mid-July to mid-August. Failed breeders may depart as early as late June. Most adults pass through the center of the continent, with smaller numbers along the Atlantic Coast and only a few down the Pacific Coast. Peak numbers pass through the n. Pacific Coast and n. Plains from mid-July to mid-August and through the s. Pacific, Gulf, and mid-Atlantic Coast in August. Adults are rare in North America after September. Arrival at wintering sites mostly by mid-September. Southbound **juveniles** depart the breeding grounds mostly in mid-August (earlier in Siberia) and begin to arrive in the n. Plains and along the n. Pacific and n. Atlantic coasts by mid- to late August. Peak numbers occur through most of the continent from about mid-September to mid-October, with stragglers lingering into November. Relatively scarce in southern regions, as these are apparently overflown by most birds. Arrival at wintering sites mostly late October.

#### MOLT

**First life year:** Molt out of juvenal plumage takes place entirely on the wintering grounds between October/November and March/April. This molt is variable but typically includes most head and body feathers, some median coverts, and sometimes tertials and central tail feathers. Some individuals may undergo a complete molt, including primaries and secondaries. Between late February and May there is a partial molt into breeding plumage involving a variable number of head and body feathers and sometimes a few coverts, tertials, and tail feathers. This molt takes place on the wintering grounds. It resembles adult breeding, but some individuals retain worn juvenal flight feathers and some coverts. Others may be indistinguishable from adults.

**Adult:** Between late February and early May there is a partial molt into breeding plumage involving most head and body feathers, some tertials, and some median coverts. This molt takes place on the wintering grounds before migration. The complete molt to nonbreeding plumage begins on the breeding grounds in late June or early July, with a variable but usually small number of head and body feathers. The molt is then suspended during migration and completed after the wintering grounds are reached from about October to February.

#### VOCALIZATIONS

Flight call is a variable low, trilled *chrrk,* similar to Baird's Sandpiper's but usually lower pitched. Display song is an eerie, rapidly repeated hooting, *hooa-hooa-hooa-hooa . . . ,* like a fanciful alien spaceship.

## Sharp-tailed Sandpiper, p. 294

***Calidris acuminata***

#### STATUS

Asian species. Uncommon to fairly common in fall and casual in spring in w. Alaska; rare in fall along Pacific Coast;

casual elsewhere. Breeds in n. Siberia. Winters primarily in Australia, with smaller numbers in New Guinea and New Zealand.

**TAXONOMY**

No subspecies recognized.

**BEHAVIOR**

A bird of fresh and saltwater marshes, grasslands, and mudflats. Breeds in moist or wet Arctic tundra with scattered raised ridges or hummocks. Also uses shrub-covered hummocks in peat hollows. During migration and winter frequents a variety of habitats, including tidal and inland mudflats, fresh and salt marshes, grasslands, and drier wetland margins. Feeds in similar fashion to Pectoral, picking food from surface of mud or grasses or probing in mud or shallow-water substrates. More likely to utilize mudflats for feeding than Pectoral. Often crouches low to ground to avoid detection when alarmed. Food source not described, but probably similar to Pectoral Sandpiper's. May be seen singly or in small to large flocks during migration and winter.

Nests from late May to mid-July.

**MIGRATION**

Long-distance migrant. Migrates across a broad front, both overland and coastal. Unlike most other shorebirds, males migrate before females.

**Spring** migration takes place between March/April and early June, with peak movements in early May. Departs wintering grounds in late March/April and arrives at breeding areas in late May/early June. Casual spring sightings in mainland North America occur primarily from late April to mid-May and in w. Alaska in late May/early June. Very few nonbreeders remain at wintering areas during northern summer months, but destination of nonbreeding migrants is unreported.

**Fall** migration takes place from late June to late November. **Adult** males depart the breeding grounds mostly in July, with females following in August. A few birds, presumably failed breeders, depart as early as late June. Casual sightings of adults in mainland North America occur mostly from late June to early August. **Juveniles** gather in large numbers at staging areas in Alaska's Yukon-Kuskokwin Delta in August/September, with a few lingering into October. Modest numbers occur elsewhere in w. Alaska at the same time. Small numbers of juveniles pass through the U.S. Pacific Coast, mostly from mid-September to early November. Peak numbers occur in se. Asia in August/September (adults and juveniles), though some still moving in November (mostly juveniles). Arrives at wintering grounds from late August to November, with peak arrival in September.

**MOLT**

**First life year:** Molt out of juvenal plumage is either partial or nearly complete and occurs primarily at wintering sites from September/December to March. Limited head-and-body molt rarely takes place before wintering grounds are reached. Commencement of heavy molt varies in timing, probably corresponding with arrival on wintering grounds. Some birds undertake an almost complete molt from December to March, including outer primaries, tail, wing coverts, tertials, and back. Other birds molt only limited head and body feathers. From late January to April there is a variable partial molt to breeding plumage. This molt includes most head and body feathers, many tertials, a few wing coverts, and maybe some tail feathers. One-year-old birds may be distinguished from adults by the retention of worn juvenal (at least inner) primaries; sometimes extensive retained juvenal plumage.

**Adult:** The partial molt to breeding plumage begins in late January and is completed by late March/early April. It includes most head and body plumage, many tertials, some wing coverts, and some tail feathers. The complete molt to nonbreeding plumage takes place between mid-July and mid-January, mostly later in migration or on the wintering grounds. Only limited molt takes place by late August, with more extensive molt of most head and body feathers by mid-September. Completion occurs after arrival on wintering grounds with back, rump, tail, wing coverts, and flight feathers being molted last. Molt is completed in w. Australia by mid-January.

**VOCALIZATIONS**

Flight call is a soft, whistled *weep* or *treeip;* often given in short, twittering sequence recalling Barn Swallow; very different from the lower, huskier Pectoral Sandpiper.

## Purple Sandpiper, p. 173

### *Calidris maritima*

**STATUS**

Fairly common but highly localized. Breeds in Arctic and subarctic regions from Banks I. in Canada east to Laptev Sea in central Siberia. Has a more northerly distribution in winter than any other shorebird. Winters from se. Labrador south along the Atlantic Coast to n. Florida as well as in s. Greenland, Iceland, and w. Europe from Norway to Spain. Rare but regular migrant through the Great Lakes region and rare winter visitor along the Gulf Coast. The world population is estimated at 65,000, with 15,000 in North America.

**TAXONOMY**

No subspecies currently recognized. Variation in size from the smallest birds breeding in James and Hudson Bays to the largest breeding in Iceland.

**BEHAVIOR**

A bird of cold weather and rocky coastlines. Hops from rock to rock, fluttering to avoid crashing waves. In flight always low to the water. Breeds on barren Arctic and alpine tundra as well as gravel riverbanks and beaches. During migration and winter strictly tied to rocky, wave-pounded coastlines. At more southern latitudes found almost exclusively on man-made rock jetties or breakwaters. Rarely seen on adjacent tidal flats. Forages mostly visually. In rocky habitats forages by walking quickly, picking prey from barnacles, mussels, rock crevices, or seaweed. On tundra picks prey off vegetation or probes shallowly into soil. Eats a variety of invertebrates, including gastropods, insects, spiders, crustaceans, and worms as well as some plant material. Feeds during both day and night. Typically seen in small flocks of 20–30 but occasionally forms groups of over 100. Roosting birds form tight packs on sheltered, sunny side of rocks.

Nest initiation primarily mid- to late June (late May in Iceland). Egg dates early June to late July. Incubation 21–22 days. Male does most incubation and tends chicks alone. One brood. Peak hatching mid-July. Young can fly 3–4 weeks after hatching. Male's display flight begins with a series of flutters and stiff-winged glides to a height of 25–80 ft. (8–24 m), followed by wide circular glides with head up and chest out as it delivers its song. The flight is completed by a rapid zigzag descent to the ground with wings held up in a "V" as terminal portion of song is delivered.

**MIGRATION**

Short- to medium-distance migrant. More northerly breeders apparently migrate southeast to winter in Greenland and Europe, while southerly breeders migrate across Quebec and Labrador to winter on the American side. Some populations (for example, s. Greenland) apparently resident.

In **spring,** wintering birds depart the Atlantic Coast between April and early June, with peak departure in early May. The very few interior migrants occur from late March to mid-June, with a peak in early May. In the Arctic, spring migration is mostly in the first half of June. Arrival at breeding sites takes place between late May and mid-June, males arriving before females. **One-year-old** birds tend to depart later than adults but only rarely remain at wintering sites through the summer.

In **fall,** southbound migration is protracted. **Adult** females depart the tundra by mid-July and head to nearby coastal areas, where they remain until early August. Males are about two weeks behind females and head south by mid-August. The first few migrants arrive in the Maritime Provinces in early August, with the first major wave in mid-October and peak numbers by mid-November. In the mid-Atlantic, first arrivals begin in late September (rarely earlier), with the first major wave in late October and peak numbers by late November or December. **Juveniles** head to coastal staging areas by mid-August and depart to join adults by mid-September. Rare fall/winter visitor in the Great Lakes region, mostly from mid-October to mid-January, and on the Gulf Coast, mostly from November to early April.

#### MOLT

Molt patterns poorly studied in North America.

**First life year:** A partial molt out of juvenal plumage begins near the breeding grounds in September, with most head and body feathers replaced by late October/late November (the time when the wintering grounds are reached). Wing coverts, tertials, and flight feathers are retained. Between about late March and early May another partial molt into breeding plumage takes place involving many head and body feathers, but juvenal wing feathers (including coverts and tertials) are retained.

**Adult:** Between late March and early May adults undergo a partial molt to breeding plumage involving most head and body feathers and some tail feathers. The complete molt to nonbreeding plumage takes place rapidly between early July and mid-September, mostly in the Arctic before migration. Females molt about 2 weeks earlier than males.

#### VOCALIZATIONS

Flight call is a low, scratchy *kweesh,* sometimes extended into a chatter. Chase call is a higher *kwi-ti-ti-dr.* Song, usually given in flight over territory, is two- to three-parted with a shrill introductory chatter, *kwi-ti-ti-ti-bli-bli-bli-bli-bli-bli,* followed by a buzzy, rising *dooreee dooreee dooreee* . . . Each element may be given in a long series. Also gives a rapid, playful *dr-dr-dr-dr-dr* . . . , sometimes in long series.

## Rock Sandpiper, p. 176

***Calidris ptilocnemis***

#### STATUS

Fairly common but highly localized in coastal areas. Breeds in the Bering Sea region from the Kamchatka and Chukotski Pens. in e. Russia east to the Seward Pen. and Kodiak I. in w. Alaska. Many populations resident. Others winter in Japan and from s. Alaska to central and rarely s. California. The world population is estimated at 100,000 to 200,000, the majority of these in North America.

#### TAXONOMY

Four subspecies generally recognized, three occurring regularly in North America. Nominate *C. p. ptilocnemis* ("Pribilof Sandpiper") breeds on the

Bering Sea islands of St. Paul, St. George, St. Matthew, and Hall, Alaska, and winters along the s. Alaskan coast, primarily at Cook Inlet, rarely south to Washington. *C. p. tschuktschorum* ("Northern Rock Sandpiper") breeds from Russia's Chukotski Pen. east to w. Alaska from Seward Pen. and St. Lawrence I. south to Bristol Bay; winters from s. Alaska to California. *C. p. couesi* ("Aleutian Sandpiper") is largely resident throughout the Aleutian Arch. east to the Alaska Pen. and Kodiak I.; rarely winters south to Washington. *C. p. quarta* ("Commander Sandpiper") is largely resident in Commander and Kuril I. off Russia's Kamchatka Pen. but has occurred in w. Alaska.

*C. p. ptilocnemis* is the largest, palest subspecies and is distinctive in all plumages. *C. p. couesi* and *tschuktschorum* are very similar and not always distinguishable, but *couesi* averages darker in all plumages. *C. p. quarta* is also very similar to the latter two subspecies but is marginally brighter above in breeding plumage and averages smaller.

**BEHAVIOR**

Like Purple Sandpiper in the Atlantic, a bird of cold weather and rocky coastlines. Breeds primarily on lowland heath tundra along the coast. Some also breed on montane subarctic tundra with low vegetation in the coastal mountains of w. Alaska. During migration and winter found on rocky headlands, gravel or cobble beaches, and, in some areas, mud or sand flats. In southern parts of wintering range more strictly tied to rocky, wave-pounded coastlines. Forages mostly visually. In rocky habitats forages by walking quickly, picking prey from barnacles, mussels, rock crevices, or seaweed. On tidal flats or tundra walks quickly and picks or shallowly probes. During the breeding season eats primarily adult and larval insects and spiders on the tundra. During the remainder of the year feeds on invertebrates, especially marine clams and snails and larval flies. Feeds during both day and night. Typically seen in small flocks of 20–30 but occasionally forms groups of several hundred.

Nest initiation primarily mid-May in south to early to mid-June in north. Egg dates early May to late July. Incubation 20–24 days. Both parents share parental duties, but female usually departs before brood fledges. One brood but will renest after failure. Peak hatching mid-June in south to early July in north. Young can fly at 20–22 days. Male's display flight involves a steep rise to a height of 30–325 ft. (9–100 m), at which point it hovers on alternating fluttering and arched wings as it delivers the first part of its song. The display ends in a steep descent with wings held in a "V" as terminal portion of song is delivered. Bird often does single-wing lift after landing.

**MIGRATION**

Short- to medium-distance migrant with movements either coastal or over water. *C. p. couesi* largely resident with only minor local movements. *C. p. ptilocnemis* migrates a short distance and exhibits marked facultative movements. *C. p. tschuktschorum* migrates up to 5,600 mi. (9,000 km) between Russia and California.

In **spring,** *tschuktschorum* departs wintering areas in California between mid-March and mid-May, with peak numbers passing through British Columbia between mid-April and early May. Arrival on the breeding grounds takes place mostly in mid- to late May. *C. p. ptilocnemis* probably departs wintering areas in March and early April, arriving at breeding sites in mid- to late April. *C. p. couesi* also returns to breeding sites in mid- to late April.

In fall, **adult** *ptilocnemis* and *tschuktschorum* depart breeding areas and move to the mainland Alaska coast to molt by mid- to late July, remaining until early October. Some *couesi* apparently move to Gulf of Alaska sites to molt in July and remain through the winter. *C. p. ptilocnemis* reaches wintering sites in the Cook Inlet area by mid-October. These birds regularly move up to 100 mi. (160 km) during severe winter cold spells, then return when temperatures rise. *C. p. tschuktschorum* reaches British Columbia by mid-September (occasionally as early as late July) and peaks there in late October/early November. California wintering areas are reached by early November. **Juveniles** of all races join adults at coastal staging areas by August or September and generally migrate with them.

**MOLT**

**First life year:** Head-and-body molt out of juvenal plumage begins by late July and is completed between September and November, before migration to wintering areas. Flight feathers, most wing coverts, and sometimes a few tertials and scapulars are retained. Between late February and early May there is a partial molt into breeding plumage involving most head and body feathers. This plumage resembles adult breeding but is distinguished by retained juvenal primaries, coverts, and some tertials.

**Adult:** Between late February and early May there is a partial molt into breeding plumage involving most head and body feathers. The complete molt to nonbreeding plumage begins in early July and is completed between September and late October at postbreeding staging areas.

**VOCALIZATIONS**

Flight call is a short, scratchy *kwit,* sometimes extended into a chatter. *C. p. ptilocnemis* gives a lower *cheet.* Chase call is a higher, softer *kwi-di-di-di.* Male's song, given in display flight over territory, is rolling and buzzy, beginning with a harsh, pulsating *di-jrrr di-jrrr di-jrrr di-jrrr* . . . delivered as bird hovers, then higher, more melodic *quida-we-quida-we-quida-we-quida-we* . . . as he descends to the ground.

## Dunlin, p. 180

### *Calidris alpina*

**STATUS**

Locally abundant and widespread, especially in coastal areas, occurring nearly throughout the Northern Hemisphere. Breeds in subarctic and Arctic regions of Alaska, Canada, e. Greenland, and from Iceland to Siberia. Winters from s. Alaska and Nova Scotia south to n. Mexico as well as in Europe, n. Africa, the Middle East, n. India, and se. Asia. Generally scarce migrant inland except through California's Central Valley, the w. Great Basin, the n. Prairies, and the Great Lakes region. The world population is estimated at 3,934,000, with 1,252,000 in North America.

**TAXONOMY**

Complex, with seven subspecies recognized worldwide. Three subspecies occur regularly in North America and a fourth reported as a vagrant. *C. a. arcticola* breeds in nw. Canada and n. Alaska north of Lisburne Pen. and winters in se. Asia. *C. a. pacifica* breeds in w. Alaska south of Lisburne Pen. and winters along the American Pacific Coast. *C. a. hudsonia* breeds in central Canada and winters along the Atlantic and Gulf coasts. *C. a. arctica* breeds in e. Greenland and winters primarily in nw. Africa. It has been reported as a vagrant along the Atlantic Coast. Other Old World subspecies include *alpina* (breeds n. Scandinavia, nw. Siberia; winters w. Europe, Mediterranean, w. India), *schinzii* (breeds se. Greenland, Iceland, Britain, s. Scandinavia; winters w. Africa), and *sakhalina* (breeds ne. Siberia; winters se. Asia).

Differences in appearance are subtle and somewhat clinal. The three North American races have brighter upperparts in breeding plumage than Old World races. *C. a. pacifica* is the largest and longest billed. *C. a. hudsonia* is very similar but tends to be more heavily streaked in breeding plumage, with a less distinct white band between the breast and belly-patch and more extensive streaking along the flanks and undertail coverts (flanks and undertail coverts also more heavily streaked in nonbreeding plumage). *C. a. arcticola* is slightly smaller and shorter billed (distinctly so in males) than *pacifica* and averages whiter on the breast and more suffused with gray above in breeding plumage. *C. a. arctica* is about 15–20 percent smaller than North American races, with a 25–30 percent shorter bill. Breeding plumage is drab with dull brownish-gold scapulars, heavy streaking on head and breast, and a smaller black belly-patch than North American races. Unlike North American races, adults molt to nonbreeding plumage on the wintering grounds, so fall migrants are in worn breeding plumage.

**BEHAVIOR**

Breeds in moist to wet tundra, often near ponds and short ridges. Staging birds in late summer use barren intertidal flats. During migration and winter, closely tied to intertidal areas, preferring extensive mudflats and sandy beaches. To a lesser degree also uses flooded fields and other seasonal wetlands. Tactile feeder that forages by probing, picking, and jabbing at substrate, often with bill open. Walks steadily during foraging attempts until prey items are found. Occasionally runs from one foraging site to the next. Eats a wide variety of invertebrates, including clams, worms, insect larvae, and amphipods. Occasionally eats plant material or small fish. Feeds during both day and night. Typically seen in large flocks of several hundred or more, both while foraging and roosting. Roosting birds form tight packs.

Nest initiation primarily early June *(pacifica)* and mid-June *(arcticola* and *hudsonia)*. Egg dates late May to early July. First-year birds nest slightly later than adults. Incubation 21–22 days. Both adults share parental duties, but female deserts male and young after about 6 days. One brood per season but will renest after failure. Peak hatching late June *(pacifica)* and early July *(arcticola* and *hudsonia)*. Young can fly 18–24 days after hatching. Display flight, performed mostly by male, involves a series of flutters and short glides on stiff arched wings at a height of 50–100 ft. (15–30 m).

**MIGRATION**

Short- to medium-distance migrant. Migrates by day or night, depending on weather conditions.

In **spring,** *arcticola* departs Asian wintering areas in March, with peak numbers passing through Japan in mid-May. Arrival on the breeding grounds takes place in late May. *C. a. pacifica* begins to depart wintering areas as early as January, with peak numbers along the central Pacific Coast in April and early May and at the Copper River Delta, Alaska, in early to mid-May. Arrival on the breeding grounds takes place in mid-May. *C. a. hudsonia* begins to depart wintering areas along the Gulf and s. Atlantic coasts in mid-March, with abrupt departure in early May. Interior migrants make direct flight to stop-over areas in n. Prairies and Great Lakes, where peak numbers occur in mid- to late May. Atlantic Coast migrants make more gradual northward movement, with peak numbers occurring in the mid-Atlantic in early to mid-May and in New England in mid- to late May. Arrival on the breeding grounds takes place in late May. In all subspecies males arrive before females. Some **one-year-old** birds remain at wintering sites through the summer, while others migrate north to breed.

In **fall, adult** *arcticola* departs breeding grounds and moves to nearby coastal staging areas from early July to early September, with a peak presence in mid-August. Most birds then move to coastal w. Alaska in September/October, where they mix with *pacifica* before departing to se. Asian wintering areas. Peak passage through Japan takes place between mid-September and late October, though some arrive as early as late July. Adult *pacifica* departs breeding grounds to nearby coastal areas, with a presence there from late June to early November and peak abundance from mid-August to early October. Most of these birds depart abruptly in early October and make single transoceanic flight to coastal areas from s. British Columbia to central California. Some of these birds gradually drift south or inland to wintering areas in Mexico and California's Central Valley. Adult *hudsonia* departs breeding grounds to staging areas along Hudson and James Bays between July and early September. Less migration through the interior than in spring. Peak numbers pass through New England and the mid-Atlantic in October and November. **Juveniles** of all three races join adults at coastal staging areas by mid-August and generally migrate south with them. However, juvenile *pacifica* often move before adults.

**MOLT**

The adult's molt to nonbreeding plumage takes place on the breeding grounds in *pacifica, arcticola,* and *hudsonia* and on the wintering grounds in *arctica.* However, a few individuals of *pacifica, arcticola,* and *hudsonia* migrate before molting extensively.

**First life year:** Head and body molt out of juvenal plumage begins in late July or early August, before juvenal plumage is fully grown. Most birds undergo moderately extensive molt before departure from the Arctic. This molt is completed by November or December after the wintering grounds are reached. Flight feathers and most wing coverts are retained. Between late February and early May there is a variable molt into breeding plumage, involving most head and body feathers. This plumage resembles adult breeding but is typically duller with fewer brightly colored scapulars. Some individuals completely lack breeding plumage and resemble nonbreeding. All first-summer birds may be distinguished by retained juvenal primaries and some coverts.

**Second life year:** For birds oversummering on the wintering grounds the complete molt to nonbreeding plumage takes place primarily between late April (of first life year) and mid-August, with primary molt beginning as early as early June. For birds summering on the breeding grounds the schedule is probably more like adult's. Between late February and May, second-year birds undergo a variable head and body molt, after which they resemble adults.

**Adult:** Between late February and early May adults undergo a partial molt to breeding plumage involving most head and body feathers and some tail feathers. Much of this molt takes place rapidly in late April. The complete molt to nonbreeding plumage begins shortly after arrival on the breeding grounds in late May or early June. Primary molt takes place gradually through the summer and is completed by mid-August to mid-September at staging areas. Most body feathers are molted between late June and mid-September *(pacifica)* or July and August *(hudsonia)*, but the molt is often completed on the wintering grounds as late as December.

**VOCALIZATIONS**

Flight call is a buzzy, even-pitched *jeeezp*, often given by nocturnal migrants. Threat call is a buzzy *droy droy droy*, similar to Semipalmated Sandpiper's but buzzier. Song is a 3–4 second buzzy, chattering trill, rising slightly then descending; often combined with a repeated cricket-frog-like *wrraah wrraah wrraah* . . . Song given while standing or in flight; given primarily on the breeding grounds but also by migrants, particularly in spring.

## Curlew Sandpiper, p. 296

### *Calidris ferruginea*

**STATUS**

Eurasian species. Rare but regular migrant (mostly adults) along Atlantic Coast; casual elsewhere. Fairly common within range. Breeds primarily in narrow band of coastal tundra across n. Siberia; has bred casually in n. Alaska (1962–72). Winters mainly from sub-Saharan Africa east to se. Asia, Australia, and New Zealand, with smaller numbers in Israel, Iraq, and occasionally w. Europe.

**TAXONOMY**

No subspecies recognized, but females of eastern population (ne. Siberia) show less dark barring on underparts in breeding plumage than western birds.

**BEHAVIOR**

A bird of fresh and saltwater wetlands. Breeds in high Arctic coastal tundra, with nest site situated on margins of boggy depressions and tundra pools. During migration and winter, prefers muddy inland and coastal wetland fringes, beaches, and tidal mudflats. Feeds while walking quickly and steadily, head down, continuously pecking and probing at surface of mud. Occasionally runs. Will also wade or swim in fairly deep water, probing like Stilt Sandpiper in submerged mud. Eats a wide variety of prey, but mostly invertebrates, including insects (adults, pupae, and larvae), crustaceans, mollusks, and worms. Plant material, especially seeds, also part of diet. Often seen in large flocks away from breeding grounds. In North America most often seen with Semipalmated or Western Sandpipers and occasionally with Dunlin. Normal flight is swift and low, though higher in migration.

Nest initiation primarily mid-June. Egg dates at least mid-June to mid-July (probably longer). Incubation period unreported but probably 21–23 days, as in similar-sized *Calidris.* Parental duties by female only, with males departing breeding grounds by late June, shortly after egg laying. One brood per season, but probably replaces eggs after loss. Peak hatching late June/early July. Fledging period unknown. Age of first breeding 2 years.

**MIGRATION**

Long-distance migrant. Two geographically separate breeding populations migrate across a broad front, including coastal w. Europe and overland across e. Europe and Siberia. Migration is rapid and generally involves long uninterrupted flights.

**Spring** migration takes place mostly between April and early June. Departs wintering areas in April and undertakes long-distance flights to arrive on breeding grounds in early June. Spring sightings along the U.S. Atlantic Coast take place from late April to mid-June, with a peak in the second half of May. Most **one-year-old** birds remain at wintering areas through the summer.

**Fall** migration takes place from late June to October. **Adults** depart breeding grounds from late June (males) to August and head straight to staging areas, usually along the coast, where they often remain for extended periods to molt. Adults then continue their journey with rapid continuous flights, arriving on wintering grounds in September. Fall sightings of adults along the U.S. Atlantic Coast take place from late June to late September, with a peak from mid-July to early August. **Juveniles** move south about a month later than adults, mostly August to October, and arrive at wintering areas primarily in September/October. Rare sightings of juveniles in North America take place primarily between mid-September and mid-October.

**MOLT**

**First life year:** The partial molt out of juvenal plumage takes place between late September/October and November/December, mostly after arrival at wintering grounds. This molt includes most head and body feathers and some tertials. Very limited body molt may take place during migration. Between early February and early May many birds replace outer primaries and often replace remaining tertials and tail feathers at same time. This molt is highly variable, with some birds replacing all primaries and others none. Between late February and May there is a partial molt to breeding plumage involving a variable number of head and body feathers. Most one-year-old birds remain on wintering grounds through the summer, and many of these skip this molt.

**Second life year:** For birds that oversummer on the wintering grounds, the complete molt to nonbreeding plumage begins late July/September and is completed from October/December. The molt starts with replacement of p1, wing coverts, and scattered feathering on head and body from late July to early September, with rest of molt completed after this time. Last juvenal inner primaries are shed in September. After this molt, second-year birds are on the same schedule as adults, with a partial molt to breeding plumage.

**Adult:** The partial molt to breeding plumage begins late January/March, is suspended during migration, then completed April/early May. It includes most head and body feathers, inner tertials, and some wing coverts and tail feathers. The complete molt to nonbreeding plumage takes place from early July to February. Limited molt may begin on the breeding grounds. After arrival at stop-over sites, extensive head and body molt takes place between late July to early September. Molt is then suspended during migration and completed at wintering sites. Head and body molt is completed by early September/mid-October. Flight feather molt takes place mostly at wintering sites, early September/early October to December/February; in some cases may begin at stop-over sites from late July and be suspended during migration.

**VOCALIZATIONS**

Flight call a low, trilled *chrreep,* with a quality like Pectoral's but with a pattern recalling Least. Songs on breeding grounds include a series of chatters, trills, and whinnies.

## Stilt Sandpiper, p. 185

***Calidris himantopus***

**STATUS**

Locally common migrant through the Great Plains in spring and fall, and along the Atlantic Coast in fall. Scarce in the West. Breeds in coastal subarctic to Arctic areas from n. Alaska east to James Bay. Winters primarily from the w. Gulf Coast south to Costa Rica and in interior South America from e. Peru and s. Brazil to n. Chile and n. Argentina. Small numbers winter locally at widely scattered sites, including inland. Estimates of the world and North American population range from 50,000 to 200,000.

**TAXONOMY**

No subspecies recognized.

**BEHAVIOR**

Breeds primarily in open marshy tundra with scattered pools. Less often uses drier tundra. Prefers coastal locations but nests farther inland during late-snowmelt years. During migration and winter prefers flooded fields, marsh pools, impoundments, and quiet lagoons. Seldom uses mudflats. Typically forages belly-deep in water, where it walks with neck stretched out and bill pointed down, occasionally picking at the surface. Also probes with dowitcher-like "sewing-machine" motion but usually with head under water and tail tilted up. More active than dowitchers. Eats a variety of invertebrates, particularly adult and larval insects, as well as snails and seeds. Usually found in flocks of a few dozen to over 1,000. Often associates with Lesser Yellowlegs and dowitchers, particularly Long-billed.

Nest initiation primarily early to mid-June. Egg dates early June to mid-July. Incubation 19–21 days. Both adults share parental duties. One brood per season but will renest after failure. Peak hatching early July. Young can fly after about 17–18 days. The male's display flight involves a mix of hovering and gliding over territory at a height of 65–200 ft. (20–60 m) as song is monotonously repeated. Display may last for several minutes and usually ends with the male swooping low over territory, then flying out of sight.

**MIGRATION**

Long-distance migrant. Primary route is through the Great Plains in spring and fall and along the Atlantic Coast in fall, with smaller numbers elsewhere. Migrates mostly at night and early morning.

**Spring** migrants depart South America between late February and mid-May, with arrival along the Gulf Coast beginning in March. Peak numbers pass through the Gulf Coast and s. Plains states in early to mid-May and through the n. Great Plains in the second half of May. Small numbers pass through se. United States, mostly from late April to mid-May. Rare in the West in spring except at Salton Sea, California, where peak numbers pass through in mid-May.

Arrival on the breeding grounds is mostly in late May to mid-June. Most **one-year-old** birds oversummer on or near the wintering grounds, though a few apparently migrate north to breed.

**Fall** migration more widespread than spring, though still most numerous through the prairies and at many sites along the Atlantic Coast. **Adults** depart the breeding grounds in late June (failed breeders), early to mid-July (females), and mid- to late July (males) and make nonstop flights to stop-over sites. Peak numbers occur at stop-over sites through much of the continent from late July to mid-August. Arrival at wintering grounds early August to early September. Southbound **juveniles** depart the breeding grounds from late July to early September. Peak numbers appear at stop-over sites through much of the continent from mid-August to mid-September. Arrival at wintering sites mostly late September.

**MOLT**

**First life year:** Molt out of juvenal plumage takes place from about mid-August to February, mostly at migratory stops. Most head and body plumage and some wing coverts are replaced by late September/October when wintering grounds are reached. Between November and February outer 1–7 primaries are replaced along with remaining head and body feathers, wing coverts, and central tail feathers. Between January and May there is a partial molt into breeding plumage involving a variable number of head and body feathers. This plumage may resemble adult breeding or may be intermediate between breeding and nonbreeding. One-year-old birds may be distinguished by worn juvenal inner primaries and outer tail feathers.

**Second life year:** For birds that oversummer on the wintering grounds the complete molt to nonbreeding plumage takes place between June and September, a month earlier than adults. Birds that migrate to the Arctic may be on a more adultlike schedule. After this molt, second-year birds are usually indistinguishable from adults. Between mid-February and mid-May there is a partial molt into breeding plumage involving many head and body feathers, sometimes 1–2 wing coverts, and central tail feathers.

**Adult:** Between mid-February and mid-May there is a partial molt into breeding plumage as in second-year birds. The complete molt to nonbreeding plumage takes place from early July to December/February. Limited head and body molt begins on the breeding grounds in early July, then heavy molt takes place at stop-over sites from about late July to mid-August. Head and body molt is completed on the wintering grounds by October. Primary molt begins in the Gulf Coast and Caribbean region in late August and is suspended, then completed by late December/February on the wintering grounds.

#### VOCALIZATIONS

Flight call is a low, muffled *pewf,* very unlike calls from other shorebirds. Calls frequently while flying and while feeding. Also gives a scratchy *krrit* and a sharp *kew-it.* Display song is a scratchy, gurgling *k-k-k-kroy-kroy-kroy-kreee-aaaw, kreee-aaaw* . . . and a more musical, rising *urrEEE, urrEEE, urrEEE* . . . , sometimes repeated for minutes on end.

## Spoon-billed Sandpiper, p. 298

***Eurynorhynchus pygmeus***

#### STATUS

Asian species. Casual migrant in w. Alaska; accidental in Alberta (two spring records) and British Columbia (one fall record). Rare, endangered, and declining within range due largely to heavy predation on the breeding grounds by Arctic Foxes and loss of key Yellow Sea staging sites. Breeds in Arctic coastal tundra in ne. Siberia. Winters coastally from se. India east to Singapore, se. China, and Hainan in small numbers. World population is estimated at 1,000 to 2,500 birds.

#### TAXONOMY

No subspecies recognized.

#### BEHAVIOR

A bird of coastal habitats. Breeds on Arctic coastal tundra, favoring grassy areas near freshwater pools or stream mouths, sparsely vegetated ridges, and sandy-banked lagoons. During migration and winter found primarily along coastlines in lagoons or muddy tidal flats. Also uses grassy or algae-covered margins of freshwater pools. Feeds mostly in shallow water or soft mud. Walks forward with head down, sweeping spatulate bill back and forth "vacuum-cleaner" style like a small spoonbill. Also picks food from surface of mud and sometimes "quivers" bill tip momentarily as if to strain out unwanted material. Probably eats a variety of invertebrates, like other small shorebirds. Usually found singly or in small groups, often with Red-necked Stint.

Nests mid-June to late July.

**MIGRATION**

Long-distance migrant. Most movement apparently along the coast.

**Spring** migration takes place primarily between April and mid-June. Most birds depart wintering areas in April and arrive at breeding areas in early to mid-June. No information available regarding where nonbreeding birds oversummer.

**Fall** migration takes place at least between July and October. Based on female's lack of participation in parental duties, some females may depart breeding grounds as early as early to mid-July after young hatch. Most **adults** depart breeding areas by early August and first arrive at staging areas in Japan, Korea, and China by early to mid-August, where some remain at least through September. **Juveniles** depart breeding areas by mid-August and pass through Japan and Korea mostly from September to mid- to late October. Arrival on wintering grounds unknown, but possibly between late August and late October.

**MOLT**

Very little information is available on molt schedule and timing.

**First life year:** Juvenal plumage held at least through October, but timing of molt undescribed. Some birds show mostly replaced head and body feathers by December.

**Adult:** The partial molt to breeding plumage apparently takes place largely between early April and early May, somewhat later than on the small stints with which Spoon-billed Sandpipers associate. In early April, when Red-necked Stint is showing a good deal of breeding plumage, Spoon-billed Sandpiper still retains extensive nonbreeding feathers. Most head and body feathers, tertials, and some inner wing coverts are replaced. The molt to nonbreeding plumage is undescribed but is presumably complete. Extensive breeding plumage is usually retained at least to mid-August, when birds reach staging areas, but it is likely molted quickly after that, as many fall migrants are in mostly nonbreeding.

**VOCALIZATIONS**

Flight call is said to be a quiet, rolled *preep* or a shrill *wheet.*

## Broad-billed Sandpiper, p. 300

***Limicola falcinellus***

**STATUS**

Eurasian species. Casual in fall on w. Aleutian Is.; accidental in New York (Aug. 1998). Uncommon to locally common within range. Breeds from Scandinavia to e. Siberia. Winters from Persian Gulf and s. Africa east to se. Asia and Australia.

**TAXONOMY**

Two subspecies recognized. Nominate *L. f. falcinellus* breeds from Scandinavia east to w. Russia and winters from Persian Gulf to w. India and s. Africa. *L. f. sibirica* breeds in e. Siberia and winters in se. Asia and Australia. *L. f. sibirica* differs from nominate by having brighter, more rufous fringes to upperparts and cinnamon wash to breast in breeding plumage. Also the upper supercilium is less well defined. In juvenal plumage buff fringes are broader in *sibirica.*

**BEHAVIOR**

A bird of inland and coastal wetlands. Breeds in subarctic montane and lowland zone. Prefers wettest, least vegetated bogs with ridges of peat mud lightly overgrown by sedges and grasses. Usually found at a variety of interior wetlands during mostly overland migration, including wet meadows, pond margins, and muddy patches in marshes. In winter prefers a variety of mostly coastal habitats, including muddy estuaries, pebbly beaches, tidal mudflats, tidal creeks, and sewage ponds. Feeding style similar to Dunlin but bill held more vertically. Walks quickly on short legs while methodically probing or picking at soft substrates. In several wintering areas forages exclusively by capturing worms and other prey by surface pecking, with probing done only on sufficiently soft mud. Often flattens itself in a snipelike fashion when threatened. Eats a variety of invertebrates from freshwater and marine habitats, including insects, crustaceans, mollusks, and marine worms. Also eats a variety of seeds. Usually seen singly or in small numbers among large stint flocks. Large flocks of several hundred may occur in migration.

Nests early June to mid-July.

**MIGRATION**

Long-distance migrant. Nominate *falcinellus* migrates mostly overland to wintering areas. Movements of *sibirica* are poorly known.

**Spring** migration concentrated and takes place between early April and mid-June, with peak movements in May. Most likely departs wintering areas in early April, based on earliest European records, and arrives in Scandinavian breeding grounds from late May to mid-June. Some nonbreeders, presumably **one-year-old** birds, remain at wintering areas through the summer, while others migrate north to summer in s. Russia. Some one-year-old birds may breed.

**Fall** migration takes place from early July to early November, with peak passage occurring August (**adults**) to September (**juveniles**). Adults depart Fenno-Scandia breeding grounds in July and arrive at wintering grounds from late July to early August. Juveniles depart in August, with a few lingering into September, and arrive at wintering areas mostly in September/October.

**MOLT**

**First life year:** The partial molt out of juvenal plumage takes place between early September/mid-October and April. Nonbreeding appearance is mostly achieved by mid-November after most head and body feathers are replaced. Remaining head and body feathers, and most wing coverts, tertials, and tail feathers, are replaced by February. Between early December and April a variable number of outer primaries are replaced, but inner primaries and secondaries are retained. Between late February and early April there is a variable partial molt to breeding plumage. The molt includes most head-and-body feathers, tertials, tail feathers, and a few wing coverts. For birds oversummering on wintering areas, molt is less complete.

**Second life year:** For birds summering on the wintering grounds the complete molt to nonbreeding plumage is as in adults, but may begin a bit earlier (July). After this molt, second-year birds are indistinguishable from adults. The partial molt to breeding plumage is as in adults.

**Adult:** The partial molt to breeding plumage begins in late February and is completed by early April. It includes most head and body feathers, often central tail feathers, and a limited number of median coverts. The complete molt to nonbreeding plumage takes place between late July/August and early November/late January. Limited head-and-body molt begins at stop-over sites in late July/August and is suspended until wintering areas are reached. Most birds arrive on wintering grounds in late August/early September with much of head, body, and tertials in nonbreeding plumage, though a few arrive in mostly breeding plumage. Remainder of head and body feathering replaced by late September/late October. Flight feather molt may begin at stop-over sites, but takes place mostly on the wintering grounds between September and November/late January.

**VOCALIZATIONS**

Flight call is a drawn-out, buzzy, Dunlinlike *jrrreeeit* with a distinct rising inflection; sometimes gives a shorter *tzit.*

## Buff-breasted Sandpiper, p. 190
### *Tryngites subruficollis*

**STATUS**

Uncommon to fairly common migrant through Great Plains in spring, with significant concentrations at various staging areas; mostly rare along Pacific Coast and uncommon along Atlantic Coast in fall. Breeds in high Arctic regions from n. Alaska eastward to central Canada. Winters primarily in pampas of s. South America, including parts of Argentina, Uruguay, and Paraguay. World population is estimated at about 15,000, but numbers are difficult to verify. This species is generally considered at risk for potential population declines.

**TAXONOMY**

No subspecies recognized.

**BEHAVIOR**

Breeds sporadically along Arctic coastal tundra from ne. Alaska to Devon I., Canada. In migration prefers short-grass areas, such as pastures, sod farms, golf courses, and meadows. Also uses wet rice fields with short growth, recently plowed or mowed agricultural areas, lakeshores, and occasionally sandy beaches, especially near sparse vegetation or wrack lines. Forages by walking quickly and deliberately, picking food from the ground or foliage, with occasional quick dashes to snatch prey. Walks with peculiar pigeonlike head bobbing motion. Eats mostly terrestrial invertebrates, especially insects and spiders, and occasionally seeds. Typically seen singly or in small groups but may gather in large flocks of several hundred to several thousand at important spring staging areas. During migration often loosely associated with other "grasspipers," including Pectoral, Baird's and Upland Sandpipers, and American Golden-Plover.

Nest initiation primarily mid-June. Egg dates from early June to late July. Incubation 23–30 days. Incubation and brood rearing by female alone. Single-brooded. Peak hatching mid- to late July. Young achieve sustained flight in about 20 days. On breeding grounds males display together in leks, often involving half a dozen or more birds, depending on available habitat. Occasionally single males will display. Elaborate courtship display involves single-wing waves to attract females to a lek, and double-wing courtship embrace that involves wing vibration, standing on toes, inflating chest, and rotating slowly. Females, sometimes up to half a dozen at once early in the season, approach to inspect striking underwing pattern. Copulation sometimes occurs when closest female(s) opens both wings and turns back on male. Copulation often broken up by competing males masquerading as females. Polygamous by nature, males will copulate with multiple females during the several days to 3-week active lek period. Females may copulate with several different males during egg laying. Males depart soon after lek activity ceases. Wing-flash displays also performed at spring staging areas and occasionally on fall migration (by adults and juveniles).

**MIGRATION**

Long-distance migrant, often covering over 18,000 mi. (28,970 km) during annual movements. Spring migration takes place between early February and early June. Unlike most other shorebirds, males migrate before females in fall.

**Spring** migration takes place between early February and mid-June. Northbound **adults** depart wintering grounds by early February, though some birds remain into late March. Northward migration in South America continues through late April. Birds arrive in Texas and Louisiana in late April/early May after flying over Gulf of Mexico. Migration in North America follows a narrow corridor through the central United States and Canadian provinces, with significant concentrations occurring at staging areas east of Houston, Texas; in York County, Nebraska; at Beaverhills Lake near Edmonton, Alberta; and possibly near Saskatoon, Saskatchewan. Males and females arrive on breeding grounds in late May at s. edge of breeding grounds, and mid-June at more remote sites. Very rare in spring away from the central flyway. Some birds apparently oversummer in n. Canadian prairies. Absence of birds on wintering grounds during austral summer suggests that **one-year-old** birds fly north, possibly stopping short of breeding grounds.

**Fall** migration takes place from mid-June through late October. **Adult** males, nonbreeders, and failed breeders depart Arctic from mid-June to early July. Females and **juveniles** depart late July to early September. Most birds generally follow same route as in spring migration, but in a wider corridor. Peak passage of adults takes place between late July and late August. Some juveniles and a very few adults follow a different route, flying east to Hudson Bay, across the Great Lakes, and through New England and n. mid-Atlantic states, before flying over the w. Atlantic to ne. South America. Peak numbers pass through the Northeast from mid-August to early October. The vast majority of these are juveniles. A small number of juveniles also occur annually in the Pacific Northwest Coast from mid-August to mid-September. Typical arrival dates on wintering grounds in Argentina is early September through mid-October.

#### MOLT

**First life year:** Molt out of juvenal plumage begins upon arrival on winter grounds (September to late October) and is completed before start of spring migration. Molt almost complete, with the exception of a few remaining juvenal feathers on upperparts.

**Adult:** Adults undergo a partial molt to breeding plumage in March/April which includes head and much of body. The complete molt to nonbreeding plumage begins in early July on breeding grounds with some head and body feathers. All molt is suspended in late July or August and completed on the winter grounds, starting in October. Flight feathers are molted last, sometimes as late as February/March.

#### VOCALIZATIONS

Mostly silent. Flight call, sometimes heard, is a short, soft *gert* or a longer, rising *grriit;* similar to Pectoral Sandpiper but much higher and softer. Males give a series of very soft *tick* or *tuck* notes during the double-wing embrace courtship display.

## Ruff, p. 301

***Philomachus pugnax***

#### STATUS

Old World species. Rare but regular migrant in Alaska, along Atlantic and Pacific coasts and through Midwest; casual through interior West. Has bred at least once in Alaska and winters occasionally along Pacific Coast, rarely elsewhere. Common within range. Breeds from Britain and Scandinavia east to e. Siberia. Winters primarily in sub-Saharan Africa, with smaller numbers through w. Europe, the Middle East, India, Indochina, and s. Australia.

#### TAXONOMY
No subspecies recognized.

#### BEHAVIOR
A bird of freshwater, grassy habitats, and coastal pools. Breeds in moist Arctic tundra as well as freshwater marshlands and grasslands in more temperate zones. During migration and winter prefers similar freshwater habitats such as marshes, rice paddies, flooded fields, and impoundments as well as sheltered coastal pools. Less often at intertidal habitats. Walks actively while foraging, pecking at surface of ground or water for food, in characteristic hunched posture. Will also run, wade in deep water, or swim in search of food. Eats a variety of invertebrates, especially on breeding grounds, but diet also includes plant material (mainly seeds or rice) outside of breeding season. Often found in large concentrations away from breeding areas (up to 1 million in w. Africa), but usually seen singly in North America.

Nest initiation at more southern breeding areas mid- to late May, early to mid-June at Arctic locations. Egg dates late April to late July. Incubation is 20–23 days. One brood per season but replacement eggs laid after loss. Parental and nest building duties by female only. Female feeds young for several days, but self-feeding afterwards. Peak hatching middle half of June at southern locations; early to mid-July in Arctic. Age of first breeding 1–2 years. Males are polygamous, exhibiting dramatic grouselike lek courtship behavior to enable copulation, after which they take no part in parental duties.

#### MIGRATION
Mostly long-distance migrant. Migrates across a broad front, with spring and fall movements using different pathways. Migrant and winter flocks often sexually imbalanced due to different migration patterns, with males wintering farther north in general. Sexes also frequently segregated in large flocks where both occur.

**Spring** migration takes place between mid-February and mid-June, with peak movement from March through mid-April. Spring passage generally occurs more easterly than fall. Departure from African wintering grounds begins in mid-February, with greatest numbers departing from March to mid-April. Arrives at breeding areas progressively from mid-April (North Sea region) to mid-June (Siberia). Large numbers of nonbreeders remain at wintering areas through the summer, even south of equator, while others migrate north to breed or to summer near breeding areas. In North America peak numbers are seen along the mid-Atlantic Coast from late March to early May and in the Upper Midwest in early to mid-May. A few, perhaps nonbreeders, may linger into June.

**Fall** migration takes place from late June to early November. **Adult** males, with no parental duties, depart breeding grounds in late June/early July. Females and **juveniles** form flocks on breeding grounds before departing by late July and August. Main movements occur overland across Europe from late July to mid-September. First adult males arrive in Africa in mid-July, but main trans-Saharan passage occurs from mid-July to mid-September. Smaller numbers of juveniles continue to arrive at wintering areas until mid-November. In North America peak numbers of adults occur along the mid-Atlantic Coast from late June to early August. Juveniles, which are much more numerous along the Pacific Coast than elsewhere, peak from late August to early October.

#### MOLT
**First life year:** The partial molt out of juvenal plumage takes place between late

September/October (rarely late July) and March. Most head and body plumage is replaced by November/February. On some birds (mostly females?) the outer two to three primaries are replaced between January and March. Between January and March (males) or February and March (females) there is a partial molt to breeding plumage involving a variable number of head and body feathers. This molt is similar to adult's but takes place slightly later and is usually less extensive. Birds that oversummer on the wintering grounds typically show only scattered breeding feathers and begin primary molt (from p1) as early as January/March. Males that reach the breeding grounds undergo an additional (supplementary) molt between mid-March and mid-May involving a variable number of head and body feathers, including fancy head tufts and neck "ruffs."

**Second life year:** For birds that oversummer on wintering grounds the complete molt to nonbreeding plumage may start as early as January/March (of first life year) with primary molt. Molt is probably completed by August/September, after which birds are on the same schedule as adults.

**Adult:** The partial molt to breeding plumage takes place between November/January and February/March (males) or December/January and March (females). This molt involves most head and body feathers as well as most tertials and some coverts and tail feathers. Males undergo an additional (supplementary) molt between mid-March and mid-May involving a variable number of head and body feathers, including fancy head tufts and neck "ruffs." In males the complete molt to nonbreeding plumage takes place between early June/early July and late August/December. Limited head-and-body molt takes place on or near the breeding grounds; then, at staging areas, mostly nonbreeding head and body plumage is acquired by late July/early August. Primary molt begins late June/early August and is completed by late August/early September at staging areas or else suspended, then completed on the wintering grounds by December (sometimes later). A few birds molt primaries entirely at the wintering grounds. This molt is more variable in females but usually takes place about 1 week later, with more extensive molt (including most or all flight feathers) taking place on the wintering grounds.

#### VOCALIZATIONS

Mostly silent. Feeding birds in winter sometimes give a series of low grunts, and a low, nasal *rret* or *rret-et* alarm call is heard rarely. Occasionally gives a shrill, rising *hoo-ee.*

## Short-billed Dowitcher, p. 194

***Limnodromus griseus***

#### STATUS

Common throughout most coastal areas of North America. Breeds in subarctic regions from s. Alaska to e. Canada. Winters in coastal areas from Oregon and Virginia south to Peru and Brazil. Much rarer migrant inland than Long-billed Dowitcher. The world and North American population is estimated at 320,000.

#### TAXONOMY

Three distinct subspecies are recognized. *L. g. caurinus* breeds along the s. coast of Alaska and Yukon Territory and winters along the Pacific Coast north to Oregon. Migrants are mostly confined to the Pacific Coast. *L. g. hendersoni* breeds in central Canada from e. British Columbia to Manitoba and probably winters along the s. Atlantic and Gulf coasts and along both coasts of Central America. Migrants are most common in the prairies but occur along the Atlantic Coast in moderate numbers, particularly in fall and from Virginia south, moving somewhat earlier than *griseus. L. g. griseus* breeds in Quebec and Newfoundland and probably winters mostly in the West Indies and coastal Venezuela and Brazil. May also winter in se. United States. Migrants are seen primarily along the Atlantic Coast, but small numbers occur through the eastern interior and moderate numbers occur along the e. Gulf Coast. Population breeding along the s. shore of Hudson Bay appears intermediate between *hendersoni* and *griseus.*

Differences in appearance among the three races are moderately distinct in breeding plumage, though apparent intergrades occur between *hendersoni* and *griseus* in the Hudson Bay region. Juvenal and nonbreeding plumages are similar and may overlap, but *caurinus* averages darkest, *hendersoni* palest, and *griseus* intermediate.

**BEHAVIOR**

Breeds primarily in the transition zone between boreal forest and subarctic tundra. Nest site is typically in sedge meadows or bogs with interspersed black spruce and tamarack. In migration and winter prefers coastal mudflats and salt marshes and adjacent freshwater pools. Seldom found at freshwater sites and seldom seen inland except at large lakeshores. Like Long-billed Dowitcher, typically feeds in tight flocks with characteristic "sewing-machine" probing motion. Walks relatively slowly while foraging, typically taking only one or two steps between probing efforts. While not feeding, typically stands with bill held down at a 45° angle. Eats a variety of aquatic and terrestrial invertebrates including annelids, snails, and insects. On the breeding grounds may also eat spiders and seeds. Forages during both day and night. Readily swims. Typically seen in tight flocks of 20–30 to several hundred.

Nest initiation primarily early June. Egg dates range from late May to mid-July. Incubation 21 days. Both parents share in incubation but male performs nearly all chick-rearing duties. Single-brooded but will renest after egg loss. Peak hatching late June/early July. Young can make short flights after about 16–17 days. In display flight gives flight calls while rising on quivering wingbeats, then song delivered on a downward glide with outstretched neck.

**MIGRATION**

Migrates during both day and night under a variety of weather conditions.

**Spring** migration takes place between March and early June. Between March and May, **adult** *griseus* depart South America and probably fly nonstop to the s. Atlantic Coast. Numbers of *griseus* peak along the mid-Atlantic Coast in mid- to late May, after which they make the nonstop flight to the breeding grounds. Most *hendersoni* depart the Gulf and Atlantic coasts early to late May and apparently fly nonstop to the breeding grounds, with only a few stopping in the interior. Arrival on the breeding grounds takes place between mid-May and early June. Peak numbers of *caurinus* pass through the West Coast in mid- to late April and depart in early May, flying directly to the s. Alaska coast, where numbers peak in early to mid-May prior to dispersal to breeding areas. Many **one-year-old** birds oversummer on or near the wintering grounds, while others probably return partway or all the way north to the breeding grounds.

**Fall** migration takes place between late June and early November. Most **adult** females depart the breeding grounds in late June or early July, soon after egg hatching, and fly directly to coastal areas. Failed breeders and nonbreeders probably head south around the same time. Breeding males depart about 2 weeks later, after young have fledged. Migrant *griseus* first appear along the Atlantic Coast in late June/early July, with peak numbers from mid-July to early August. First arrivals in South America occur in early August.

Migrant *griseus* occur in small numbers along the e. Gulf Coast and only rarely in interior Northeast. Southbound adult *hendersoni* first appear along the s. Atlantic Coast in early July and peak in mid- to late July. Numbers of *hendersoni* equal or outnumber *griseus* from Virginia south but become much less common to the north, where they make up about 10–12 percent of the fall migrant dowitchers in New Jersey. Small numbers of adult *hendersoni* appear at interior eastern and prairie sites in July and August. Along the Gulf Coast adult *hendersoni* first appear in mid-July and peak in early August. Southbound adult *caurinus* migrate almost entirely down the coast with peak numbers in the Pacific Northwest from late June to mid-July. **Juveniles** depart the breeding grounds in early to mid-August, heading primarily toward coastal areas. Peak numbers appear in the Pacific Northwest in August and early September and are rare there after October. Along the mid-Atlantic Coast juveniles first appear in early August and peak in the second half of the month. Numbers diminish by mid-September, though some birds linger well into the fall. Arrival at wintering sites continues through late October/early November.

**MOLT**

**First life year:** Molt out of juvenal plumage takes place between mid-September and February and involves most head and body plumage, some tertials and coverts, most tail feathers, and sometimes outer two primaries. Head and body is in mostly nonbreeding by October, but tail and flight feather molt is usually completed by January/February. This molt usually begins on migration and is completed on the wintering grounds, but many birds hold full juvenal plumage well into October. Between late February and May there is a highly variable molt to breeding plumage involving head and body feathers, some wing coverts and tertials, central tail feathers, and occasionally outer primaries. Some individuals replace as many feathers as adults and are not readily distinguishable from adults in the field. In other birds, particularly those summering near wintering grounds, the molt is more restricted and may be skipped altogether. First-summer birds may be distinguished from adults by retained worn juvenal inner primaries.

**Second life year:** For birds that oversummer south of the breeding grounds, the complete molt to nonbreeding plumage takes place primarily from June to August, more than a month earlier than in adults. Other birds may be on a more adultlike schedule. After this molt, second-year birds are on the same schedule as adults.

**Adult:** Between late February and May adults undergo a partial molt to breeding plumage involving almost all body feathers, some wing coverts, and central tail feathers. There is variation between subspecies in the number of wing coverts replaced, with *hendersoni* and *caurinus* typically replacing more coverts than *griseus.* The complete molt to nonbreeding plumage takes place between mid-July and late September/early October. Limited head-and-body molt takes place on the breeding grounds in males, but females begin molt at stop-over sites. Heavy head-and-body molt takes place at stop-over sites from mid-July to mid-August and is typically completed by mid-August to mid-September, after the wintering grounds are reached. Primary molt takes place mostly between early August and late September/early October, when birds are at or near the wintering grounds. Some birds retain nearly immaculate breeding plumage until the wintering grounds are reached.

**VOCALIZATIONS**

Flight call is a low, rapid *kiu-tu-tu,* typically three to four notes; similar to Lesser Yellowlegs but much faster; pitch variable though always lower pitched than Long-billed Dowitcher. Feeding flocks typically silent though may occasionally utter a flight call or a single *tu* note. Flight song is a three-part, gurgling *drri-dididi-drrrree-drrrroi,* repeated several times and often preceded by several flight calls; given primarily on the breeding grounds but occasionally given by migrants between May and July.

## Long-billed Dowitcher, p. 201

***Limnodromus scolopaceus***

**STATUS**

Common migrant throughout much of the West and South at both coastal and interior sites. Less common and more localized in the Northeast. Breeds in high Arctic regions of e. Siberia, n. Alaska, and nw. Canada. Winters from British Columbia and New Jersey south to Mexico and n. Central America. Winters along the coast at northerly latitudes, inland to the south. The world population is estimated at 500,000.

**TAXONOMY**

No subspecies recognized.

**BEHAVIOR**

Breeds primarily in coastal tundra. Nest site is typically in wet grassy or sedge meadows near extensive flats and freshwater ponds. During migration and winter prefers freshwater pools, flooded fields, and lakeshores. In coastal areas regularly uses sheltered lagoons, protected salt marsh pools, and adjacent freshwater ponds. Uses tidal flats in some regions but generally prefers freshwater sites when available. Like Short-billed Dowitcher, typically feeds in tight flocks with characteristic "sewing-machine" probing motion. Walks relatively slowly while foraging, typically taking only one or two steps between probing efforts. While not feeding, typically stands with bill held down at a 45° angle. While resting, stands more upright than Short-billed Dowitcher due to more front-heavy proportions. Eats a variety of aquatic and terrestrial invertebrates, including worms, crustaceans, bivalves, amphipods, and insects along with seeds and other plant material. Forages during both day and night. Readily swims. Typically seen in tight flocks of 10–20 to several hundred.

Nest initiation primarily late May. Egg dates range from late May to late July. Incubation 20 days. Female assists male with incubation at first, then leaves male to complete incubation and raise brood. Single-brooded. Peak hatching late June/early July. Young can fly after about 15–30 days. In display flight male rises to a height of 50–70 ft. (15–21 m) and sings as he hovers on quivering wings. Also gives abbreviated song during pauses in territorial chase.

**MIGRATION**

Migrates during both day and night.

**Spring** migration takes place between early February and late May. Peak numbers pass through Texas in mid- to late April; through California, the Great Basin, and Midwest (and rarely the Northeast) between late April and early May; and through the Pacific Northwest and s. Alaska in early to mid-May. Arrival on the breeding grounds takes place in mid- to late May. Unlike Short-billed Dowitcher, virtually all **one-year-old** birds return north to the Arctic.

**Fall** migration takes place between early July (rarely late June) and December. **Adult** females depart the breeding grounds mostly in early to mid-July, soon after egg hatching. Males depart about 2 weeks later, after young have fledged. Migrants first appear in interior Alaska in early July and reach staging areas throughout the continent, including the Pacific, Atlantic, and Gulf coasts, by mid-July, with peak numbers of adults between early August and early September. More numerous in the Northeast in fall than spring, but always less common there than Short-billed Dowitcher. **Juveniles** depart the breeding grounds in early to mid-August, heading toward nearby coastal areas. The first juveniles typically arrive in the Pacific Northwest by mid-August, with peak numbers from late September to mid-October. On the Atlantic Coast juveniles first appear by mid-September (exceptionally late August) and peak in October, with later migrants lingering into December. On the Gulf Coast juveniles arrive by late September/early October, with peak numbers by late October and November.

### MOLT

**First life year:** Molt out of juvenal plumage takes place between early September and February and involves most head and body plumage, some tertials and coverts, and sometimes outer 2 primaries. Head and body is in mostly nonbreeding by November, but flight feather molt usually completed by January/February. This molt often begins earlier than in Short-billed Dowitcher, so that most migrants show at least some replaced scapulars when they first arrive at southern latitudes. Between late February and May there is a partial molt to breeding plumage involving head and body feathers, many wing coverts and tertials, and some tail feathers. Unlike Short-billed Dowitcher, virtually all first-year birds acquire a breeding plumage similar to adult's. First-year birds may be distinguished from adults by retained worn juvenal primaries.

**Adult:** Between late February and early May adults undergo a partial molt to breeding plumage involving almost all head and body feathers, tertials, many wing coverts, and central tail feathers. The complete molt to nonbreeding plumage takes place between mid-July and October, primarily at coastal and interior stop-over sites. Primary molt begins in late July and is completed by mid-September (note that Short-billed Dowitcher molts primaries only at or near coastal wintering areas and begins molt slightly later). Only limited body molt takes place before mid-August, but head molts early, resulting in distinctive gray-headed, salmon-bellied look in late summer. Heavy body molt takes place mostly between mid- to late August and early October.

### VOCALIZATIONS

Flight call is a sharp, whistled *peep* or *pseep*, delivered singly or in a long series when agitated. Sometimes confused with Short-billed Dowitcher when delivered in a series, but always higher pitched than calls of Short-billed. Flight calls regularly given by standing and flying birds and regularly given by feeding flocks. Male's flight song is a buzzy chatter, *pipipipipipi-chi-drrr,* repeated several times; similar to Short-billed but higher and with last element descending; given by standing or flying birds on the breeding grounds and, to a lesser degree, during migration in late spring and early fall.

## Jack Snipe, p. 306

***Lymnocryptes minimus***

**STATUS**

Eurasian species. Probably casual visitor but easily overlooked due to secretive habits. Recorded from w. Alaska (Apr.), Washington (Sept.), California (Nov.), and Labrador (Dec.). Uncommon throughout range. Breeds across n. Eurasia from Norway, Sweden, and Finland east to Siberia. Winters from w. and s. Europe south to n. Middle East, central Africa, and se. Asia.

**TAXONOMY**

No subspecies recognized.

**BEHAVIOR**

A secretive bird of vegetated freshwater marshes and wet meadows. Breeds in open boreal and deciduous marshes and bogs. During migration and winter prefers inland freshwater marshes and soft muddy pond edges, often with dense vegetation. Feeds "snipelike", walking slowly and methodically probing mud with relatively long bill. When feeding, often rocks body rhythmically up and down. When alarmed may run through dense vegetation in rail-like fashion, or may freeze in place, using camouflage to hide its presence. Flushes only on very close approach (1 meter), then flies weakly away before dropping back into cover a short distance away. Eats primarily adult and larval insects, mollusks, worms, and plant material. Usually solitary, but occasionally found in small groups.

Nests late April to late July.

**MIGRATION**

Short- to long-distance migrant. Migrates across broad front, with late fall and early spring movements.

**Spring** migration takes place between February and mid-May, with peak movement in April. Departs southernmost wintering areas in February, but mainly March/April. Arrives at some w. European breeding areas in March, but most from mid-April to mid-May (later arrivals in n. Siberia). Records of birds in North Sea countries in July/August suggest that some nonbreeders remain at wintering areas during summer months.

**Fall** migration takes place from late August to early December. **Adults** and **juveniles** remain on or near breeding grounds to molt during August/September and typically depart these areas in September. Peak numbers occur in migration through Europe south of Baltic from mid-September to mid-November. Arrival at northern part of wintering range occurs from late August onwards. Arrival at southern part of wintering ranges does not occur until November/December.

**MOLT**

**First life year:** The molt out of juvenal plumage is unknown but is presumably similar to adult's fall molt, possibly taking place later.

**Adult:** The partial molt to breeding plumage begins in February on wintering grounds and is completed by May. Molt includes most head and body feathers and some tail feathers. The complete molt to nonbreeding plumage is not well known but apparently takes place between July and November, mostly on the breeding grounds. Primary molt presumably takes place from July to late August, prior to migration. Head, body, and tail molt probably begins on the breeding grounds and is completed during migration by November.

**VOCALIZATIONS**

Mostly silent away from breeding grounds. When flushed, rarely gives a quiet, harsh *gah* or *gatch*.

## Common Snipe, p. 307

### *Gallinago gallinago*

#### STATUS

Primarily Eurasian species. Regular migrant in w. Aleutians, where it has bred and been recorded in winter. Rare or casual migrant elsewhere in w. Alaska. Common throughout much of the Old World. Breeds from Iceland, the Azores, and Europe east to Siberia and China. Winters from Europe and Africa east to Japan and Malaysia.

#### TAXONOMY

Two subspecies recognized. Nominate *G. g. gallinago* breeds across most of range from Europe to Siberia and has occurred in Alaska. *G. g. faeroeensis* breeds in Iceland and the Faroes, Orkney, and Shetland Is., casual to Greenland. Compared to nominate, *faeroeensis* is slightly brighter tawny overall, with more diffuse pattern on breast.

#### BEHAVIOR

Like Wilson's Snipe.

Nest initiation primarily early April in southern areas to mid-May in the Arctic. Egg dates late March to mid-August (mostly mid-May to early June at high latitudes). Incubation 18–21 days by female alone. One brood per season but replacements laid after egg loss. Late clutches are probably by one-year-old birds. Peak hatching early May in southern regions to early to mid-June at high latitude. Male cares for first two chicks, female for second two. Young can fly 19–20 days after hatching. Winnowing display flight similar to Wilson's.

#### MIGRATION

Short- to medium-distance migrant.

**Spring** migration takes between late February and mid-May, with peak passage through Europe in March/April. Breeding areas may be reached between early April and mid-May, depending on region.

**Fall** migration takes place between July and November, with peak numbers passing through Europe in September/October. May arrive in sub-Saharan Africa by early September, though not common there until October. No information on differences in timing of **adults** and **juveniles,** but they are presumed to migrate together.

#### MOLT

**First life year:** The partial molt out of juvenal plumage takes place between July and December and includes most head and body feathers, tail, tertials, and most wing coverts. Molt begins on breeding grounds in July, often before juvenal plumage is fully grown. Peak molt of body feathers takes place in late August/September and wing coverts in September/October. Molt is completed on the wintering grounds. Between February and May many body feathers, a few coverts, and from one to all tail feathers are molted again. This molt may begin on the wintering grounds but takes place largely during spring migration.

**Adult:** Between February and May many body feathers, a few coverts, and from one to all tail feathers are molted again. This molt may begin on the wintering grounds but takes place largely during spring migration. The complete molt to nonbreeding plumage takes place between July and November/December, mostly on the breeding grounds. Primary molt begins on or near the breeding grounds in July (usually p1–3), then is completed at stopover sites or occasionally on the wintering grounds after suspensions. Body molt begins in July and is completed simultaneously with renewal of p10.

#### VOCALIZATIONS AND DISPLAY SOUNDS

Vocalizations very similar to Wilson's. Winnow much lower pitched, hollower, and more humming than Wilson's due to broader outer tail feathers.

## Wilson's Snipe, p. 207

***Gallinago delicata***

**STATUS**

Common and widespread in North America. Breeds through most of Canada and the n. United States and winters from s. Alaska and Massachusetts south to n. South America, though southern edge of range unclear due to confusion with South American Snipe *(G. paraguaiae).* Found just as frequently inland as along coast. The world and North American population is estimated at 2 million.

**TAXONOMY**

No subspecies recognized but recently split from the Old World Common Snipe *(G. gallinago).*

**BEHAVIOR**

Breeds in a variety of wetlands, including marshes, bogs, and willow/alder swamps. Requires small clearings with soft rich soil for feeding and scattered clumps of dense vegetation, trees, or posts for perching. During migration and winter prefers damp areas with vegetative cover, including marshes, wet fields, and ditches. Generally secretive, using camouflaged plumage to hide from predators. Often not seen until flushed, when it explodes into a zigzag flight, uttering flight call. Forages by probing deeply in mud or shallow water, bill often remaining submerged while swallowing prey items. Generally forages in a small area, walking little. Occasionally employs foot stomping or bouncing to find prey. Eats mostly insect larvae but also crustaceans, earthworms, and mollusks. Feeds during both day and night but primarily crepuscular and regularly seen flying to and from foraging areas at dawn and dusk. Typically seen singly or in small flocks of a dozen or more. Sometimes gathers in larger numbers in prime habitat.

Nest initiation primarily mid-May in southern areas, mid-June at high latitudes. Egg dates range from early May to early August. First-year birds nest slightly later than adults. Incubation 18–20 days, by female alone. One brood per season but will renest after failure. Peak hatching early June in southern regions to mid-July at high latitude. Male cares for first two chicks, female for second two. Young can fly 19–20 days after hatching. During display flight, flies in wide circles at about 500 ft. (152 m) or more, then periodically dives at a 45° angle and "winnows." Display most common in evening but may be performed at any time of day or night, mostly by male.

**MIGRATION**

Short- to medium-distance migrant. Migrates mostly at night.

**Spring** migration takes place between late February and late May. Departure from southern wintering areas takes place between late February and April. Peak migratory wave passes through the Gulf Coast in March, through much of the West, Midwest, and Middle Atlantic in April, and through s. Canada in late April and May. Arrival at breeding sites varies geographically, with southerly breeding sites being occupied by early March but Arctic regions not until late May.

**Fall** migration takes place between mid-July and December. Timing of **adult** departure from breeding sites varies geographically. The first few fall migrants, probably failed breeders, arrive at nonbreeding sites along the central Pacific Coast in mid-July, the mid-Atlantic Coast in late July, and the Gulf Coast in early to mid-August. Peak movements pass through the Rocky Mts. in September, lowland areas of the West from mid-September to mid-October, and much of the East from mid-October to late November. Close proximity of breeding birds in the Rocky Mts. accounts for earlier movements in the West. Arrives in South America by late October. Little information on differences in timing of adults and **juveniles,** but they are presumed to migrate together. Juveniles tend to gather in loose aggregations of up to 100 at age 6 weeks (mid-July to late August, depending on region).

**MOLT**

**First life year:** The molt out of juvenal plumage begins between September and late October (at about 3½ months of age) and is completed by November or December. This molt may begin on migration but is completed on the wintering grounds and involves body feathers as well as some coverts, tertials, and tail feathers. Between January and April a variable number of body and tail feathers are molted again. This molt takes place largely or entirely on the wintering grounds.

**Adult:** Between January and April a variable number of body and tail feathers are replaced, producing breeding plumage. This molt takes place largely or entirely on the wintering grounds. The complete molt to nonbreeding plumage may take place largely on the breeding grounds or may be completed at stop-over or wintering sites. It begins during or just after incubation in June or July and is completed by October/November.

**VOCALIZATIONS AND DISPLAY SOUNDS**

Flight call is a tearing *skaaip,* frequently given after being flushed as well as during long overhead flights, including nocturnal migration; resembles the sound of a boot being withdrawn from a muddy path. On the breeding grounds, perched and low-flying birds give a monotonous *kup-kup-kup-kup-kup* . . . or a two-syllable *chip-a-chip-a-chip-a-chip-a* . . . , apparently in alarm and or aggression. In display flight produces a nonvocal "winnowing" sound created by air rushing past outer rectrices of spread tail, modulated by the beating wings. The resulting sound, which functions as a song, is a fast series of low whistles, gently rising in pitch and increasing in volume toward the end, much like the song of Boreal Owl. Winnowing performed primarily on the breeding territory but also by spring migrants near the breeding grounds.

## Pin-tailed Snipe, p. 309

***Gallinago stenura***

**STATUS**

Asian species. Accidental in w. Aleutians (two spring records). Locally common to uncommon within range. Breeds from nw. Russia east across Siberia. Winters from India through most of se. Asia.

**TAXONOMY**

No subspecies recognized.

**BEHAVIOR**

A bird of damp marshlands and drier grassy habitats. Breeds in grassy marshes, moist open meadows, bogs, and near small muddy ponds in low Arctic or boreal taiga zones. During migration and winter prefers freshwater marshes, flooded fields, and tidal marshes. Tolerates drier habitats than Common Snipe, such as harvested fields, pastures, and dry grass meadows. Behavior similar to Common Snipe, but feeds with more picking than probing. Feeds largely on invertebrates, mainly insects and their larvae, mollusks, and earthworms, with smaller intake of seeds and crustaceans. Escape flight heavier and steadier than Common Snipe. Usually solitary or in small groups, but larger loose flocks of up to 60 birds possible on wintering grounds where food is plentiful.

Nests late May to early July.

**MIGRATION**

Long-distance migrant. Migrates overland across a broad front to wintering areas in se. Asia.

**Spring** migration takes place primarily from March/early April to June, with peak movement in April. Earliest departure from s. wintering areas in February, but most departures occur in March and first half of April. Arrival at southern breeding areas in May, early June at northern locations.

**Fall** migration takes place from late July to late October. Early departure from Siberian breeding grounds occurs in late July, but majority of birds leave in August and September. Arrives at wintering areas from late August to late October, with peak numbers occurring in second half of September and October. **Juveniles** probably arrive later than **adults.**

**MOLT**

**First life year:** The partial molt out of juvenal plumage takes place between November and early April and involves most head and body feathers as well as tail and wing coverts. Most juvenal wing coverts are replaced by midwinter, although some are retained until following spring. Tail feathers are replaced from December to February. Timing and extent of partial molt to breeding plumage is undescribed.

**Adult:** The extent and timing of the partial molt to breeding plumage is undescribed. The complete molt to nonbreeding plumage takes place between August and January. Primary molt is descendant, with from 1 to 3 (sometimes 8) outer primaries replaced on the breeding grounds in August/September, then suspended and completed on the wintering grounds in November/December. Body molt begins after inner primaries are replaced and is completed in about 3 weeks.

**VOCALIZATIONS**

Mostly silent away from breeding grounds but occasionally gives a high-pitched, ducklike *squak* when flushed; softer, higher and less raspy than Common or Wilson's Snipe.

## Eurasian Woodcock, p. 310

***Scolopax rusticola***

**STATUS**

Eurasian species. Formerly casual vagrant to e. North America from Newfoundland to Alabama, mostly in late fall and winter. Most records were in 19th century, coinciding with population increase. Fairly common within normal range. Breeds from Norway and the Canary Is. east to se. Russia, n. China, and n. Japan. Winters from the British and Canary Is. east to s. Japan and the Philippines.

**TAXONOMY**

No subspecies are recognized.

**BEHAVIOR**

A primarily upland species. Breeds in extensive damp woodland with dense undergrowth and adjacent clearings or fields. During migration and winter sometimes ventures into drier, brushier habitats. Very secretive, using camouflaged plumage to hide from predators. Often not seen until flushed, when it bursts into zigzag flight. Forages by probing deeply in mud or damp soil and by picking food off the surface. Eats mostly earthworms and insects but also a variety of other invertebrates, small vertebrates, and some plant material. Feeds mostly in forest during the day in spring and summer, mostly in fields at night in fall and winter. Regularly seen flying to and from foraging and roosting areas at dawn and dusk. Typically seen singly.

Nests early March to early September.

**MIGRATION**

Partial to intermediate-distance migrant. Migrates strictly at night.

**Spring** migration takes place between late February and mid-May, with peak movement in late March and early April. Most breeders are on territory by early April.

**Fall** migration generally begins in early September and extends to late November, with peak movement from early October to mid-November. Timing variable, dictated largely by the onset of frost. Late season movements may occur in harsh weather.

#### MOLT

**First life year:** The partial molt out of juvenal plumage takes place from late July to September/October. This molt involves most head and body feathers, inner secondaries, tail feathers, and some coverts. Some may molt tail in winter. Spring molt undescribed. First-year birds may be distinguished from adults by retained juvenal primaries, outer secondaries, and some coverts.

**Adult:** Between February and May there is a partial molt to breeding plumage involving some head and body feathers. The complete molt to non-breeding plumage takes place from late July to October/December. Flight feather molt is usually completed by late September, before migration.

#### VOCALIZATIONS

Flight call is a harsh, snipelike *aatsh,* sometimes rapidly repeated. On takeoff, wings produce a swishing sound quite different from American.

### American Woodcock, p. 210

***Scolopax minor***

#### STATUS

Common but secretive in e. North America. Breeds from s. Manitoba and s. Newfoundland south to ne. Texas and s. Georgia. Winters from central Oklahoma and s. Connecticut south to central Texas (rarely ne. Mexico) and s. Florida. The world and North American population is estimated at 5 million.

#### TAXONOMY

No subspecies are recognized.

#### BEHAVIOR

A primarily upland species. Breeds in damp second-growth forest, forest openings, overgrown fields, and bogs. Requires small clearing or field for display and nighttime roost. During migration and winter forages in damp woods during the day and in adjacent damp fields at night. During severe cold seeks warm sunny forest edges and roadsides where soil may be less solidly frozen. Very secretive, using camouflaged plumage to hide from predators. Often not seen until flushed, when it bursts into flight on twittering wings. Forages by probing deeply in mud or damp soil. Generally forages in a small area, walking little. Eats mostly earthworms but also a variety of invertebrates, small vertebrates, and some plant material. Feeds primarily during the day in spring and summer, also at night during winter. Regularly seen flying to and from foraging and roosting areas at dawn and dusk. During severe cold snaps or after ice or snow storms, may forage along sunny field edges where exposed, unfrozen soil may be found. Walks with rhythmic rocking motion to body. Typically seen singly, though large numbers may gather in prime habitat along coast during fall migration.

Nest initiation primarily early February in South, late April in North. Egg dates range from early January to mid-July. First-year birds nest slightly later than adults. Incubation 20–21 days. Female alone incubates and raises young. One brood per season but will renest after failure. Peak hatching early March in South, mid- to late May in North. Young can fly 18–19 days after hatching. Breeds in first year. During courtship display male calls from the ground, then spirals high in the air on twittering wings to a height of 200–300 ft. (61–91 m). As the bird zigzags back to the ground, the wings produce three to six sets of descending twitters. Display is performed mostly at dusk and dawn, to a lesser degree at night.

#### MIGRATION

Short- to intermediate-distance migrant. Migrates strictly at night.

**Spring** migration takes place from mid-January to mid-April. Departure from southern wintering areas takes place between mid-January and late February. Peak migration takes place through the Midwest and mid-Atlantic Coast from late February to late March. Northern breeding areas reached between mid-March and mid-April.

**Fall** migration takes place between August and mid-December. Local dispersal from breeding grounds begins in August, but southward migration generally begins in early September. Migration timing variable, depending on harshness of weather. Peak movements usually take place between mid-October and early to mid-November. Arrival at southernmost wintering areas takes place from late October to mid-December. Differential timing between **adults** and **juveniles** poorly known due to difficulty of aging, but timing probably similar. Adult males winter farther north than females and juveniles.

#### MOLT

**First life year:** The partial molt out of juvenal plumage begins mid-July/early August and ends in November. This molt involves most head and body feathers, inner two to nine secondaries, tail feathers, and some coverts. Flight feather molt ends in October, before migration. There is no molt in spring. First-year birds may be distinguished from adults by retained juvenal primaries, outer secondaries, and some coverts.

**Adult:** No molt in spring. The complete molt to nonbreeding plumage takes place from late June to November. Flight feather molt is completed by October, before migration.

#### VOCALIZATIONS

Call, given by male primarily during display, is low, nasal *beent.* This call is given singly, often with long intervals between calls. Sounds made during display flight are nonvocal, produced by air rushing past narrow outer primaries. Display is performed on the breeding territory and also by spring migrants near the breeding grounds. Flushed birds also produce a distinctive wing whistle (slightly higher pitched in males due to narrower outer primaries).

## Wilson's Phalarope, p. 213

### *Phalaropus tricolor*

#### STATUS

Fairly common through most of West. Uncommon and local in the East. Widespread breeder in western states and provinces, with smaller numbers in a narrow band across Great Lakes Region. Isolated nestings have been recorded in many areas outside usual breeding range. Virtually all adults stage in huge flocks at hypersaline/alkaline lakes of w. North America before migrating to similar wintering habitats in South America, primarily highland lakes of Peru, Chile, Bolivia, and nw. Argentina. Uncommon migrant, particularly in fall, along Atlantic Coast. North American and world population estimated at 1.5 million.

#### TAXONOMY

No subspecies are recognized.

#### BEHAVIOR

Breeds in shallow grassy wetlands of interior North America. Most terrestrial of the three phalarope species. During migration and winter prefers mainly open, shallow-water habitats, particularly hypersaline lakes and alkaline ponds.

Also in winter, mudflats and open-water habitats of high Andean salt lakes and freshwater marshes in s. South America. Rarely reported at sea, lacking developed salt glands and lobed toes of other phalaropes. Typically forages in shallow water, where it spins to create vortex, drawing aquatic invertebrates to surface. To the same end sometimes follows ducks, particularly Blue-winged Teal and Shoveler, to feed on nutrients they stir up. In shallow water may feed by rapidly scything bill sideways through water, like Lesser Yellowlegs. Also actively feeds on land, chasing and plucking prey from surface of mud or shallow pools, typically with head held low. Erratic movements often disturb other feeding shorebirds. Eats a variety of invertebrates, particularly brine shrimp and flies, as well as seeds of aquatic plants. Also consumes gravel for digestion. May be seen in large flocks of several hundred to several thousand at selected spring and fall staging areas. Otherwise seen singly or in small groups.

Pair bond established after female competition for available males, with mobile flocks centered around single male. Nest initiation by female, primarily early to late May. Egg dates range from early May to mid-July, with peak dates late May/early June. Incubation averages 23 days. Role reversal is practiced by pair, with male assuming all incubation and brooding duties. Female typically deserts male shortly after clutch completion, often then competing for additional males. Peak hatching latter half of June. Courtship occurs during migration and throughout breeding season, and involves a variety of behaviors, including female-female aggression.

**MIGRATION**

Somewhat long-distance migrant, crossing equator from w. North American breeding sites to wintering grounds in South America. Virtually entire population winters in Southern Hemisphere, with small numbers occasionally in s. California and s. Texas.

**Spring** migration takes place between mid-March and early May. Northbound migrants depart wintering grounds in mid-March and move through highlands of South America. Migration continues through Central America or across Gulf of Mexico. Peak migration through w. intermountain region and s.-central Plains takes place late April/early May, with about 90 percent of population passing through Cheyenne Bottoms in Kansas in nondrought years. Arrival at breeding areas takes place late April to early May. **One-year-old** birds return to breeding grounds, but percentage that breed is unknown.

**Fall** migration takes place from mid-June to early November. **Adult** females arrive at staging areas in w. United States by mid-June and depart from late July through mid-August. Adult males average 2 weeks later. Southbound adults undergo a fast nonstop flight of over 50 hours from staging areas in w. North America to coastal w. South America. Almost no adult records from Central America and n. South America. **Juveniles** move more slowly and across a broader range overland in w. and central United States, with smaller numbers in e. United States, continuing south through Mexico to n. South America. Peak numbers of juveniles pass through the Atlantic and Pacific coasts in early to mid-August, with most gone by mid-September (rarely lingering to mid-October). Juveniles first arrive in South America in late August. Adults and juveniles migrate south through Andes, arriving at wintering areas from late September to early November.

**MOLT**

**First life year:** Molt out of juvenal

plumage is partial or complete and takes place between mid-July/early August and January/April. Most head and body feathers are replaced rapidly by mid-August with flight feathers, wing coverts, tail, and tertials retained. Remaining body and wing feathers are replaced on the wintering grounds from November to January/April. Some retarded birds skip or only partially complete wing molt. Between February and May there is a partial molt to breeding plumage, including most head and body plumage as well as tail feathers. Retarded individuals may skip this molt and molt directly to nonbreeding plumage over the following summer (of second life year).

**Adult:** Between February and May adults undergo a partial molt to breeding plumage, including all body plumage and tail feathers. The complete molt to nonbreeding plumage takes place between mid-June/July and January. Females molt 2–4 weeks earlier than males, beginning soon after arrival at staging areas (mid-June to early July). Males begin limited head-and-body molt while brooding young, then begin heavy molt, including some flight feathers, at staging areas. By mid-August most or all body feathers, some tail feathers, and some primaries are replaced. Flight feather molt for both sexes is suspended in late July/mid-August and completed on the wintering grounds from late September/mid-November to mid-January.

#### VOCALIZATIONS

Generally quiet away from the breeding grounds. Flight call a low, nasal, froglike *werpf,* given mostly on the breeding grounds. On the breeding grounds also gives a higher, repeated *emf emf emf emf* . . . or *luk luk luk luk* . . . as well as a purring call. Soft gurgling calls are given at staging area roosts.

## Red-necked Phalarope, p. 217

### *Phalaropus lobatus*

#### STATUS

Common visitor to nearshore pelagic waters of Atlantic and Pacific Oceans as well as along the Pacific Coast and, locally, at interior West staging areas. Uncommon migrant through e. Great Lakes and along Atlantic Coast; rare elsewhere. Breeds in low Arctic or subarctic habitats throughout North America, Europe, and Asia. Winters at sea, primarily off w. coast of South America, the Arabian Sea, waters off Indonesia, and possibly in w. South Atlantic. Winters locally elsewhere, including a few in s. California. No estimate of world population, but total Canadian population estimated at greater than 2 million. Recent decline in fall at historic Bay of Fundy staging area from several million to zero individuals is alarming.

#### TAXONOMY

No subspecies recognized.

#### BEHAVIOR

A bird of open-water bodies, from small breeding ponds to large lakes and open oceans. Flies rapidly, usually low to the water, and alights on surface with amazing ease and quickness. Breeds in wet Arctic tundra and subarctic freshwater marshes and bogs. During migration found in a variety of locations, including nearshore pelagic waters, coastal ponds, gravel and sand flats, and virtually all types of inland water bodies, especially large lakes, impoundments, and sewage ponds. Mostly aquatic and constantly in motion, feeds by jabbing and picking tiny invertebrates from or just below water's surface, gliding across the water in a restless zigzag path. Spins in shallow water to create vortex, which draws aquatic invertebrates to surface. More terrestrial than Red but less so than Wilson's Phalarope, it will infrequently walk at water's edge. Food consists mostly of small invertebrates and aquatic larvae of flies. At sea prefers copepods and other marine invertebrates, which it finds along lines of seaweed or other flotsam. Typically seen in small active flocks but occasionally gathers in large groups numbering in the tens of thousands at favored stop-over or wintering areas. Sometimes forms mixed flocks with Red Phalarope, especially far offshore.

Nest initiation varies from mid-May at warmer sites to mid-June in colder regions. Egg dates range from mid-May to late July. Incubation 18–21 days, depending upon region. Role reversal is practiced by pair, with male assuming all incubation and brooding duties. Brighter plumaged female rarely returns to nest after last egg is laid. Hatching dates vary with region, from mid-June in southern areas to mid-July in northerly breeding grounds. Young able to fly after 16–17 days, but adept fliers by 20–22 days. Typically monogamous, but females may have multiple mates when available. Some renesting males may pair with different female. Females typically initiate pair bond.

**MIGRATION**

Long-distance migrant, moving primarily over nearshore pelagic waters, with smaller numbers overland. Less often over deep pelagic waters than Red Phalarope. Stormy weather with onshore winds may deposit larger numbers along the coast or inland.

**Spring** migration takes place between mid-March and early June, primarily along coast (in the West) or offshore, though many fly overland. Birds wintering off w. South America probably depart by mid-March. Peak migration takes place along the U.S. Pacific Coast from mid-April to late May and through the interior West and offshore Atlantic in May. More numerous at interior sites in s. Canada than farther south, peaking in mid- to late May. Arrival on breeding grounds takes place from mid-May to early June. Most **one-year-old** birds return to breeding grounds.

**Fall** migration takes place between mid-June and November. Southbound **adult** females depart the breeding grounds from mid-June to mid-July, shortly after egg laying. Failed breeders and attending males follow between early July and mid-August. Movement to wintering grounds is mostly coastal or pelagic, but significant numbers migrate partially overland. Migrant adults first appear along the Pacific Coast in late June/early July and peak in late July and August. Much smaller numbers occur through the interior West at the same time. Along the Atlantic Coast adults are generally very uncommon except offshore, where they first appear in mid-July and peak in late August. **Juveniles** first appear along the Pacific Coast in mid-August, with peak numbers in September and early October and a few lingering into early November. Juveniles are more widespread inland and more regular along the Atlantic Coast than adults, with peak numbers occurring from early September to early October.

**MOLT**

**First life year:** The molt out of juvenal plumage takes place between September/November and late winter/early spring. Fall migrants hold full or extensive juvenal plumage, with only a scattering of scapular or mantle feathers replaced by late October/November. Molt to first winter occurs mainly between November and late winter/early spring, with head and body plumage and most coverts replaced. Flight feather molt variable, with some birds apparently replacing all primaries and others none. Between February and June (mostly April/May) some birds undergo a partial molt to breeding plumage involving most head and body feathers and rarely a few tertials. Birds oversummering on the wintering grounds apparently skip this molt.

**Adult:** Between February and June (mostly April/May) there is a partial molt to breeding plumage involving most head and body feathers and rarely a few tertials. This molt probably begins on the wintering grounds and is completed during spring migration. The complete molt to nonbreeding plumage takes place between mid-June and April. Limited head-and-body molt takes place on the breeding grounds from mid-June to mid-July. Most head-and-body molt takes place at stop-over sites in July and August, so most adults have a nonbreeding appearance by September. A few primaries may be molted in July, but flight feather molt is suspended, then completed on the wintering grounds from from late September/late November to early February/early April.

**VOCALIZATIONS**

Flight call is a hard, squeaky *pwit* or *kit,* lower and huskier than Red Phalarope. Flocks often very vocal and cumulative sound is a twittering chorus. On breeding grounds most vocalizations involve variations of flight-call-like notes, including a long series of notes. Other calls during nest building and courtship include a series of sharp, buzzy whistles, *wedu wedu wedu* or *wewewewewe.*

## Red Phalarope, p. 221

### *Phalaropus fulicarius*

**STATUS**

Common but highly pelagic outside breeding season and seldom found in nearshore waters or on land. Breeds along the Arctic coast through much of Northern Hemisphere. Winters over deep temperate and tropical waters, with major concentrations in the Humbolt Current off Peru and Chile and the Benguela Current off w. Africa. In North America moderate numbers winter regularly off California, particularly in the California Current, and off the s. Atlantic Coast, particularly in the Gulf Stream. Migrants seen regularly in deep water off both coasts. Regular in small numbers along the Pacific Coast in fall. Rare elsewhere. World population estimated at 5 million, with between 1 and 2.5 million in North America.

**TAXONOMY**

No subspecies recognized.

**BEHAVIOR**

Highly pelagic species. Breeds in marshy coastal tundra, where it frequents shallow edges of tundra ponds and meltwater zones. Breeds sparingly farther inland. Away from breeding grounds occurs in deep pelagic waters near plankton-rich oceanic upwellings. Less often on inshore waters, coastal bays and ponds, or interior lakes. Clearly an aquatic species, swims with amazing buoyancy, as if barely touching the water, and flies skillfully over rough surf, swirling high to avoid waves.

Feeds visually while swimming, picking prey from surface and occasionally submerging head or upending to secure prey. Like other phalaropes, spins in water to create vortex, drawing nutrients to the surface. Also forages on foot when it walks slowly, picking prey from surface of mud or water. At sea eats mainly marine copepods, amphipods, fish eggs, and larval fish, which it finds along lines of seaweed and other flotsam. On breeding grounds diet includes adult and larval insects and crustaceans.

Typically seen in small to moderate flocks of a few to several hundred individuals, but may gather by the thousands at favored feeding areas.

Sometimes forms mixed flocks with Red-necked Phalarope.

Nest initiation timing related to snow

cover, with earliest snow-free areas usually available in Alaska in early June and in Far North by mid- to late June. Egg dates range from late May to late July. Incubation 19 days. Role reversal is practiced by pair, with male assuming all incubation and brooding duties. Brighter plumaged female rarely returns to nest after last egg is laid. Although mostly monogamous, some males may pair with another female during renesting. Peak hatching occurs early July in Alaska, mid-July in Far North. Young are able to fly after 18 days. Mating system somewhat promiscuous, with females initiating most courtship activity. Courtship involves numerous displays, including aggressive behavior between pair and intruders. Pair bond is short-lived, with female abandoning male shortly after he begins incubation.

**MIGRATION**

Long-distance migrant between Arctic breeding grounds and temperate to tropical oceans, mostly over open pelagic waters. Occasionally driven to coast, rarely inland, during stormy weather with strong onshore winds. Most coastal sightings occur past peak times in late spring and late fall. Occasional "wrecks" of hundreds, mostly juveniles, have been seen along the Pacific Coast, primarily from October to December.

**Spring** migration takes place between late March and mid-June. Probably leaves wintering areas mostly late March/early April. In the Pacific, peak migration takes place off California from mid-April to early May and off the Pacific Northwest and Alaska in mid- to late May. In the Atlantic, peak migration takes place off North Carolina in April, off New England in early to mid-May, and through Hudson Strait, Canada, in mid-June. Arrival on breeding grounds mostly early to mid-June. Many **one-year-old** birds apparently oversummer at sea, such as a few present off California throughout the summer. None oversummer off North Carolina.

**Fall** migration takes place between late June and early January. **Adult** females, failed breeders, and nonbreeders start movements to sea in late June, with remaining males departing mid-July to early August. Peak numbers of adults occur off the Pacific Coast and off e. Canada from about mid-July to mid-September. Adults are much less common farther south in the Atlantic and are only occasionally seen off North Carolina in August and September. Most **juveniles** depart the breeding grounds in early August and stage in nearshore areas until early September before continuing out to sea. The first juveniles arrive off the Pacific Coast in late August and peak from early to mid-September to late October, with occasional large movements into early December. In the Atlantic, peak numbers occur off New England from early September to mid-October and off North Carolina mostly in December.

**MOLT**

**First life year:** The partial molt out of juvenal plumage takes place slowly from August/September to December and includes complete replacement of head and body feathers. Molt begins early, so most fall migrants show some to many nonbreeding scapulars even early in the season (compared to Red-necked, which often shows full juvenal plumage well into the fall). Subsequent molt through the winter and spring poorly known. Birds that oversummer at sea apparently retain juvenal wing and tail feathers. Birds that return to the Arctic may replace primaries over the winter and undergo a head-and-body molt to breeding plumage in spring, like adults.

**Second life year:** Poorly known. Birds

returning to the Arctic seem to be on the same schedule as adults. For birds that oversummer at sea there is apparently a partial molt to breeding plumage from May to August, including replacement of some head and body feathers as well as juvenal primaries.

**Adult:** The partial molt to breeding plumage takes place at sea between mid-March and early June and includes head, body, tertials, and possibly a few tail feathers. Most molt takes place late in this time period, with only 50 percent of birds in full breeding by early May and many still in nearly full nonbreeding as late as mid-May. The complete molt to nonbreeding plumage takes place from early July/early Aug to late September/March. It begins on the breeding grounds and is completed either during migration or on the wintering grounds. Head and body plumage is usually mostly nonbreeding by late August/mid-September. Primary molt begins in early to mid-July and is either completed by late September, prior to migration, or is suspended (with outer one to three primaries old) and then completed on the wintering grounds November/December to March.

#### VOCALIZATIONS

Flight call is a sharp, whistled *psip* or *pseet* often given in rapid succession; not unlike calls of Long-billed Dowitcher or Arctic Tern; higher, clearer, and more metallic than Red-necked Phalarope. Alarm call is a variable, two-syllable *pu-dip*, or a clearer, more emphatic, drawn-out *chuu-IT.* Advertisement call, given on the breeding grounds, is a buzzy, Dunlinlike *zheeep.*

## COURSERS AND PRATINCOLES (FAMILY GLAREOLIDAE)

### Oriental Pratincole, p. 311

***Glareola maldivarum***

#### STATUS

Asian species. Accidental in w. Alaska in spring (two records). Summer breeder in se. Asia, from ne. Mongolia and s. Manchuria south to India, Burma, Thailand, and s. China. Largely resident, but nomadic in India. E. Asian migrants winter in Indonesia and n. Australia. Population in e. Asia/Australasia estimated at 75,000.

#### TAXONOMY

No subspecies recognized.

#### BEHAVIOR

A swallowlike shorebird of open country, with aerial feeding habits. Breeds colonially in dry, open, flat habitats with scattered vegetation, often adjacent to fresh or alkaline wetlands. During migration also uses reservoirs, desert oases, and coastlines. Resident populations are nomadic by nature, dispersing after breeding to food-rich areas. Highly gregarious, feeds in large flocks on swarming insects, mostly at dawn and dusk. During feeding flight performs graceful aerial maneuvers reminiscent of Black Tern. Also feeds on ground in a ploverlike manner, often running or lurching forward after flying or grounded insects. Eats mostly insects, especially beetles, locusts, grasshoppers, and crickets.

Migratory population nests mostly from May to July, while resident birds in India probably nest much earlier.

#### MIGRATION

Partial migrant. Populations in India are mostly resident/nomadic, augmented by migrants in winter. Nomadic birds follow heavy rains and the resulting insect emergence. Birds that breed in e. Asia are mostly migratory, but se. Asia and

Indochina also support sedentary populations.

**Spring** migration takes place from about late February to May. Departure from main wintering areas in n. Australia begin in late February, with major exodus in late March/early April. First arrives in Thailand in late February, and migrants occur in se. China from early March to May. Most northern breeding areas are occupied by April and May.

**Fall** migration takes place from July to late November. Peak numbers of migrants pass through ne. China from July to September, through Japan in August/September, and through se. China in September/October. Wintering areas in n. Australia are reached by late September/October and are fully occupied by late November/December. No information on the differential timing of **adults** and **juveniles.**

### MOLT

**First life year:** The molt out of juvenal plumage is usually complete and takes place from August/early September to late February. The molt begins on the breeding grounds with inner primaries (usually p1–5), some tail feathers and wing coverts, and a variable number of head and body feathers. Molt is suspended during migration and completed on the wintering grounds between October and February. Some birds retain a few secondaries, tertials, and wing coverts. From February to April there is a partial molt to breeding plumage involving a variable number of head and body feathers. The molt is similar to that of adults but is usually less extensive, with a variable amount of nonbreeding and sometimes juvenal plumage retained.

**Adult:** The partial molt to breeding plumage takes place on the wintering grounds between late October and March and includes head and some body plumage. The complete molt to nonbreeding plumage takes place from early July/early August to December/February. This molt begins on the breeding grounds in early July/early August with replacement of most head and body plumage, innermost primaries, and central tail feathers by about August/September. Molt is suspended during migration and is completed after wintering areas are reached from October/November to December/March.

### VOCALIZATIONS

Flight call is a sharp, somewhat nasal, ternlike *kewp* or *kik* and a higher, chattering *kuw-ik-ik*. Also gives a soft *to-wheet.*

APPENDIX

GLOSSARY

BIBLIOGRAPHY

PHOTO CREDITS

INDEX

# APPENDIX

### Answers to Quiz Questions in Species Photos

*Caption number is located at beginning of answer.*

**BLACK-BELLIED PLOVER**

**3 (page 30)** A juvenile at center right is showing checkered upperparts and streaked breast. A nonbreeding adult at center left looks much more softly patterned. A molting adult behind the juvenile is showing the last remnants of breeding plumage. Other species present include Red Knot (left) and Short-billed Dowitcher (center).

**11 (page 33)** The best way to sort out a flying group like this is to start by picking out the largest and smallest birds. The Black-bellied Plovers are the largest. There are four very small birds in the photo—three along the bottom and another in the upper right. These are "peeps," and although it is admittedly difficult to identify them with certainty, they are too pale brown above and pale-headed for Least but just right for Semipalmated (and there are virtually no Westerns in New Jersey in spring). The next size up from Semipalmated are birds showing reddish backs, bold wing stripes, and (visible on at least two birds) black belly patches. These are Dunlin. Then there are three different species of a similar "medium" size. Two birds near the upper right are identifiable as dowitchers by their dark overall upperparts with white "cigar" up the back and pale trailing edge to the wings. If you focus on wing shape and pattern, you'll find another dowitcher on the far left. Although the dowitchers are not identifiable from this view, only Short-bills occur regularly in New Jersey in May. Left of the two dowitchers is a similarly sized bird with weak wing stripes and a pale (but not white) rump. This description matches only Red Knot. At least two other knots are visible in the tight pack at top center. In the middle of that pack is another medium-sized bird showing a bold white wing stripe, a white shoulder patch, and a bold rump or tail pattern. This is a Ruddy Turnstone.

**AMERICAN GOLDEN-PLOVER**

**9 (page 37)** The center left bird is the Black-bellied.

**WILSON'S PLOVER**

**2 (page 48)** Some of the most obvious differences are the larger head and bill, slightly larger overall size, and paler upperparts. Also notice a slightly different face pattern, with white wider over the forehead. Though not that obvious in this photo, the legs are pinkish in color, not yellow-orange as in Semipalmated.

**SEMIPALMATED PLOVER**

**9 (page 53)** This bird is a juvenile, identifiable by thin pale fringing to the coverts and an overall fresh look.

PIPING PLOVER

**7 (page 56)** Orange legs and pale face with prominent dark eye rule out Snowy Plover. Snowies show grayish to dull pink legs and a dark "mascara" mark behind the eye. They also look a bit stubbier bodied, bigger headed, and longer legged.

KILLDEER

**3 (page 59)** To the right of the Killdeer is an American Golden-Plover and behind it are two Black-bellied Plovers. In the back right is a Buff-breasted Sandpiper.

MOUNTAIN PLOVER

**5 (page 63)** American and Pacific Golden-Plovers have gray underwings and are never so white on the belly.

AMERICAN OYSTERCATCHER

**4 (page 65)** Three birds at back left are juveniles, showing dark eyes and dark bill tips. Although it's hard to actually make out the eyes and bill tips on those birds, a quick look at all the other birds reveals glowing yellow eyes and bills with orange all the way to the tips.

**7 (page 66)** First the non-shorebirds: there are four Black Skimmers and three Common Terns. The shorebird at bottom left is at an odd angle but is showing the distinctive Ruddy Turnstone wing and head patterns. The only other shorebird species are visible at either side of the tip of the sharp Common Tern's upper wing. To the left is a smallish bird with a moderately long bill, heavy chest, and some blackish blobs on the belly. This is a molting adult Dunlin. A second Dunlin is visible toward the center of the photo. To the right of the tern is a slightly larger, very long billed bird with a gray hood, smudgy flanks, and an evenly patterned underwing—a dowitcher. Two other dowitchers are visible below (one showing the distinctive white "cigar" up its back). Which dowitcher is hard to say for certain, but based on the relatively pale tails, pale heads, and lack of a contrasting white patch on the leading edge of the underwing, they appear to be Short-bills. Note that the oystercatcher at bottom right is probably a first summer. It is showing extensive dark on the bill yet has yellow eyes (duller than adult's). It also has adultlike primaries, which are broader and less pointed than juvenal primaries (compare the juvenile at center left).

BLACK-NECKED STILT

**5 (page 73)** This is a good example of how general color distribution can be a good field character. The bird at left is a Long-billed Dowitcher, showing a reddish cast to the belly but duller neck and breast. The bird on the right is a Short-billed, showing a reddish cast to the breast but whiter belly.

AMERICAN AVOCET

**4 (page 75)** The dowitcher at center front has that fat-bodied, "swallowed a grapefruit" look that is diagnostic of Long-billed when visible. Others are at less favorable angles, but all appear to have the same shape. All are probably Long-bills.

GREATER YELLOWLEGS

**3 (page 79)** A second Lesser is facing head-on behind the juvenile Greater. A second juvenile Greater is only partially visible, two birds behind the side-on Lesser. The large size, brownish cast, and prominent tertial spots are enough to identify and age it.

LESSER YELLOWLEGS

**4 (page 84)** The Lesser Yellowlegs is peeking out from behind the rightmost Stilt Sandpiper. Note its short, straight, dark bill, slim neck, and small head.

**8 (page 86)** In body size, the dowitcher looks marginally larger, though the yellowlegs stands much taller.

**EASTERN WILLET**

**3 (page 91)** The Eastern Willet is the back right bird.

**SPOTTED SANDPIPER**

**7 (page 103)** The bird in the back is another Spotted Sandpiper. Note the same shape and color pattern. The leg color looks different only because the bird in the foreground has muddy legs.

**WHIMBREL**

**5 (page 109)** The godwit looks fatter bodied and less attenuated, with buffier overall color.

**LONG-BILLED CURLEW**

**6 (page 113)** Although the yellowlegs look small, they are standing next to one of the largest shorebirds in the world. Both are Greaters. Note the long, sturdy, subtly upturned bills and heavy, sculpted bodies with thick chests.

**MARBLED GODWIT**

**7 (page 122)** Marbled has a fatter, rounder body than Hudsonian and lacks that species' trim athletic look.

**BLACK TURNSTONE**

**3 (page 129)** One Ruddy is standing side-on at center left. The other is at left facing away (notice its pale wing coverts and brighter orange legs).

**SURFBIRD**

**3 (page 132)** Most of the other birds are Black Turnstones, looking a bit smaller than Surfbird and very dark (too dark for anything else) on upperparts, head, and breast. Two other birds, one at center and one near back right, are smaller than turnstones with extensive rufous above and finely streaked flanks. These are Rock Sandpipers.

**RED KNOT**

**3 (page 135)** The Western Sandpiper is at center left, out of focus. The Dunlins to its left and right are slightly larger, slightly darker above, and much darker across the breast.

**SANDERLING**

**6 (page 141)** The Sanderlings are the big pale birds with rotund body shape and very white breasts. The Dunlins are similar in size to Sanderlings (here looking marginally smaller) and are much darker and browner above, with extensive dark across the breast. The Western is like a miniature Dunlin, but slightly paler above and much whiter on the breast.

**15 (page 144)** The three smallest birds with thinner bills (top center, center right, and head only in extreme upper left) are Semipalmated Sandpipers; the bright red, black, and white bird just above center is a Ruddy Turnstone; three large birds with salmon-colored faces and underparts are Red Knots; the long-winged, long-billed bird with a black breast patch (just below and a little left of center) is a Dunlin. Note the obvious smaller size, but also smaller head and bill and reduced white upperwing stripe of Semipalmated Sandpiper; also larger, bulkier body of Red Knot.

**SEMIPALMATED SANDPIPER**

**10 (page 148)** This species was named for partial webbing between its middle and outer toes, a trait that is perfectly visible on this chick. Western Sandpiper has a similar foot structure.

**WESTERN SANDPIPER**

**12 (page 154)** Visible here is a rufous crown and cheek and a whitish, streaked supercilium. Western and Stilt Sandpiper both come close, but the fact that the rufous is brighter on the crown than cheek is unique to Western.

**LEAST SANDPIPER**

**9 (page 159)** The back bird is another

Least. Although out of focus, structure and color pattern perfectly match the closer bird. Note especially the stubby rear end and brown overall coloration, including a brown-hooded appearance. Even the yellow legs are visible.

**13 (page 160)** The Least is at front left and is identifiable by its very small size, small head, dark brown coloration, and compact frame with stubby rear end. It is identifiable as a juvenile by its relatively bright coloration, neatly patterned wings, and scattered white spots on the upperparts (each lower scapular and greater covert has a white outer tip). The "peep" at far right is a Semipalmated. It is noticeably larger than Least and is drab gray-brown above with scattered black spots, which are retained breeding scapulars. Those scapulars lack rufous, which rules out Western. The bird next to it is a Western, showing extensive fresh, pale gray nonbreeding upperparts—much more nonbreeding plumage than a Semipalmated would show in New Jersey. The bird behind the Least is another Western, this one showing some of the distinctive breast spotting of breeding plumage. It is also showing wing covert molt, something Semipalmated never shows away from the wintering grounds (the pale spot on the wing is created by whitish greater covert bases, normally concealed by median coverts, which are missing on this bird). The bird at far left is another Semipalmated. Notice its similarity to the bird at far right. The three Semipalmated Plovers are distinctive with their black breast-bands, smooth brown upperparts, and stubby, orange-based bills. The remaining birds are at least as large as the plovers and look fat and pale. These are Sanderlings.

### WHITE-RUMPED SANDPIPER

**5 (page 162)** Behind the White-rump is a Dunlin in lovely breeding plumage. To the left is a bird nearly equal in size to the Dunlin, with heavy-looking body, relatively heavy bill, and uniform pale rufous head, breast, and upperparts. This is a Sanderling. To the right is a bird that looks similar to the White-rump but slightly smaller and shorter winged, with much whiter flanks and a drab gray-brown color above. This is a Semipalmated Sandpiper.

### BAIRD'S SANDPIPER

**9 (page 168)** The Baird's is at center and looks larger and much longer winged than Least, with a paler, buffier overall color. It has black legs, though that difference is less obvious at long range.

### PECTORAL SANDPIPER

**7 (page 171)** This is an adult male in worn breeding plumage. Males are much larger than "peep," sometimes four times the weight. This one isn't quite that big but is still very large and bulky. Worn breeding plumage shows a simple pattern of dark-centered, pale-edged scapulars and coverts that are moderately pointed (due to wear) and not so neatly arranged as in a juvenile.

### PURPLE SANDPIPER

**6 (page 175)** The most obvious difference between Purple and Dunlin is in color. Purple is much darker, with more extensive streaking down the flanks. Purple is also fatter, thicker necked, and bigger headed, with shorter, heavier legs and a shorter, heavier bill. Its bill-base and legs are dull orange.

### ROCK SANDPIPER

**11 (page 179)** The Rock Sandpipers are the smallest birds. The Black Turnstones are the black-headed birds. The only one left is the Surfbird (right of center at the top). It shows a grayish hood, dirty (streaked) body, stubby bill, and a white "rump" (actually the uppertail coverts).

**DUNLIN**

**6 (page 182)** In this case, the Dunlins look slightly larger. Both species are somewhat variable in size, and either may look slightly larger or smaller. Compare this photo with photo #5 in the Sanderling account.

**15 (page 184)** Because this photo was taken in Florida, the right guess would be *hudsonia*. This is supported by the streaked flanks and lack of a white highlight above the belly patch.

**17 (page 184)** The smallest birds (7) are Western Sandpipers, looking like miniature Dunlins but with shorter bills, whiter breasts, and slightly paler overall coloration. The upper left bird is a Sanderling. It is very close in size to a Dunlin and has a glowing white head, relatively short bill, and bold wing stripe. Another Sanderling head is visible at the left edge of the frame.

**STILT SANDPIPER**

**5 (page 187)** The background bird is a Lesser Yellowlegs. Notice the longer, bright yellow legs and shorter, straighter bill. Body size looks similar here, which may be an illusion. Usually Lesser Yellowlegs looks slightly larger (as in photo #4).

**11 (page 189)** Stilt Sandpipers have slightly pointed tails that lack the barring of a yellowlegs. Legs are a duller olive coloration.

**SHORT-BILLED DOWITCHER**

**6 (page 196)** The rightmost bird in full view is a dowitcher with its head tucked. Notice the same face, breast, and flank pattern as the other dowitchers. The two small birds in front with a similar back color to dowitchers are Dunlins. Notice their long, slightly drooped black bills, short legs, and dirty gray wash across the head and chest. Behind the right Dunlin is a Sanderling, looking similar in size to Dunlin but with white head and shorter, straight bill. Mostly hidden in the flock are some dowitcher-sized birds with dowitcher-like face patterns, but they are paler overall with short bills. These are Red Knots. Standing taller than all the other birds are several Black-bellied Plovers with their big blocky heads and short thick bills. Some still show remaining bits of breeding plumage.

**17 (page 199)** Most of these are *griseus*, with one molting *hendersoni* in the foreground and at least two brighter *hendersoni* in the back.

**23 (page 200)** The other bird is a Dunlin, showing its long drooped bill, bold wing stripe, and reddish back.

**LONG-BILLED DOWITCHER**

**5 (page 203)** The Short-billed is on the left, looking slimmer, paler, and flatter backed.

**WILSON'S PHALAROPE**

**10 (page 216)** The Stilt Sandpiper is at bottom center and looks slightly larger than the others, with longer legs and bill and barred underparts contrasting with white underwing. The smallest bird at upper right is a Least Sandpiper. It looks very small, stubby bodied, and small headed, with a dark hood and extensive dark markings along the leading edge of the underwing. At bottom right is a slightly larger, more elongated "peep" with a larger, paler head and whiter underwing. This is a Semipalmated. The remaining birds are White-rumped Sandpipers. They look intermediate in size between the Semipalmated and Wilson's Phalaropes. They have long wings, grayish hoods with some gray spilling down the flanks, and white rumps contrasting with dark tails (visible on a few).

**SPOON-BILLED SANDPIPER**

**3 (page 299)** The Spoon-billed Sandpiper is the bird at center facing head-on.

### Full Page Photo Information

*Page number precedes caption information.*

**Pages ii–iii — Spotted Sandpiper:** juvenile, Aug., N.Y.

**Page 27 — Black-bellied Plovers with Short-billed Dowitchers:** Aug., N.Y.

**Page 28 — Buff-breasted Sandpiper:** June, Alaska

**Page 31 — Black-bellied Plover:** nonbreeding, Dec., Fla.

**Page 36 — American Golden-Plover:** breeding male, June, Churchill, Manitoba

**Page 39 — American Golden-Plover:** juvenile, Oct., N.Y.

**Page 57 — Piping Plover:** breeding male, June, N.J.

**Page 67 — American Oystercatcher:** adult, Jan., Fla. — Ring-billed Gull stole freshly shucked conch from the Oystercatcher's bill by hovering over it as soon as it went to wash the sand off the food. After three successful thefts of food by the gull, the Oystercatcher flew off with its next meal to avoid another robbery.

**Page 77 — American Avocet:** breeding female, Apr., Calif.

**Page 81 — Greater Yellowlegs:** mostly nonbreeding, Mar., Tex.

**Page 114 — Long-billed Curlew:** adult transition to breeding, Mar., Calif.

**Page 117 — Hudsonian Godwit:** breeding male, June, Churchill, Manitoba

**Page 121 — Marbled Godwit:** Jan., Calif.

**Page 138 — Red Knot:** breeding adults with Ruddy Turnstones, May, N.J.

**Page 140 — Sanderling:** juvenile, Sept., N.J.

**Page 149 — Semipalmated Sandpiper:** molting adult, late Aug., N.J.

**Page 186 — Stilt Sandpiper:** juvenile, Aug., N.Y.

**Page 193 — Buff-breasted Sandpiper:** juvenile, Aug., N.J.

**Page 195 — Short-billed Dowitcher:** juvenile, Aug., N.Y.

**Page 212 — American Woodcock:** Jan., N.J.

**Page 224 — Red Phalarope:** breeding female, June, Alaska

**Page 225 — Common Redshank:** May, Portugal

**Page 229 — Northern Lapwing:** Aug., England, on foggy day

**Page 238 — Common Ringed Plover:** juvenile, Aug., England

**Page 287 — Little Stint:** breeding adult, May, England

**Page 305 — Ruff:** breeding male, May, Finland

**Page 318 — Red Knots:** June, N.J.

### Captions for Impression Photos in Domestic Section

**Page 29 — Black-bellied Plover:** adult breeding (right) with two first-summer birds, May, N.J.

**Page 34 — American Golden-Plover:** molting adults, late Aug., N.J.

**Page 40 — Pacific Golden-Plover:** nonbreeding, Feb., Calif.

**Page 44 — Snowy Plover:** Dec., Calif.

**Page 47 — Wilson's Plover:** adult female on nest, Apr., Tex.

**Page 50 — Semipalmated Plover:** adults and juveniles, late Sept., N.J.

**Page 54 — Piping Plover:** adult and juvenile (foreground) with Semipalmated Plovers and Sanderling, Aug., N.J.

**Page 58 — Killdeer:** three birds in foreground with birders in background, Nov., N.J.

**Page 62 — Mountain Plover:** winter flock in plowed field, Feb., Calif.

**Page 64 — American Oystercatcher:** Adults, June, N.J.

**Page 68—Black Oystercatcher:** Adult, May, Alaska
**Page 71—Black-necked Stilt:** adult males and females, Apr., Tex.
**Page 74—American Avocet:** winter flock with Long-billed Dowitchers, Feb., Calif.
**Page 78—Greater Yellowlegs:** nonbreeding, Nov., Tex.
**Page 83—Lesser Yellowlegs:** Apr., Tex.
**Page 87—Solitary Sandpiper:** Sept., N.J.
**Page 90—Eastern Willet:** adult with chick, July, Conn.
**Page 95—Western Willet:** winter flock with Red Knots and Whimbrel, Feb., Fla.
**Page 98—Wandering Tattler:** Feb., Calif.
**Page 101—Spotted Sandpiper:** early Apr., Fla.
**Page 104—Upland Sandpiper:** juvenile, Aug., Colo.
**Page 107—Whimbrel:** late Apr., N.J.
**Page 111—Long-billed Curlew:** Feb., Calif.
**Page 115—Hudsonian Godwit:** molting adults with Laughing Gulls and Lesser Yellowlegs, Sept., N.J.
**Page 119—Marbled Godwit:** in mixed flock with both dowitchers, Black-bellied Plover, and Western and Least Sandpipers, Apr., Tex.
**Page 123—Ruddy Turnstone:** breeding males and females with Laughing Gull and Semipalmated Sandpiper, late May, N.J.
**Page 128—Black Turnstone:** winter flock, Feb., Calif.
**Page 131—Surfbird:** with Black Turnstone, late Mar., Calif.
**Page 134—Red Knot:** flock with Black-bellied Plovers and Dunlin, late Oct., N.J.
**Page 139—Sanderling:** Sept., N.J.
**Page 145—Semipalmated Sandpiper:** with Dunlin, early June, N.J.
**Page 151—Western Sandpiper:** large migratory flock, May, Alaska
**Page 156—Least Sandpiper:** nonbreeding, Feb., Calif.
**Page 161—White-rumped Sandpiper:** Aug., N.J.
**Page 165—Baird's Sandpiper:** juvenile, Aug., N.Y.
**Page 169—Pectoral Sandpiper:** molting adults, Aug., N.J.
**Page 173—Purple Sandpiper:** with Ruddy Turnstone, Mar., N.J.
**Page 176—Rock Sandpiper:** winter flock, Feb., Alaska
**Page 180—Dunlin:** nonbreeding, Jan., N.J.
**Page 185—Stilt Sandpiper:** molting juveniles, Sept., N.J.
**Page 190—Buff-breasted Sandpipers:** Sept., N.J.
**Page 194—Short-billed Dowitcher:** July, N.J.
**Page 201—Long-billed Dowitcher:** with Dunlin, Apr., Tex.
**Page 207—Wilson's Snipe:** Apr., N.J.
**Page 210—American Woodcock:** Jan., N.J.
**Page 213—Wilson's Phalarope:** molting to breeding plumage, Apr., Nev.
**Page 217—Red-necked Phalarope:** Sept., N.C.
**Page 221—Red Phalarope:** Dec., Calif.

# GLOSSARY

**accidental:** One or very few records and thought unlikely to occur again for many years.

**attenuated:** Slender and elongated at rear end, or gradually tapered. For example, White-rumped Sandpiper has a more attenuated shape than Semipalmated Sandpiper.

**auriculars:** "Cheeks" or ear coverts; feathers that cover the sides of the head below the eye.

**axillaries (axillars):** The "wingpits." These feathers cover the underside of the base of the wing. For example, Black-bellied Plover has black axillaries in all plumages.

**casual:** Multiple records, and thought likely to occur again in the near future.

**common:** Always or almost always encountered in appropriate habitat and season, usually in moderate to large numbers.

**coverts** ("wing coverts"): In the broad sense, rows or groups of relatively small feathers that overlap the bases of the primaries, secondaries, and tail feathers, both on the upper side and under side (see topography illustration, pp. 14–15). In this guide, we most often use the term *coverts* (or *wing coverts*) as a collective shorthand term for all secondary coverts on the upper wing, including greater, median, and lesser secondary coverts. Because these feathers are referred to often, such a shorthand term is useful. On a shorebird's folded wing, usually the greater and median coverts are most visible. Coverts of the primaries, tail, and underwing are referred to much less frequently, so their full names are always used (for example, primary coverts, underwing coverts, undertail coverts, uppertail coverts).

**crown:** The top of the head extending from the bill to the back of the skull.

**culmen:** The upper edge of the bill.

**decurved:** Curved downward, as in a curlew's bill.

**dimorphism:** The existence, among one species, of two distinct forms that differ in one or more characteristics such as size, structure, or coloration. In some shorebirds, males and females differ in appearance, so are considered sexually dimorphic.

**edge:** The side of a feather, excluding the tip (compare with *fringe*).

**egg dates:** Dates that live eggs may be found from earliest laying to latest hatching.

**eye-ring:** A ring of often contrastingly pale feathers around the eye (compare *orbital-ring*).

**fairly common:** Usually encountered in appropriate habitat and season, though generally not in large numbers.

**Fenno-Scandia:** The four countries in n. Europe including Norway, Sweden, Denmark, and Finland.

**flanks:** The sides of the body from the leading edge of the folded wing to the vent.

**fringe:** The entire margin of a feather, including sides and tip (compare *edge*).

**gape:** The base of the bill between the upper and lower mandibles.

**grasspiper:** An informal name for any shorebird that regularly uses grassland habitats such as Buff-breasted or Upland Sandpipers.

**greater coverts:** Used in this book to indicate the largest, rearmost row of coverts that directly overlap the secondaries; also called greater secondary coverts. The large feathers overlapping the primaries are the *greater primary coverts.*

**lek:** A gathering of males of one species, held before and during the breeding season, in which birds display and spar for the purpose of attracting mates. This behavior is well known among several species of grouse. Among shorebirds, this behavior is exhibited by Ruff and Buff-breasted Sandpiper.

**lesser coverts:** Several rows of small coverts between the *median* and *marginal coverts.*

**leucism:** A rare genetic abnormality, usually found only in females, in which pigment is diluted. A leucistic bird will appear uniformly washed out, sometimes nearly pure white, but will have normally pigmented eyes. Leucism differs from albinism, in which pigment is lacking in part or all of the bird.

**lores:** The area between the eye and bill.

**lower scapulars:** Two or three rows of larger *scapulars,* closer to the wing than the *upper scapulars.*

**mantle:** The upper back between the *scapulars* and *nape.*

**marginal coverts:** Tiny feathers on the leading edge of the wing, bordering the *lesser coverts.* These are seldom visible on the folded wing.

**median coverts:** The second row of coverts that overlap the bases of the greater coverts.

**nape:** The back of the head and neck.

**Nearctic:** Arctic and temperate regions of the New World.

**orbital-ring:** A very narrow ring of bare skin around the eye (compare *eye-ring*).

**Palearctic:** Arctic and temperate regions of the Old World.

**peep:** Any of the seven smallest members of the genus *Calidris,* including Least, Semipalmated, and Western Sandpipers, along with Little, Long-toed, Red-necked, and Temminck's Stints. The American term analogous to *stint.*

**primaries:** The outer flight feathers attached to a bird's "hand" or outer wing. On the folded wing, tips of the outer primaries often extend past the *tertials* (see *primary projection*).

**primary coverts:** Feathers covering the base of the primaries.

**primary projection:** The degree to which the *primaries* extend past the *tertials* on a bird's folded wing. For example, White-rumped Sandpiper has a longer primary projection than Semipalmated.

**race:** See *subspecies.*

**rare:** Occurs annually, or nearly so, in very small numbers.

**rectrix(-ces):** Tail feather(s).

**remex(-iges):** Flight feather(s), including the primaries, secondaries, and tertials.

**rump:** The lower back between the uppertail coverts and the trailing edge of the wings; on standing birds, the rump is generally covered by the folded wings.

**scapulars:** The relatively large feathers

covering the base of the upper wing. On a standing bird these feathers generally cover the largest portion of the upperparts (see *lower scapulars* and *upper scapulars*).

**secondaries:** The middle flight feathers attached to a bird's "forearm" or inner wing. These are generally completely hidden on a standing bird.

**secondary coverts:** See *coverts.*

**species:** A somewhat arbitrary concept subject to much debate. Generally speaking, a species is any population whose members share close similarities in structure, behavior, voice, and DNA, and that is largely or entirely isolated reproductively. In reality, because the forces of evolution are constantly reshaping each population, a continuum exists between well defined species and poorly defined subspecies with no clear-cut division.

**split supercilium:** On some small *Calidris,* a narrow pale branch splits off of the *supraloral* and forms a division between the central and lateral portions of the crown. This upper branch is referred to as the "upper supercilium."

**stint:** An Old World term analogous to *peep.*

**subspecies:** Any recognizably distinct population that does not meet the criteria of a full species. Subspecies are generally identified by geographic distribution; variations or color morphs that occur within a population do not constitute a separate subspecies. Also called *race.* (see *species*).

**subterminal:** Refers to a marking just before the margin of a feather. A *subterminal line* parallels and "highlights" the feather fringe. Other subterminal markings are usually just inside the feather tip.

**supercilium:** The "eyebrow"; a pale line above the eye separating the cheeks and cap.

**supraloral:** The forward part of the *supercilium* directly above the *lores*; generally referred to separately when the rear supercilium is inconspicuous or absent.

**tarsus:** The "lower part of the leg" (actually the foot).

**tertials:** The innermost flight feathers attached to a bird's "upper arm," which is very short. These relatively long feathers are situated beyond the *scapulars* and function to cover the *primaries* on the folded wing. Because they are relatively large and conspicuous, their pattern is often useful for identification.

**tibia:** The "upper leg," often largely concealed in smaller species.

**uncommon:** Occurs regularly in small numbers, though easy to miss.

**undertail coverts:** The rearmost body feathers on the underside.

**underwing coverts:** Rows of small feathers on the underwing, informally referred to as "wing linings." These coverts may be split into the same groups (lesser, median, and greater primary and secondary coverts) as coverts on the upper wing.

**upper scapulars:** Usually two rows of smaller *scapulars*, between the mantle and the lower *scapulars.*

**upper supercilium:** The upper branch of a "split supercilium."

**uppertail coverts:** The rearmost body feathers on the upperside (beyond the rump).

**vent:** The area between the legs.

**wing point:** The degree to which *primaries* project beyond the tail.

# BIBLIOGRAPHY


Armstrong, R. H. *Guide to the Birds of Alaska.* Anchorage, Alaska: Northwest Publishing Company, 1983.

Baicich, P. J. *A Guide to the Nest, Eggs, and Nestlings of North American Birds.* 2d ed. San Diego: Academic Press, 1997.

Barter, M. A. "Shorebirds of the Yellow Sea: Importance, Threats, and Conservation Status." *Wetlands International, Global series 9,* Canberra, Australia: International Wader Studies 12, 2002.

Beaton, G.; P. W. Sykes, Jr.; and J. W. Parrish, Jr. *Annotated Checklist of Georgia Birds.* 5th ed. Atlanta: Georgia Ornithological Society, 2003.

Bent, A. C. *Life Histories of North American Shore Birds, Part One.* New York: Dover Publications, 1962.

———. *Life Histories of North American Shore Birds, Part Two.* New York: Dover Publications, 1962.

Chandler, R. J. *The Facts on File Field Guide to North Atlantic Shorebirds.* New York: Facts on File, 1989.

Chartier, B. *A Birder's Guide to Churchill.* Colorado Springs: American Birding Association, 1994.

Cleveland, N. J.; S. Edie; G. D. Grieef; G. E. Holland; R. F. Koes; J. W. Maynard; W. P. Neily; P. Taylor; and R. Tkachuk. *Birder's Guide to Southeastern Manitoba.* Winnipeg: Manitoba Naturalists Society, 1988.

Clover, K. J. *Stokes Field Guide to Bird Songs, Western Region.* New York: Time Warner Audio Books, 1999.

Cramp, S., and K.E.L. Simmons, eds. *The Birds of the Western Palearctic.* Vol. 3. Oxford, U.K.: Oxford University Press, 1983.

Cuthbert, C. W.; J. I. Horton; M. W. McCowan; B. G. Robinson; and N. G. Short. *Birder's Guide to Southwestern Manitoba.* Brandon, Man.: Brandon Natural History Society, 1990.

de Schauensee, R. M., and W. H. Phelps, Jr. *A Guide to the Birds of Venezuela.* Princeton, N.J.: Princeton University Press, 1978.

Dister, D. C.; J. W. Hammond; R. Harlan; B. F. Master; and B. Whan. *Ohio Bird Records Committee Checklist of the Birds of Ohio.* Columbus: Ohio Department of Natural Resources, 2002.

Downer, A., and R. Sutton. *Birds of Jamaica: A Photographic Field Guide.* New York: Cambridge University Press, 1990.

Edison, J.; M. Malone; R. Ruisinger; R. O. Russell; J. Tweit; R. Tweit; and D. Yetman. *Davis and Russell's Finding Birds in Southeast Arizona.* Tucson: Tucson Audubon Society, 1995.

Elliott, L. *Stokes Field Guide to Bird Songs, Eastern Region.* New York: Time Warner Audio Books, 1997.

Elphick, C.; J. B. Dunning, Jr.; and D. A. Sibley, eds. *The Sibley Guide to Bird Life and Behavior.* New York: Alfred A. Knopf, 2001.

Fussell, J. O., III. *A Birder's Guide to Coastal North Carolina.* Chapel Hill: University of North Carolina Press, 1994.

Grant, P., and K. Mullarney. *The New Approach to Identification.* Hunstanton, England: Witley Press, Ltd., 1989.

Harrell, B. E., ed. *The Birds of South Dakota: An Annotated Check List.* Vermillion, S.D.: South Dakota Ornithologists' Union, 1978.

Hayman, P.; J. Marchant; and T. Prater. *Shorebirds: An Identification Guide to the Waders of the World.* Boston: Houghton Mifflin, 1986.

Hilty, S. L., and W. L. Brown. *A Guide to the Birds of Colombia.* Princeton, N.J.: Princeton University Press, 1986.

Holt, H. *A Birder's Guide to Southern California.* Colorado Springs: American Birding Association, 1990.

———. *A Birder's Guide to the Rio Grande Valley of Texas.* Colorado Springs: American Birding Association, 1992.

———. *A Birder's Guide to the Texas Coast.* Colorado Springs: American Birding Association, 1993.

———. *A Birder's Guide to Colorado.* Colorado Springs: American Birding Association, 1997.

Howell, S.N.G., and S. Webb. *A Guide to the Birds of Mexico and Northern Central America.* New York: Oxford University Press, 1995.

Iliff, M. J.; R. F. Ringler; and J. L. Stasz. *Field List of the Birds of Maryland.* Baltimore: Maryland Ornithological Society, 1996.

Johnson, S. R., and D. R. Herter. *The Birds of the Beaufort Sea.* Anchorage, Alaska: BP Exploration (Alaska), Inc., 1989.

Johnston, D. W., compiler. *A Birder's Guide to Virginia.* Colorado Springs: American Birding Association, 1997.

Jones, H. L., and A. C. Vallely. *Annotated Checklist of the Birds of Belize.* Barcelona, Spain: Lynx Edicions, 2000.

Jonsson, L. *Birds of Europe: With North Africa and the Middle East.* London: Christopher Helm, 1992.

Kain, T., ed. *Virginia's Birdlife: An Annotated Checklist.* Lynchburg: Virginia Society of Ornithology, 1987.

Kaufman, K. *A Field Guide to Advanced Birding.* Boston: Houghton Mifflin, 1990.

———. *Birds of North America.* Boston: Houghton Mifflin, 2000.

Lewington, I.; P. Alstrom; and P. Colston. *Rare Birds of Britain and Europe.* London: HarperCollins, 1991.

Lockwood, M. W. *Birds of the Texas Hill Country.* Austin: University of Texas Press, 2001.

Mlodinow, S. *Chicago Area Birds.* Chicago: Chicago Review Press, 1984.

———, and M. O'Brien. *America's 100 Most Wanted Birds.* Helena, Mont.: Falcon Press, 1996.

Morrison, R.I.G., and R. K. Ross. *Atlas of Nearctic Shorebirds on the Coast of South America.* Vol. 1. Ottawa: Canadian Wildlife Service, 1989.

———. *Atlas of Nearctic Shorebirds on the Coast of South America.* Vol. 2. Ottawa: Canadian Wildlife Service, 1989.

Morrison, R.I.G.; R. E. Gill; B. A. Harrington; S. Skagen; G. W. Page; C. L. Gratto-Trevor; and S. M. Haig. "Estimates of Shorebird Populations in North America." *Canadian Wildlife Service Occasional Paper no. 104.* Ottawa: Environment Canada, 2001.

Mullarney, K.; L. Svensson; D. Zetterstrom; and P. J. Grant. *Collins Bird Guide.* London: HarperCollins, 1999.

Murin, T., and B. Pfeiffer. *Birdwatching in Vermont.* Hanover, N.H.: University Press of New England, 2002.

National Geographic Society. *Field Guide to the Birds of North America.* 4th ed. Washington, D.C.: National Geographic Society, 2002.

Paulson, D. *Shorebirds of North America: The Photographic Guide.* Princeton, N.J.: Princeton University Press, 2005.

———. *Shorebirds of the Pacific Northwest.* Seattle: University of Washington Press, 1993.

Peterson, R. T., ed. *Eastern/Central Bird Songs.* Ithaca: Cornell Lab of Ornithology, 1990.

———. *Western Bird Songs.* Ithaca: Cornell Lab of Ornithology, 1992.

———, and V. M. Peterson. *A Field Guide to the Birds of Eastern and Central North America.* 5th ed. Boston: Houghton Mifflin, 2002.

Peyton, L. J. *Bird Songs of Alaska.* Ithaca: Cornell Lab of Ornithology, 1999.

Pierson, E. C.; J. E. Pierson; and P. D. Vickery. *A Birder's Guide to Maine.* Camden, Me.: Down East Books, 1996.

Pizzey, G. *A Field Guide to the Birds of Australia.* Princeton, N.J.: Princeton University Press, 1980.

Poole, A., P. Stettenheim, and F. Gill, eds. *The Birds of North America: Life Histories for the 21st Century.* Washington, D.C.: American Ornithologists Union, 2002.

Pranty, B. *A Birder's Guide to Florida.* Colorado Springs: American Birding Association, 1996.

Prater A. J.; J. H. Marchant; and J. Vuorinen. "Guide to the Identification and Ageing of Holarctic Waders." *BTO Guide 17.* Tring, England: British Trust for Ornithology, 1977.

Pratt, H. D.; P. L. Bruner; and D. G. Berrett. *A Field Guide to the Birds of Hawaii and the Tropical Pacific.* Princeton, N.J.: Princeton University Press, 1987.

Raffaele, H.; J. Wiley; O. Garrido; A. Keith; and J. Raffaele. *A Guide to the Birds of the West Indies.* Princeton, N.J.: Princeton University Press, 1998.

Raynes, B., and D. Wile. *Finding the Birds of Jackson Hole.* Jackson, Wyo.: Darwin Wile, 1994.

Richards, A. *Shorebirds: A Complete Guide to Their Behavior and Migration.* New York: Gallery Books, 1988.

———. *Shorebirds of North America.* New York: Gallery Books, 1991.

Ridgely, R. S. *A Guide to the Birds of Panama.* Princeton, N.J.: Princeton University Press, 1976.

Righter, R., and G. A. Keller. *Bird Songs of the Rocky Mountain States and Provinces.* Ithaca: Cornell Lab of Ornithology, 1999.

Roberson, D. *Monterey Birds: Status and Distribution of Birds in Monterey County, California.* 2d ed. Carmel, Calif.: Monterey Peninsula Audubon Society, 2002.

Robertson, W. B., Jr., and G. E. Woolfenden. *Florida Bird Species: An Annotated List.* Gainesville: Florida Ornithological Society, 1992.

Roche, J. C. *All the Bird Songs of Britain and Europe.* Norfolk, England: Wild-Sounds and Sittelle; Rue des Jardins, France, 1993.

Rosair, D., and D. Cottridge. *Photographic Guide to the Shorebirds of the World.* New York: Facts on File, 1995.

Sample, G. *Bird Songs and Calls of Britain and Northern Europe.* London: HarperCollins, 1996.

Sibley, D. *Sibley's Birding Basics.* New York: Alfred A. Knopf, 2002.

———. *The Sibley Guide to Birds.* New York: Alfred A. Knopf, 2000.

———. *The Birds of Cape May.* Bernardsville: New Jersey Audubon Society, 1997.

Slater, P., P. Slater, and R. Slater. *The Slater Field Guide to Australian Birds.* Willoughby, Australia: Lansdowne-Rigby Publishers, 1986.

Sonobe, K., and J. W. Robinson, eds. *A Field Guide to the Birds of Japan.* Tokyo: Wild Bird Society of Japan, 1982.

Stallcup, R. *Ocean Birds of the Nearshore Pacific.* Stinson Beach, Calif.: Point Reyes Bird Observatory, 1990.

Steele, M., ed. *A Birder's Guide to Eastern Massachusetts.* Colorado Springs: American Birding Association, 1994.

Stiles, F. G., and A. Skutch. A Guide to the Birds of Costa Rica. Ithaca, N.Y.: Cornell Univ. Press, 1989.

Svingen, D., and K. Dumroese, eds. *A Birder's Guide to Idaho.* Colorado Springs: American Birding Association, 1997.

Taylor, R. C. *A Birder's Guide to Southeastern Arizona.* Colorado Springs: American Birding Association, 1995.

Tomkovich, P. S. "Breeding Biology of the Great Knot, *Calidris tenuirostris* 13." *Bulletin of Moscow Society of Naturalists,* vol. 106, part 4, 2001.

Veit, R. R., and L. Jonsson. "Field identification of smaller sandpipers within the genus *Calidris.*" *American Birds* 38, 1984.

Walsh, J.; V. Elia; R. Kane; and T. Halliwell. *Birds of New Jersey.* Bernardsville: New Jersey Audubon Society, 1999.

White, M. *A Birder's Guide to Arkansas.* Colorado Springs: American Birding Association, 1995.

# PHOTO CREDITS

**Goran Altstedt/Windrush Photos:** 296 top
**Lee Amery:** 87 bottom, 211 top
**Yuri Aztukhin:** 178 middle, 233 top, 250 top, 277 top left, 291 top, 291 bottom right
**Nigel Blake:** 217 bottom, 231 top, 243 top left, 243 bottom, 244 bottom, 301 bottom right
**Don Bleitz:** 268 top left
**Lysle Brinker:** 315 bottom
**Richard Brooks/Windrush Photos:** 252 top, 289 bottom, 300 bottom right
**P. A. Buckley:** 207 bottom
**Milo Burcham/www.miloburcham.com:** 69 bottom, 70 top, 70 bottom right, 196 top
**Daniel CK Chan:** 220 bottom, 247 top, 249 bottom, 250 bottom, 252 bottom, 255 middle, 265 bottom right, 275 bottom, 282 bottom
**Richard J. Chandler:** 230 top left, 274 top, 304 left
**Robin Chittenden:** 230 top right, 243 top right, 246 top, 256 top, 257 bottom, 277 top right, 316 top left, 316 top right
**Lyann Comrack:** 128 bottom
**Richard Crossley/www.crossleybirds.com:** iii, 18 top left. 18 top right, 18 second left, 18 second right, 18 bottom right, 19 top right, 19 third right, 19 bottom left, 29 top, 29 bottom, 30 top, 30 middle, 30 bottom, 31, 32 top, 32 bottom, 33 middle, 33 bottom, 34 top, 34 bottom, 37 middle right, 37 bottom, 38 middle, 38 bottom, 43 middle left, 43 middle right, 45 top, 45 bottom, 46 middle, 46 bottom left, 48 top, 48 middle, 48 bottom, 50 top, 51 bottom left, 51 bottom right, 52 middle, 52 bottom, 53 top, 53 middle, 54 top, 55 top, 55 middle, 56 top, 56 third, 56 bottom, 58 top, 59 top, 59 middle, 59 bottom right, 60 bottom, 61 top, 61 bottom, 62 top, 63 bottom, 64 top, 64 bottom, 65 bottom, 66 top right, 66 bottom, 72 top, 73 middle left, 73 middle right, 74 bottom, 75 top left, 76 bottom, 78 bottom, 79 top, 79 middle, 80 middle, 80 bottom, 84 top, 84 bottom, 85 bottom right, 86 bottom, 87 top, 88 top, 89 top, 89 middle, 89 bottom, 90 bottom, 91 bottom, 92 top, 92 bottom, 93 middle left, 96 top, 96 middle, 97 bottom, 98 top, 101 bottom, 102 top, 102 middle left, 102 bottom, 103 middle, 103 bottom left, 103 bottom right, 105 bottom left, 106 middle left, 107 top, 108 top, 108 middle, 109 top, 109 middle, 109 bottom left, 109 bottom right, 110 top right, 110 middle right, 110 bottom right, 111 top, 112 top, 112 middle, 112 bottom, 113 top, 113 middle, 113 bottom left, 113 bottom right, 115 top, 115 bottom, 116 top, 118 middle, 118 bottom left, 118 bottom right, 120 top, 122 top, 122 middle right, 122 bottom, 123 top, 124 top left, 124 top right, 124 middle, 124 bottom, 125 top, 125 bottom, 127 top, 127 middle, 127 bottom right, 128 top, 129 bottom, 134 top, 135 top, 135 middle, 137 top, 137 bottom left, 137 bottom right, 139 top, 141 top left, 141 top right, 141 middle, 140, 142 top, 142 middle, 142 bottom, 143 middle, 144 middle, 145 top, 146 top, 146 middle right, 149, 150 bottom, 151 bottom, 152 top, 152 middle left, 152 middle right, 152 bottom

left, 152 bottom right, 153 top, 153 middle, 155 top, 155 middle, 155 bottom, 156 top, 157 middle left, 157 middle right, 157 bottom, 158 top, 159 top, 159 middle, 159 bottom, 160 top, 160 middle, 160 bottom left, 161 top, 162 top left, 162 middle, 163 middle, 164 bottom, 166 top, 166 middle, 167 top, 168 middle, 168 bottom, 169 top, 171 bottom, 172 middle, 173 top, 174 top left, 174 top right, 174 bottom, 175 top, 175 middle, 175 bottom left, 180 bottom, 181 top left, 181 top right, 181 middle, 181 bottom, 182 middle, 182 bottom, 183 top right, 183 bottom right, 184 top right, 184 bottom, 185 top, 185 bottom, 186, 187 top, 189 bottom, 190 top, 191 top, 192 bottom left, 193, 196 bottom, 197 top left, 197 top right, 198 bottom, 199 middle, 200 top, 200 middle left, 200 bottom, 202 top, 202 middle, 202 bottom, 203 bottom, 205 bottom, 206 middle, 206 bottom, 207 top, 208 top, 208 middle right, 208 bottom, 209 top, 209 bottom left, 211 bottom right, 212, 213 top, 215 top, 216 top, 216 bottom, 217 top, 219 bottom, 225, 227 top, 227 middle, 227 bottom left, 228 bottom left, 228 bottom right, 229, 230 bottom, 231 bottom right, 237 top left, 237 top right, 238, 239 bottom left, 239 bottom right, 240 top, 240 middle, 240 bottom, 241 bottom, 245 top left, 245 bottom, 246 bottom, 249 top, 253 top, 253 bottom left, 256 bottom left, 256 bottom right, 259 top, 260 middle, 262 top inset, 263 top, 263 bottom left, 263 bottom right, 264 bottom, 268 bottom left, 274 bottom left, 275 top, 276 top right, 276 bottom right, 277 middle, 280 bottom, 284 bottom, 292 middle right, 302 middle, 304 bottom right, 308 top inset, below, 308 bottom inset, 312 top, 317 middle, 317 bottom left, 317 bottom right

**Mike Danzenbaker/www.avesphoto.com:** 16, 40 bottom, 53 bottom, 62 bottom, 63 top left, 63 middle, 68 bottom, 69 top right, 70 bottom left, 71 bottom, 82 middle, 88 bottom left, 94 middle left, 100 top, 100 bottom left, 107 bottom, 114, 130 top, 130 bottom left, 130 bottom right, 131 top, 131 bottom, 132 middle, 132 bottom left, 133 bottom, 166 bottom left, 177 top right, 178 top, 179 top left, 198 top left, 199 top, 218 middle, 221 top, 223 bottom left, 226 bottom, 228 top, 231 bottom left, 234 top, 239 top right, 244 top, 251 top, 253 bottom right, 254 top, 255 top, 258 top, 262 top, 264 top, 266 bottom right, 267 bottom left, 267 bottom right, 269 top, 278 top, 283 bottom left, 285 bottom left, 286 middle right, 288 top, 288 bottom left, 289 top, 290 top, 291 middle right, 291 bottom left, 292 top right, 292 bottom left, 293 top left, 293 top right, 298 top, 298 bottom, 300 top left

**Don Des Jardin:** 44 top, 88 top inset, 111 bottom, 213 bottom, 220 top, 313 bottom

**Paul Doherty/Windrush Photos:** 257 top right, 259 middle right

**Linda Dunne:** 168 top, 214 top

**Scott Elowitz:** 105 bottom right, 106 middle right, 106 bottom, 129 middle, 191 bottom, 216 middle, 271 bottom, 286 bottom, 295 top

**Ian Fisher/Windrush Photos:** 310 top

**Paul J. Fusco:** 82 bottom, 84 middle, 88 bottom right, 90 top, 126 middle, 150 middle, 154 middle, 154 bottom left, 190 bottom, 204 top, 268 bottom right, 270 top, 279 bottom

**Hubert Gallagher:** 54 bottom, 285 top left

**James A. Galletto:**195

**Brian M. Guzzetti:** 100 bottom right, 176 top, 177 middle, 179 bottom

**Brandon Holden:** 172 bottom

**Julian Hough:** 93 middle right, 165 bottom, 204 bottom

**Steve N. G. Howell:** 49 middle, 166 bottom right, 222 middle, 223 bottom right

**Barry Hughes/Windrush Photos:** 251 bottom

**Kwong Kit Hui:** 42 bottom, 110 top left, 110 middle left, 232 bottom right, 234 upper right, 235 bottom, 247 bottom, 252 mid-

middle, 76 middle, 77, 95 top, 97 top, 97 middle, 99 top, 99 bottom, 120 bottom, 121, 122 middle left, 132 bottom right, 136 top right, 146 middle left, 148 top, 161 bottom, 164 top, 170 top left, 183 middle left, 184 top left, 200 middle right, 201 bottom, 215 middle, 220 middle, 224, 279 top, 286 top, 286 middle left, 302 bottom

**Alan Murphy:** 49 top

**Michael O'Brien:** 93 top, 95 bottom, 96 bottom, 171 top right, 177 top left, 199 bottom left, 199 bottom right, 206 top left, 236 top, 236 bottom left, 248 bottom left, 293 bottom right

**Jari Peltomaki/Windrush Photos:** 265 bottom left, 288 bottom right

**Simon Perkins:** 316 bottom

**Richard Revels/Windrush Photos:** 297 top

**Wayne Richardson:** 49 bottom, 173 bottom

**Jim Rosso:** 313 top

**Robert Royse:** 317 top

**Bill Schmoker:** 167 middle

**Yoshimitsu Shigeta:** 307 bottom left, 308 top inset, above, 309 top inset

**Brian E. Small/www.briansmallphoto.com:** 178 bottom

**Brian Sullivan:** 41 top, 44 bottom, 133 top, 175 bottom right, 176 bottom, 179 top right, 214 bottom right

**Ruth Sullivan:** 278 bottom left

**David Tipling/Windrush Photos:** 245 top right, 248 top, 287, 303 top, 303 bottom, 305, 306 top, 306 bottom, 311 bottom right

**Ray Tipper:** 41 middle, 232 top left, 237 bottom right, 241 middle, 258 bottom, 262 bottom, 284 top, 285 bottom right, 290 bottom, 299 top, 308 bottom

**Arnoud B. van den Berg:** 273 top, 273 bottom, 310 bottom, 314 top left, 314 top right

**Shuji Watanabe:** 227 bottom right, 241 top, 255 bottom left, 259 bottom left, 266 top, 266 bottom left, 285 top right, 292 top left, 294 top, 294 bottom, 307 top, 307 bottom right, 311 bottom left

**Chris Wood:** 236 bottom right

**Steve Young/Windrush Photos:** 222 top right

**Jim Zipp/www.jimzipp.com:** 196 middle left, 209 bottom right, 270 bottom

I N D E X